T0366305

Perturbing Material-Components on Stable Shapes

How Partial Differential Equations Fit into the Descriptions of Stable Physical Systems

Martin Concoyle Ph.D.

Order this book online at www.trafford.com
or email orders@trafford.com

Most Trafford titles are also available at major online book retailers.

Printed in the United States of America.

ISBN: 978-1-4907-2369-3 (sc)
ISBN: 978-1-4907-2372-3 (e)

Trafford rev. 01/10/2014

www.trafford.com

North America & international
toll-free: 1 888 232 4444 (USA & Canada)
fax: 812 355 4082

Copyrights

These new ideas put existence into a new context, a context for both manipulating and adjusting material properties in new ways, but also a context in which life and creativity (practical creativity, ie intentionally adjusting the properties of existence) are not confined to the traditional context of "material existence," and material manipulations, where materialism has traditionally defined the containment of material-existence in either 3-space or within space-time.

Thus, since copyrights are supposed to give the author of the ideas the rights over the relation of the new ideas to creativity [whereas copyrights have traditionally been about the relation that the owners of society have to the new ideas of others, and the culture itself, namely, the right of the owners to steal these ideas for themselves, often by payment to the "wage-slave authors," so as to gain selfish advantages from the new ideas, for they themselves, the owners, in a society where the economics (flow of money, and the definition of social value) serves the power which the owners of society, unjustly, possess within society].

Thus the relation of these new ideas to creativity is (are) as follows:

These ideas cannot be used to make things (material or otherwise) which destroy or harm the earth or other lives.

These new ideas cannot be used to make things for a person's selfish advantage, ie only a 1% or 2% profit in relation to costs and sales (revenues).

These new ideas can only be used to create helpful, non-destructive things, for both the earth and society, eg resources cannot be exploited to make material things whose creation depends on the use of these new ideas, and the things which are made, based on these new ideas, must be done in a social context of selflessness, wherein people are equal creators, and the condition of either wage-slavery, or oppressive intellectual authority, does not exist, but their creations cannot be used in destructive, or selfish, ways.

This book is dedicated to my wife M. B. and to my mom and dad

This new book has much material similar to two old books originally put onto scribd.com, 2013, as well as new material (put, m concoyle, into the search-bar at the scribd.com website)
The old books are:

Physical description based on the properties of stability,
geometry, and consistency:
Short essays which are: simple, "clear," and direct
Presented to the Joint math meeting San Diego (2013)
and
Introduction to the stability of math constructs;
and a subsequent: general, and accurate, and
practically useful set of descriptions of the observed stable material systems (2013)
By Martin Concoyle Ph. D.

Notes:

Double spaces can mean a sudden new direction of the discussion without a new paragraph title. The *'s represent either favorites (of the author) or (just as likely) indecision and questions about (logical) consistency. Information and discussion about ideas is not a monolithic endeavor pointing toward any absolute truth, the wide ranging usefulness, in regard to practical creativity, might be the best measure of an idea's truth, it is full of inconsistencies and decisions about which path to follow (between one or the other competing ideas) are either eventually made or the entire viewpoint is dropped, but this can occur over time intervals of various lengths.

The marks, ^, associated to letters, eg a^2, indicates an exponent.

The marks, *, in math expressions can have various math meanings, such as a pull-back in regard to general maps which can, in turn, be related to differential-forms, defined on the map's domain and co-domain (or range), but in this book it usually denotes the "dual" differential-form in a metric-space of a particular dimension, eg in a 4-dimensional metric-space the 1-forms are dual to the 3-forms and the 2-forms are self-dual, etc.

The main idea of thought (or of ideas) is that it is about either sufficiently general and sufficiently precise descriptions based on simple patterns, or it is about developing patterns (of description) which lead to particular practical creativity, or to new interpretations of observed patterns, or to directions for new perceptions.

This book is an introduction to the simple math patterns which can be used to describe fundamental, stable spectral-orbital physical systems (represented as discrete hyperbolic shapes, ie hyperbolic space-forms), the containment set has many-dimensions, and these dimensions possess macroscopic geometric properties (where hyperbolic metric-space subspaces are modeled to be discrete hyperbolic shapes). Thus, it is a description which transcends the idea of materialism (ie it is higher-dimensional, so that the higher-dimensions are not small), and it is a math context can also be used to model a life-form as a unified, high-dimension, geometric construct, which generates its own energy, and which has a natural structure for memory, where this construct is made in relation to the main property of the description being, in fact, the spectral properties of both (1) material systems, and of (2) the metric-spaces, which contain the material systems, where material is simply a lower dimension metric-space, and where both material-components and metric-spaces are in resonance with (or define) the containing space.

Partial differential equations are defined on both (1) the many metric-spaces of this description and (2) on the lower-dimensional material-components, which these metric-spaces contain, ie the laws of physics, but their main function is to act on either the, usually, unimportant free-material components (so as to most often cause non-linear dynamics) or to perturb the orbits of the, quite often condensed, material which has been trapped by (or is defined within) the stable orbits of a very stable hyperbolic metric-space shape.

It could be said that these new ideas about math's new descriptive context are so simple, that some of the main ideas presented in this book may be presented by the handful of diagrams which show these simple shapes, where these diagrams indicate how these simple shapes are formed, and folded, or bent, so as to form the stable shapes, which can carry the stable spectral properties of the many-(but-few)-body systems . . . . , where these most fundamental-stable-systems have no valid quantitative descriptions within the math context which is determined by the, so called, currently-accepted "laws of physics," (ie the special set of partial differential equations which are associated to the, so called, physical laws) . . . . so that the diagrams of these stable geometric shapes are provided at the end of the book.

This new measurable descriptive context is many-dimensional, and thus, it transcends the idea of materialism, but within this new context the 3-dimensional (or 4-dimensional space-time) material-world is a proper subset (in a subspace which has 3-spatial-dimensions),

The, apparent, property of fundamental randomness (in a currently, assumed, absolutely-reducible model of material, and its reducible material-components) is a derived property, but now in a new math context, in which stable geometric patterns are fundamental,

The property of spherically-symmetric material-interactions is shown to be a special property of material-interactions, which exists (primarily, or only) in 3-spatial-dimensions, of Euclidean space, wherein inertial-properties are to, most naturally, be described,

It is a descriptive context which is both reductive . . . . , (to some sets of small material-components, but elementary-particles are most likely about components colliding with higher-dimensional lattice-structures, which are a part of the true geometric context of physical description) . . . , and unifying in its discrete descriptive contexts (relationships) which exist, between both a system's components, and the system's (various) dimensional-levels (where these dimensional-levels are particularly relevant, in regard to understanding both (1) the chemistry of living systems, and (2) the functional organization of living systems),

But most importantly, this new descriptive language (new context) describes the widely observed properties of stable-physical-systems, which are composed of various dimensional-levels and of various types of components and interaction-constructs, so that this new context provides an explanation about both (1) "how these systems form" and (2) "how they remain stable," wherein, partial differential equations, which model material-interactions, are given a new: context, containment-structure, organization-context, interpretation, and with a new discrete character, whose function is to perturb the material-interaction structure of material contained in metric-spaces where these metric-spaces possess very stable shapes.

It provides a (relatively easy to follow, in that, the containment set-structure for these different-dimensional stable-geometries are simple dimensional relations) 'map' "up into a higher-dimensional context (or containment set) for existence," wherein some surprising new properties of existence can be modeled, in relation to our own living systems, which are also to be modeled as higher-dimensional constructs, and this map of "the dimensions of existence" can shed-light onto our own higher-dimensional structure (as living-systems), and the relation which these living systems have to both existence, and to the types of experiences into which we (the living-systems) may enter (or possess as memory) (or within which we might function), but where because any idea about higher-dimensions is difficult to consider, and is relatively easy to hide and ignore these higher-dimensions, especially, if we insist on the idea of materialism.

Foreword

Considerations fundamental to "descriptive knowledge"

One needs math patterns which are associated to both:

(1) a conceptually geometric, basis "upon which a very simple (and new) math construct is founded," where the construct is used to describe both material properties and the properties of existence, so that math descriptions can be both

   (a) used practically, and
   (b) used in a consistent methodical manner, to calculate the observed properties of (fundamental) measurable systems, eg the stability of general nuclei and the stability of solar systems,

and

(2) new interpretations of the traditional authoritative old math patterns, so as to follow "new math directions" based both on new interpretations and new contexts for (the old) math patterns.

The measurable properties of stable, definitive, spectral-orbital physical-systems (from nuclei to solar systems) cannot be described using the math constructs of:

1. non-linearity,
2. indefinable randomness, and
3. local sets of linear relations which relate a function's values to its domain values which are not continuously commutative (with perhaps one exceptional domain-point) and

4. defining (arbitrary) convergences on (or within) a continuum, nor can they be described with any mathematical description which is associated to
5. the idea of materialism.

Rather, an

Alternative set of (math): constructs, contexts, interpretations, containment sets, organization of math patterns, and quantitative structures are needed.

In other words:

Godel's incompleteness theorem states that "precise language is limited in regard to the patterns which it can describe," is an idea which might be best interpreted to mean that "if one cannot build what one wants with an existing (or the traditional, or the authoritative) measurable language of math, then build a new precise language," where the math patterns must have a coherent relation to both reliable measuring processes, and (to) controllable system-coupling processes.

Thus we are forced to decide between:

(1) Are physical system's only to be based on non-linearity and indefinable randomness, and thus physical systems are simply too complicated to describe?

or

(2) Does the existence of stability imply that "Godel's incompleteness theorem" should be correctly interpreted, and different types of descriptive language structures (including new assumptions, new interpretations, and new contexts) should be considered.

However, as wage-slaves we must (we are forced to) chose (1), which fits into the narrowly defined categories of commerce and the related narrowly defined interests of the investment class, who also want great (or absolute) control over all knowledge and its associated creativity within society.

Nonetheless,

When one looks out to the geometric and spectral patterns of space and material one must ask in all generality:

"What is the shape and dimension of space?"

"What patterns of geometry, size, 'material,' and spectral-values can be detected (perceived), and at what dimension, or from what subspace, or from what well known property, do they emanate, or inter-connect?"

One must consider that the correct model of life, in all likelihood, has a dimension which exceeds the dimension which the narrow viewpoint of materialism seems to demand

There are new ways in which to consider the math patterns which are needed to be able to describe the observed stable properties of the most prevalent and fundamental of physical systems, and they are multi-dimensional patterns which depend most significantly on the very stable math patterns of the discrete hyperbolic shapes, ie both the bounded and unbounded discrete hyperbolic shapes, which exist from (at least) 2-dimensional hyperbolic metric-spaces, to 10-dimensional hyperbolic metric-spaces, all modeled as discrete hyperbolic shapes, so that all of these shapes can be contained within an 11-dimensional hyperbolic metric-space and organized so as to allow for valid descriptions of the observed stable properties of the most prevalent and fundamental of physical systems.

It could be said that these new ideas about math descriptive context are so simple that the main ideas presented in this book are presented by the handful of diagrams about these simple stable discrete hyperbolic shapes, and how they are folded, and how they interact, which are provided in the diagrams at the end of the book.

Contents

PART II
THE SPEECH BY CONCOYLE, AT THE SAN DIEGO MATH CONFERENCE (2013), ABOUT USING GEOMETRIZATION TO ESTABLISH A NEW DESCRIPTIVE CONTEXT FOR THE PHYSICAL WORLD

PART III
NEWER MATERIAL

Part I

Preface

(Various essays about stable math constructs, material interactions, and finite quantitative structures)

This book is an introduction to the use of simple math patterns to describe fundamental, stable spectral-orbital physical systems (as discrete hyperbolic shapes), the containment set has many-dimensions and these dimensions possess macroscopic geometric properties (which are discrete hyperbolic shapes), thus it is a description which transcends the idea of materialism (it is higher-dimensional), and it can also be used to model a life-form as a unified, high-dimension, geometric construct, which generates its own energy, and which has a natural structure for memory, in relation to the main property of the description being spectral properties of both material systems and of the metric-spaces which contain the material systems, where material is simply a lower dimension metric-space.

The math descriptions, about which what this book is about, are about using math patterns within measurable descriptions of the properties of existence which are: stable, quantitatively consistent, geometrically based, and many-dimensional, which are used to model of existence, within which materialism is a proper subset.

In regard to the partial differential equations which are used to describe stable "material" systems they are: linear, metric-invariant [ie isometry (SO, as well as spin, and translations) and unitary (SU, Hermitian invariant [finite dimensional]) fiber groups], separable, commutative (the coordinates remain globally, continuously independent), and solvable, ie controllable.

The metric-spaces, of various dimensions and various metric-function signatures [eg where a signature is related to R(s,t) metric-spaces] have the properties of being of non-positive constant curvature, where the coefficients of the metric-functions (symmetric 2-tensors) are constants.

That is, the containment sets and "material systems" are based on (or modeled by) the simplest of the stable geometries, namely, the discrete Euclidean shapes (tori) and the discrete hyperbolic shapes (tori fitted together), where the discrete hyperbolic shapes are very geometrically stable and they possess very stable spectral properties.

One can say that these shapes are built from "cubical" simplexes (or rectangular simplexes).

Both the (system containing) metric-spaces and the "material" systems have stable shapes of various dimensions and various metric-function signatures, where material interactions are built around the structures of discrete Euclidean shapes (sort of as an extra toral component of the interacting stable discrete hyperbolic shapes), within a new dimensional-context for such material-interaction descriptions, and there are similar interaction constructs in the different dimensional levels. The size of the interacting material "from one dimensional level to the next" is determined by constant multiplicative factors (defined between dimensional levels) which are (now) called physical constants.

Furthermore, the basic quantitative basis for this description, ie the stable spectra of the discrete hyperbolic shapes, forms a finite set. The quantitative structure is, essentially: stable, quantitatively consistent, and finite.

This descriptive construct can accurately, and to sufficient precision, and with wide ranging generality, describe the stable spectral-orbital properties of material systems of all size scales, and in all dimensional levels. It is a (linear, solvable) geometric and controllable description so it is useful in regard to practical creativity.

The many-dimensions allow for new high-dimension, well organized, controllable models of complicated systems, such as life-forms. These ideas provide a "map" to help envision these geometric structures.

These new ideas are an alternative to the authoritative (and overly-domineering) math patterns used by professional math and physics people which are based on non-linearity, non-commutativity, and indefinable randomness (the elementary event spaces do not have a valid definition), where these are math-patterns, which at best, can only describe unstable, fleeting patterns, which are unrelated to practical creative development, and whose measured properties can only be related to feedback systems (whose stability depends on the range of validity of such a system's differential equation).

That is, it is a math construct which is not capable of describing the stable properties of so many fundamental (relatively) stable physical systems, eg nuclei, where within this authoritative descriptive context it is claimed that these stable fundamental physical systems are "too complicated" to describe.

There are many social commentaries, in this book, this is because such a "new context of containment, in regard to measurable descriptions," which possesses so many desirable properties, one would think that such a descriptive language should be of interest to society. But inequality, and its basis in arbitrary (and failing) authority, and the relation which this authority has to extreme violence (in maintaining its arbitrary authority, and in maintaining a social structure (as

Mark Twain pointed-out) which is based on: lying, stealing, and murdering) have excluded these new ideas from being expressed within society.

People have been herded, and tricked, into wage-slavery, where deceiving people is easy with a propaganda system which allows only one authoritative voice, and that one-voice is the voice of the property owners (with the controlling stake), and the people are paralyzed by the extreme violence which upholds this social structure, where this extreme violence emanates from the justice system, and whereas the political system has been defined as "politicians being propagandists" within the propaganda system (politicians sell laws to the owners of society, for the selfish gain of the politicians, and then the politicians promote those laws on the media).

Note

These essays span 2004 to 2012 and some old ideas (along these same lines) expressed around 2004 may not be ideas considered correct by myself today (2012), but I have not re-edited them.

Ideas are worth expressing, and the development of ideas can have interesting histories, and old ideas can be re-considered.

Today there exist experts of "dogmatic authority" who are represented in the propaganda system (as well as in educational institutions which also serve the interests of the owners of society, ie the modern day Roman Emperors) as being "always correct," yet they fail to be able to describe the stability of fundamental physical systems, and their descriptions have no relation to being related to practical creative development, since they are descriptive constructs based on probability and non-linearity, and deal with systems made-up of only a few components which possess unstable properties.

"Peer review" checks for the "dogmatic purity" of its contributors. However, such a situation in science and mathematics does not express the (true) spirit of knowledge. Knowledge is related to "practical" creativity and knowledge is about equal free-inquiry with an "eye on" "what one wants to create." In this context knowledge should be as much about re-formulating, and re-organizing technical (precise) language as about learning from the current expressions of knowledge [with its narrow range of creativity associated to itself].

A community of "dogmatically pure" scientists and mathematicians is not about knowledge, but rather about the power structure of society (a society with a power structure which is essentially the same as that of the Holy-Roman-Empire, ie fundamentally based on extreme violence) and the scientists and mathematicians are serving the owners of society (the new Emperors) by competing in a narrow dogmatic structure of authority, so as to form a hierarchical array of talent to be selected from by the Emperor, and then used within the narrowly defined ranges of creativity, about which the owners of society want attended.

That is, scientists function in society as elite wage-slaves for the owners of society, and they are trained experts, similar to trained lap-dogs.

This structure of knowledge is the opposite of "valid" knowledge, which should be related to a wide range of creative efforts, by many people, expressing many diverse interests.

It is good to express a range of ideas.

Whereas, "being correct" is associated to a "false," or at best, limited "knowledge," which serves "the owners of society" and is mostly used to express in the propaganda system, the (false) idea, that people are not equal. It is this type of idea which the Committee on Un-American Activities should investigate, since the US Declaration of Independence states that all people are equal, and this should be the basis for US law, and not: property rights, and minority rule, which is the same basis as for Roman-Empire Law.

Chapter 0

Succinct-explanation

This paper is about explaining a simple math construct . . . , within which existence is best contained in regard to both accuracy and practically useful descriptions concerning the patterns of existence . . . , in a succinct manner.

First separate one's academic and supposedly scholarly mind from the math constructs of non-linearity and fundamental (indefinable) randomness, ie let go of general relativity, particle-physics, and quantum physics as well as all the speculative theories derived from these failing representations of physical law, eg string theory, etc. The properties of randomness and spherical symmetry, as well as other observed patterns etc, can be recovered or re-interpreted within a (relatively new (2002)) simpler math construct.

The conditions of both containment and structure that allow functions . . . , (which model the measured values of "physical system properties," where the system is contained in the function's domain space and the system is stable) . . . , to be valid measures of a system's properties, and still have properties of stability (either very stable, or relatively stable) are the rigid geometric conditions of the globally solvable shapes, ie the circle-spaces (related to right rectangular simplexes as fundamental domains), mainly the "discrete hyperbolic shapes."

One could say these ideas are about "taking the very simplest forms of description, based on very stable geometry, as far as one can go."

Functions are used, in math, in a context of measurement, and relating functions to quantitatively consistent mathematical frameworks of containment sets, ie a physical system is contained in a coordinate domain space and its measurable properties are represented either as functions or as (physically measurable) operators [or sets of differential equations] defined on the domain space.

This context of measurement is modeled either locally linearly as a derivative operator (classical physics) or as a set of differential equations so as to define measured values as spectral values (when based on the idea of randomness this is quantum physics).

Classical physics (geometry)

When measurement is modeled by derivatives (or partial differential equations) which define local vector relationships as matrices, which are defined on the local coordinate vectors, where the vector values are associated to linear function-approximation values through differentiation.

The solution function is to be uniquely defined by an (almost) inverse integral operator (inverse to the derivative) and a specified set of function values on the domain space.

The diagonal matrix conditions required of the many-variable descriptive context (of containment) . . . . , so as to have the properties of solvability, quantitative consistency, and stability [ie linear, metric-invariant, and separable (partial) differential equations which model the measured values of the system] . . . , must be global (fill the entire coordinate space) in order to have a stable solution function (or set of functions).

Quantum (or spectral) physics

Metric-invariant, linear sets of partial differential equations, which define spectra, also need to be separable in order to be solved.

However, are there an infinite number of spectral equations, or "Can a finite number of spectral equations be used to determine the properties of measurable, physical systems?"

Assume that stable shapes are the basis for physical descriptions of physical systems which possess very stable, and apparently, controllable properties

This very constraining requirement on stability and quantitative consistency requires that the shapes of the coordinates, as well as the controllable solution functions, possess the shapes of the circle-spaces, ie the very stable discrete hyperbolic shapes (which also possess very stable spectral properties) and the discrete Euclidean shapes (which form a geometrically consistent link (connection) in space between the stable, but interacting, material components [modeled as discrete hyperbolic shapes]). Furthermore, spectral systems can also be placed in the context of the very stable shapes of some circle-spaces and the spectral set can be defined to be finite in this new geometric context.

The metric-spaces which contain the material are also modeled as discrete hyperbolic shapes, whose standard-sizes, in regard to material components, are determined by constant multiplicative factors defined between dimensional levels, where these constant multiplicative factors manifest as physical constants.

Spherical shapes and other non-linear patterns are too complicated to be practically useful descriptions

The spherical shapes cannot be used, since they are non-linear, and when perturbed from their spherical shapes, they are both quantitatively-inconsistent and unstable, while (on the other hand) the discrete hyperbolic shapes are rigid and stable geometric structures.

The basics and the simple math patterns which can be related to these fundamental properties of description

In the context of physical description, there is: measuring, geometry, spectra, and changes.

The changes of measurable macroscopic properties can be described with geometry and a local linear model in regard to measuring a function's values. In this context derivatives define local vector (or local linear quantitative) relationships, where these properties are also assumed for force-fields, where the force-fields get represented as differential-forms.

Either stable shapes or sets of harmonic functions

However, the underlying stable "spectral" properties of physical systems, can be related to either sets of differential equations, so as to define a set of "harmonic" (oscillatory [periodic]) functions, or sets of stable discrete hyperbolic shapes, defined over a many-dimensional containment space, where each dimensional level (also) possesses a discrete hyperbolic shape, all contained within an 11-dimensional hyperbolic metric-space.

Very stable geometries

The fundamental set of quantitatively consistent, and stable shapes are the circle-spaces, eg discrete hyperbolic shapes, which define a globally solvable structure.

Does stability imply stable geometries as a basis? (Yes! (?))

Because fundamental spectral properties of physical systems are stable, and thus apparently emerge from a controllable description, and since randomness and non-linearity cannot describe these stable spectral patterns, then the discrete hyperbolic shapes must be their basis (or at least they are worth trying), and they best fit into a many-dimensional containment set of coordinates, where each dimensional level defines (or is defined by) a macroscopic, discrete hyperbolic shape, and the relative sizes of stable systems changes between dimensional levels (due to the existence of physical constants).

In such a many-dimensional context a geometric relation between forces and spectra becomes more natural, ie the geometric type (of the "free" materials contained in a dimensional level) is very rigid and can form (or condense) into rigid shapes, analogous to shapes which possess "holes."

The idea of stable spectra is implicit in a many-dimensional context, where each dimensional subspace (as well as all the stable material components) identifies a discrete hyperbolic shape, as well as each dimensional level being associated to a constant multiplicative factor. These discrete hyperbolic shapes define very stable spectral properties.

Are observers aware of the holes which are (naturally) a part of their containment space?

{The observer defines a dimensional level where it is assumed that no holes exist (or it is assumed that there are no metric-space orbital constructs, or that no holes are considered to exist) since the observer is in a lattice, however, the observer cannot detect the lattice, since the topology is open-closed and the (orbital) shape (of the material-containing metric-space) is assumed bounded, but this bounded-ness is also not detected (or is difficult to detect), since the lattice can be considered to expand-out, in relation to larger sized metric-spaces at higher dimensional levels, so that the distinction (of being contained within a higher-dimensional system) depends on detecting the spectral values which are consistent with the higher-dimensional metric-space, but this would be related to either stable material spectra or to stable orbital structures for condensed material, the holes in our metric-space manifest as the stable orbits of the planets}.

Material interactions

In a metric-space, which contains lower-dimension material components, the material components are discrete hyperbolic shapes, and these discrete hyperbolic material components have spatial-separation inter-relationships with other material components, where the structure of separation (within a very rigid geometric construct) is mediated by discrete Euclidean shapes, but this requires a many-dimensional containment set-up (see diagrams).

A finite spectral set as a basis for measurable descriptions

The many-dimensional context has a dimensional bound in hyperbolic space due to the properties of discrete hyperbolic shapes which exist in different dimensions, where this bound was identified by D Coxeter, ie the last discrete hyperbolic shape exists as a 10-dimensional shape (which happens to be an unbounded shape), thus the dimension of the hyperbolic containment set is 11-dimensions.

The last bounded discrete hyperbolic shape is a hyperbolic 5-dimensional shape, which is contained in a hyperbolic 6-dimensional space.

This set of dimensional relationships of discrete hyperbolic shapes allows for the definition of a finite spectral set upon which the stable geometries and spectral sets for the entire set of discrete hyperbolic shapes, which define the spectral-orbital and other geometric structures of material and space which some given 11-dimensional hyperbolic metric-space can possess. The constraint of any description to this set of spectral values is due to a need for the existence of a discrete hyperbolic shapes in such a containing space to be resonant with the spectra contained in the 11-dimensional containment hyperbolic metric-space,

where each different subspace is associated to a discrete hyperbolic shape, so that the collective set of spectral values for these subspace shapes defines a finite spectral set (since there are upper and lower bounds on the sizes of the bounded discrete hyperbolic shapes associated to each dimensional level and to each subspace in the 11-dimensional containing space, so that in the context of discreteness this defines a finite set, based on the bounded spectra which can be associated to, and inclusive of, the first five (hyperbolic) dimensional levels) with which all the material components, and all the metric-spaces must be resonant.

Chapter 1

Discrete derivative operator, and a model for a finite spectral set

The physical systems from nuclei, to general atoms, to molecules, to crystals, to solar systems: are all stable; but there is no valid descriptive context which is sufficiently general and sufficiently precise for anyone to believe the descriptive contexts which are currently used to try to describe these systems' stable properties. The current viewpoint for their descriptive contexts are (briefly) non-linearity and (indefinable) randomness, but the stable order which is observed, cannot be described, in a precise enough manner and in a general enough context, based on (most often linearly modeled, eg "regular" linear quantum physics) randomness, and the quantitatively inconsistent non-linear partial differential equation constructs or ideas (eg particle-physics or general relativity, and all other theories which are built on these two constructs).

On the other hand, the stable definitive properties of these spectral systems suggests both a stable geometric and a subsequent controllable basis for description exists, as a means for a sufficiently precise of a descriptive model, or understanding, (for providing a precise description of) these properties. To understand why stable uniform spectral systems are so common.

Math is valid only in a context of very simple geometries and simple quantitatively consistent constructs.

Namely, linear, invertible, and consistent with geometric measures of the system containing coordinate space for a metric-function which has constant coefficients.

(and where the set structure for invertibility is 1-to-1 and onto, and this set structure results in commutative matrix multiplication (in regard to both domain space coordinates and other measurable vector properties which functions can represent), where this means that the local linear matrix representation (relating local measures of functions) is a diagonal matrix, and this property must exist continuously on the entire coordinate space).

Such a local matrix structure for a partial differential equations often implies that such an equation is solvable, and when solvable then they are also controllable (by controlling boundary and initial conditions).

Only this math structure allows for the stable definitive properties associated to quantum systems with fixed component numbers for the system, eg atomic number and atomic weight etc.

If this is true then one must consider that:
The coordinate shapes of the system containing space is to be based on circles and lines.

This is consistent with the complex-number system which can contain the algebraic solution structure to polynomial equations as well as with the complex-number geometry of complex-analysis.

Furthermore, these shapes are the discrete isometry shapes of "cubical" simplexes for the fundamental domains of both the discrete Euclidean shapes and discrete hyperbolic shapes.

Discrete hyperbolic shapes are stable and have well defined discrete spectral properties.

And furthermore;

I. Derivatives become discrete operators associated to discrete structures of:

   (1) action-at-a-distance models associated to discrete spatial displacements of material components,
   (2) Weyl-angular-transformations between toral components of discrete hyperbolic shapes,
   (3) the definition of physical constants which are constant multiplicative factors defined between dimensional levels.

II. The existence of stable spectral sets in physical systems implies a many-dimensional structure to existence, that is, simply-connected implies potential functions, but non-simply-connected geometry implies (stable) spectral-orbital properties.

III. Quantitative descriptions can be based on finite spectral sets, which determine the spectral-orbital properties of both metric-spaces and material systems.
(ie and quantitative descriptions in regard to operators defined on function spaces do not need to be based on a containing set of quantities which form a continuum).

Dimensional levels, ie the different metric-spaces of different dimension and different signature . . . , [ie signature is related to R(s,t) where s and t are the spatial and temporal subspaces of the s+t=n-dimensional metric-space], . . . , are stable discrete shapes (discrete Euclidean shapes and discrete hyperbolic shapes) of various sizes, where size changes (where size changes can be associated to discrete shapes) occur at the discrete changes of dimensional levels, ie where the size of interacting material systems changes due to multiplication by constants which exist between dimensional levels, eg physical constants: c, h, G, etc.

Material systems are the same shapes as the metric-spaces, but material shapes have a dimension which allows them to be contained in an adjacent one-dimension-higher metric-space, furthermore, the sizes of a metric-space depends on discrete multiplicative factors which can multiply the higher adjacent dimensional levels, so material in 4-space is the size of solar systems and not the size of atoms. This can manifest as the values of physical constants, eg h, G, etc. Thus, one can hypothesize that gravity is 2-dimensional where inertia is related to the size of a 1-loop, and electromagnetism is 3-dimensional, where charge is a 2-dimensional discrete hyperbolic shape, etc.

The stable shapes are the discrete hyperbolic shapes.

Thus, the finite spectral set is defined as the set of discrete hyperbolic shapes which are defined for all the different dimensional subspaces of a hyperbolic 11-dimensional over-all containing space [for these discrete hyperbolic shapes (which model both the metric-spaces and material systems, and which are) associated to the set of discrete hyperbolic shapes (and multiplicative constant factors) associated to this set of subspaces (or independent dimensional levels) of the 11-dimensional hyperbolic space].

Note: The last hyperbolic dimensional-level which has bounded discrete hyperbolic shapes is (and includes) hyperbolic dimension-5.

There are also discrete Weyl-angular-transformations defined between toral components of discrete hyperbolic shapes, and these toral components are (or can be) simultaneously multiplied by a constant, so as to define the shapes of orbital envelopes for stable material and metric-space systems (or geometries), and this discrete structure for angular changes and multiplying by constants a set of particular toral components of discrete hyperbolic shapes, so as to keep metric-invariance for each dimensional level.

Note: The orbital shapes can be related to particular "rectangular" lattices associated to discrete hyperbolic shapes, thus not needing constant multiplicative factors defined between toral components.

Each metric-space (dimension and signature) has a physical property associated to itself, so the shape both defines a type of material, and it also identifies pairs of opposite metric-space states associated to a geometric (spatial or temporal) property of the metric-space, a property (or state) which is related to the particular type of material. Such a set of opposite metric-space states allows for the definition of a spin-rotation between the pairs of opposite metric-space states in a metric-space, where the opposite states are contained within the real and pure-imaginary subsets of the complex coordinates, ie so the coordinates become C(s,t) complex-coordinates leading to unitary fiber groups (as well as spin groups as fiber groups).

Material interactions are mediated by Euclidean space-forms which possess the property of action-at-a-distance, and a set of discrete spatial displacements determined for each discrete time interval defined by the spin-rotation period between the metric-space's opposite metric-space states. The spatial displacements are related to 2-forms defined on the Euclidean toral interaction shapes, and these 2-forms are, in turn, related to the geometry of the fiber group where this geometric relation is used to determine the nature of the spatial displacements of the positions of the interacting material components.

For odd-dimensional levels, there are (there can exist) discrete hyperbolic shapes (bounded or unbounded) whose genus is odd. These discrete hyperbolic shapes, since hyperbolic space contains the property of charge, if the orbital-flows (ie the faces of the cubical simplexes) of this shape are all occupied then this shape would have a charge imbalance and this would lead to oscillation which pushes opposite metric-space states together (ie matter and anti-matter states) and would generate its own energy. This is a model of radioactivity, and when associated to a spectral-memory, which can exist within a maximal torus of the system's (unitary) fiber group, this can determine a simple model of life controlled by a coherent, relatively stable, energy-generating, geometric structure which possesses a memory.

There are infinite-extent discrete hyperbolic shapes, which can exist at all dimensional levels, in which discrete hyperbolic shapes are defined.

Discrete hyperbolic shapes can be ten-hyperbolic-dimensions, so the natural containing space is an 11-dimensional hyperbolic metric-space.

Discrete hyperbolic shapes can be bounded shapes up to and including five-hyperbolic-dimensions.

That is, in the context of discrete hyperbolic shapes the infinite extent geometries can be subordinated to the bounded geometries which define a stable, finite spectral set, whose values determine the set of possible patterns for stable shapes and stable properties within an 11-dimensional hyperbolic containing space.

Thus, there can be defined "a finite spectral set" for an 11-dimensional hyperbolic containing space by defining fixed, bounded, discrete hyperbolic shapes for the different subspaces of the same dimension, which are all the various hyperbolic metric-spaces for the 11-dimensional hyperbolic containment set.

Chapter 2

Current problems in math and physics can be resolved using the geometry of circle spaces placed in higher-dimensions

Descriptions of stable math patterns based on a finite quantitative set, where the finite quantitative set represents the spectra which defines all existence vs. Quantization of unstable patterns contained in a continuum [quantization defined in a vaguely defined structure of there being distinguishable features of assumed patterns (which may not exist) and which have no stability, where the properties of these patterns depend on a continuum (a set which is "too big")]

The derivative locally relates function-values to (linear) coordinate values in a linear way. Such a linear relation is the only quantitatively consistent way in which to relate a measured property (ie a function's values) to ruler-like set of measures of the system's containing space (or domain space).

The coordinate space identifies both a metric-space, with locally linearly consistent geometric measures, (ie the coordinate functions [coordinate curves] need a local linear structure when they are differentiated) and a continuum (based on measures of length), where the continuum's properties (measures of nearness) can now be identified with the rational numbers. That is, general real numbers are excluded though particular real-numbers, such as "pi," may be used.

Only linear relations between measuring sets for the description of a system contained in a metric-space of constant curvature (really non-positive constant curvature for metric-functions whose coefficients are only constants) whose (local linear) matrix representations (between the two sets of measured values (function-values and coordinate-values)) are commutative in the global coordinate space. But this means using, for the coordinate curves, only the lines and the circles (as the quantitative sets for measuring). The commutative local linear matrices (so that the matrices are commutative all along the global coordinate shapes) is called a parallelizable shapes which is orthogonal, or can be called separable (or geometrically separable). Where such a math pattern is needed to solve a linear partial differential equation because these math patterns are quantitatively consistent (and logically consistent) and they identify actual stable (math) patterns (or stable physical systems) and such linear partial differential equations can be solved (or if they can be solved) and (subsequently) controlled.

The stable spectral-orbital properties of: general nuclei, general atoms, molecules, crystals, (Note: BCS failed when its critical temperature was exceeded by high-temperature super-conductivity) and the solar system . . . . , {as well as mysteries such as dark matter, dark energy, etc. which presently have no valid descriptive context (though peer review acts as if there is already a valid description which only dogmatic authorities can uncover)} . . . . , but the stability of the spectral-orbital properties, of both quantum systems as well as the solar system, implies a stable geometric and mathematically controllable context.

The stable shapes, or the quantitatively consistent shapes, of coordinates are circles and lines. (proof by induction)

This can be thought of as the content of the complex numbers solving polynomial equations as well as the circular and disc geometries so prevalent in complex geometry (or complex analysis).

It is not clear that the Riemann-sphere has that much value in regard to identifying stable measurable and practically useable math patterns since spheres are non-linear and when deformed they are very unstable.

The "circle spaces" are related to the discrete Euclidean and hyperbolic shapes which form from fundamental domains which are based on the cube (or rectangle). These discrete shapes are based in the classical Lie groups associated to the real metric-spaces of R(s,t) where s is the dimension of the spatial subspace and t is the dimension of the temporal subspace of an [s+t=n] n-dimensional metric-space, and the complex-coordinates, C(s,t), where each dimension of C(s,t), is represented as a complex-number plane.

Such a coordinate space composed of circles can be placed in a many-dimensional context, such that the toral components of a discrete hyperbolic shape can be given definite angular relation to one another, just as the Weyl-subgroup changes angular relationships between the maximal tori, which together fill (or cover) the entire Lie group.

When one moves . . . . both between dimensional levels and when toral components have the angular relations changed within a discrete hyperbolic shape . . . . there is also a discrete multiplication by a constant value (of the dimensional level or the toral component) affecting both size of material interactions within a dimensional level and the orbital geometry (eg forming concentric circular envelopes of orbital stability) of a discrete hyperbolic shape, where discrete hyperbolic shapes model both metric-spaces and material systems where a material shape fits into an adjacent higher-dimensional metric-space.

Furthermore, the discrete spatial displacements of the distinguished points of interacting material systems, which have the structures of discrete hyperbolic shapes, is part of the material-interaction process (along with discrete multiplication by constants and discrete angular relations between toral components) also (or all together) implies that the derivative is a discrete operator, not an operator which defines a continuum or an operator which depends for its existence on a continuum, or an operator specifically associated to function spaces, (though the derivative acts on functions and its action defines a new function).

The structures with which the derivative mostly deals, are associated to the idea of measuring, and measuring takes place between the (mostly-independent) dimensional levels of existence.

When placed in a many-dimensional context, up to hyperbolic dimension-11, where bounded hyperbolic space-forms end at dimension-5 (there are 5-dimensional discrete hyperbolic shapes which are bounded, but there are not 6-dimensional discrete hyperbolic shapes which are bounded). Then such an 11-dimensional hyperbolic metric-space, as an over-all containing space for existence, would allow for a finite set for spectral-orbital values for the bounded discrete hyperbolic shapes which compose such a math construct, so that, this finite spectral-orbital set determine the sizes, geometries, and spectra of the stable material systems (or equivalently hyperbolic metric-spaces), which are contained within such a containing space, where the geometric-spectral sets, would be controlled by the number of subspaces of any particular dimension (from hyperbolic-dimensions, 1, 2, 3, 4, 5) contained in the 11-dimensional hyperbolic space.

What is a derivative?
Can it only be defined within a continuum, or is it best defined within a continuum?
or
Is a derivative best defined in a context of discrete math constructs?

Stability and quantitative consistency

Quantities are about identifying a uniform unit of measuring and counting, thus identifying the arithmetic operation of addition (and subtraction). In this construct one needs consistency and stability, so that one can measure in a reliable manner.

The other arithmetic operation of multiplication (and division) is about changing size-scales in a consistent and reliable manner, and this is done by multiplying each member of a quantitative set by the same multiplicative constant. Furthermore, multiplication can also change number-type, eg length x width = area, where area is a different number types than is length.

In experience, measuring is a function-value representing a measurable property of a system contained within a system-containing space, so that the function's values must be related to the coordinate values of the system containing space, ie the domain space, (with the exception of being multiplied by a constant factor).

To measure a function there needs to be a local linear relation between the function-values and its domain space, where a function's value changes by means of a local linear relation to the domain values (does this require a continuum?).

Since it is believed that the measured values (ie the values of the function) exist, then such a local linear relation should be solvable, or invertible, and this requires a one-to-one and onto set structure between the set of function values and the set of domain (or containing coordinate space) values. The solution would result in the function's formula to be given in the variables of the system containing space (the domain space), and this means that the local linear measuring process is invertible and so the function's values have the same dimension as the domain's coordinates and that the matrix representation of the linear approximation of the function's values be diagonal, ie the matrices representing the vector function's values are commutative, in regard to matrix multiplication (and inversion).

If the function is a 1-form, then this vector structure is satisfied, but if a 1-form is integrated along a curve this results in a 0-form, due to its invertibility, but the 0-form is only a valid function if it is single valued, and this requires that the domain space to not have any holes in its shape.

The descriptions of stable, measurable patterns (ie function's values) are very dependent on the differential-forms defined in the coordinate space.

Geometric measures on the coordinates of a domain space based on metric-functions (measuring length of curves on coordinates) where the metric-functions are symmetric matrices and thus they always have coordinates in which they are diagonal, ie linear and invertible.

A locally linear relation which is invertible everywhere (or its matrix representation is diagonal and its matrix multiplication is commutative [and is one-to-one and onto]), it is continuously diagonal, can always (or usually (?)) be solved.

Thus we have that a linear relation which is invertible globally (everywhere within the domain space) can be solved, and it must be consistent with local geometric measures on the domain space.

This is the type of global shape which coordinates composed of lines and circles can accommodate.

However, when one move from a space which has no holes up to one higher dimension, the new space may have holes in its shape, and thus functions do not exist which are single valued, so this needs to be interpreted in a new manner.

The new, higher-dimension, space with holes in its shape represents a space which has many spectral-orbital shapes associated to itself.

This needs to be done in a quantitatively consistent manner and this requires stable shapes built around the simple shapes of lines (or line segments) and circles.

These are the discrete Euclidean shapes and the discrete hyperbolic shapes, but the discrete Euclidean shapes are not stable in regard to their spectral values, whereas the discrete hyperbolic shapes are very stable in regard to fixed spectral values.

That is, this is the argument (or the interpretation of patterns) which supports the idea that existence must be many-dimensional since the world is full of stable spectral-orbital structures. Spectral sets imply discrete isometry shapes, in particular, stable spectra require (or imply) stable discrete hyperbolic shapes, and it also requires that existence have a higher-dimensional structure. But materialism, and material interactions, as well as the values of physical constants can work together to hide the appearance of the higher-dimensions from our perception, yet dark matter can be interpreted to mean that material systems the size of the solar-system form the natural-size for material systems in 4-dimensional hyperbolic space.

Higher-dimensions

In a many-dimensional construct, based on the different dimensional levels being discrete hyperbolic shapes, there is the question of there being a free system versus a system only being determined by an orbital structure defined around holes in the space's shapes.

There is the idea that during interactions one moves from a system being free, to a component which is (becomes a) part of a new system which is a spectral-orbital structure in a higher-dimensional level.

Material interactions are about a process which involve different dimensional levels. So that there are no longer simply (n-1)-forms related by inversion to n-shapes (which are dual to n-forms). One can now identify a spectral-orbital n-set (n-shape) in an (n+1)-domain space (for an n-boundary defined on an (n+1)-space which has a shape, eg cubical-shapes (or simplexes) as fundamental domains).

Thus there can be both interactions between free components, and there can also be highly ordered and very stable spectral-orbital shapes, which result from changing the dimensional context of either the material components or the "component containing" metric-space. But when one changes between dimensional levels then there can also be changes in sizes of the material components, which are contained within a (the) metric-space "of any particular dimensional level," and this is a result of multiplying by fixed constant factors, which we call physical constants.

(1) Spectra imply dimensional levels
(2) Stability and quantitative consistency requires that the shapes for both material components or metric-spaces (at a particular dimensional level) by discrete hyperbolic shapes

There are infinite-extent discrete hyperbolic shapes, which can exist at all dimensional levels, in which discrete hyperbolic shapes are defined.

Discrete hyperbolic shapes can be ten-hyperbolic-dimensions, so the natural containing space is an 11-dimensional hyperbolic metric-space.

Discrete hyperbolic shapes can be bounded shapes up to and including five-hyperbolic-dimensions.

Thus, there can be defined "a finite spectral set" for an 11-dimensional hyperbolic containing space by defining fixed, bounded, discrete hyperbolic shapes for the different subspaces of the same dimension, which are all the various hyperbolic metric-spaces for the 11-dimensional hyperbolic containment set.

That is, in the context of discrete hyperbolic shapes the infinite extent geometries can be subordinated to the bounded geometries which define a stable, finite spectral set, whose values determine the set of possible patterns for stable shapes and stable properties within an 11-dimensional hyperbolic containing space.

Thus there are:

11 chose 1; 1-dimensional discrete hyperbolic shapes for the 1-dimensional hyperbolic metric-spaces

11 chose 2; 2-dimensional discrete hyperbolic shapes for the 2-dimensional hyperbolic metric-spaces
But the 1-spectra of the 2-shapes can be added to the spectra of the 1-spectra,

Etc to

11 chose 5; 5-dimensional discrete hyperbolic shapes for the 5-dimensional hyperbolic metric-spaces
But the 4-spectra of the 5-shapes can be added to the spectra of the 4-spectra, and the 3-spectra of the 5-shapes can be added to the 3-spectra set, etc . . . .

In this context a bounded discrete hyperbolic shape, which is a component, of dimension 5 or less, which is contained in this 11-dimensional containing space, must be in resonance with the spectral set of the same dimension, which is defined above.

Chapter 3

The derivative as a discrete operator

In order to get to an interesting, and useful, math context many of the math structures, which are part of the new context, must be simplified so as to quickly see how the different math patterns fit-together so that the global context of the descriptive construct becomes evident (so the basic idea of the descriptive construct can be seen).

The new descriptive context (the new math construct) is based on a set of bounded and unbounded spectral-orbital-shapes (discrete hyperbolic shapes) forming a set of dimensional layers (like an onion) [where, for example, a lower-dimensional unbounded orbital-shape may be bounded by a higher-dimensional (bounded) orbital shape, within which the lower dimensional shapes is contained], where all the stable shapes, either free or part of a dimensional-level construct (or dimensional-level shape), are based on the very stable discrete hyperbolic shapes.

There can be constant multiplicative factors between dimensional levels, ie physical constants (h, G, etc), as well as between toral components of the same dimension orbital shape, as well as there being discretely defined fixed angles which are formed in a fixed manner between toral components of a same dimension orbital shape (modeled as Weyl-transformations).

The lower-dimension (hyperbolic) orbital-shapes fit into higher-dimension hyperbolic orbital-shapes, in various ways.

The free orbital shapes, of the same dimension, interact with one another, and this interaction process is mediated by Euclidean tori.

All of these discrete actions . . . . of:

1. multiplication by a constant

2. angular changes by discrete angles and
3. discrete spatial displacements due to material interactions . . . . , are mediated by derivatives and action-at-a-distance toral components, which also mediate such changes in orbital shapes.

This is fundamentally a statement about (or contrasting) different math constructs;

Namely, are both materialism, and the continuum (which is believed to be needed to develop the idea of a derivative), (and algebra,) and convergences . . . . , which are all needed to develop the ideas of both

1. indefinable randomness and
2. non-linearity,

. . . . . , to be the main constructs (or the main vehicle [construct and context]) of accurate, and practically useful precise measurable descriptions?

or

Is a finite spectral set associated to (discrete) stable geometric and stable quantitative structures in a (macroscopic) many-dimensional containment space, ie material and metric-spaces are essentially the same type of a construct, to be the main vehicle of accurate, and practically useful precise measurable descriptions of the stable definitive spectral-orbital properties of many very fundamental physical systems which exist at all size scales [nuclei to solar-systems]?

It is a new context which affects the structure and interpretations of both operators and functions.

It should be taken seriously because the (current, 2012) professional mathematicians and scientists are not using math constructs which are relevant to either accurate descriptions of general systems or [do their math constructs have any relevance] to practical useful developments.

Today (2012), practical development is all about developing the ideas of 19th century classical physics. Eg electronics, thermodynamics, and the model that "rates of reactions" depend on "probabilities of component" collisions and the model of probabilities of component collisions has a very limited relation to controlling the properties of physical systems, yet there are many fundamental physical systems with very stable and definitive spectral-orbital properties, where such stability implies a geometric and controllable context for these stable systems.

A new (discrete) geometric context for: math, quantum physics, and the solar system

The derivative can be developed as a discrete operator, in contrast to the derivative as an operator which both depends on and defines a containing space which is a continuum.

The list of new properties of the new construct {the new properties are listed second}:

1. Materialism vs. many-dimensional stable-shapes (which are macroscopic shapes, but still hidden),
2. Continuum vs. discreteness (based on a finite spectral set),
3. Quantitative inconsistency, and logical inconsistency, brought about by non-linearity and "vaguely defined ideas about randomness" (ie indefinable randomness) as being the central math constructs to be used for describing the observed, measurable properties of the world, (to be used to quantize the world, in an arbitrary manner) vs. Stability and consistency and reliable measuring contexts.

The derivative is an operator (one of many such operators) which acts on functions (as opposed to algebraic operators which act on numbers), ie and it was invented because there is a need to turn curved shapes of the graphs of functions and coordinates into local linear (constant) quantities which define local vectors [as well as transformations on vectors] in order to use algebraic constructs with which to apply to local quantitative descriptions, in regard to the derivative and integral operators, [to relate measurable properties defined within a continuum to algebraic structures which locally deal with (many) numbers in a discrete manner]

However, in order for a derivative to act as a discrete operator on functions then it is best if the functions, themselves, to also become discrete shapes.

This allows the all-important property of communtativity . . . . , (a necessary condition for solvable stable controllable systems, and this is the type of information which is useful at a practical level, one might say that they are the only real patterns whose measurable properties math constructs can describe) . . . . , to be directly related to the properties of the functions of a function space.

For example, so that the operator of "multiplying by functions" does not disturb the commutative property of function-operators, as much as it now does (since the operator functions will possess simple geometries whose "geometric fit" into a set of functions can be better determined.

That is, for functions (which are now allowed to be too general functions), the commutativity of the set of operators, ie a commutative algebraic structure, would be easier to discern and realize if the functions were also all discrete shapes (eg best if they are discrete Euclidean shapes and discrete hyperbolic shapes).

As a discrete operator it (the derivative) may also introduce multiplication by a constant, in a discrete manner, and still be consistent with metric-invariance (where metric-invariance is a property associated to a continuum, but a "continuum" whose lower and upper bounds always exist, as a result of a fundamentally discrete math construct), where metric-invariance on a "continuum" is one of the stable contexts assumed for the new math constructs.

As a discrete operator

In the new math constructs, there are stable metric-spaces and stable material systems, which are both discrete hyperbolic shapes, where the dimension of the material and the dimension of its containing metric-space differ "in dimension" by at least one-dimension (ie differ . . . . by a dimension of at least 1-dimension).

The derivative of a material interaction defines an interaction in a dimension above the dimension of the metric-space within which the material interaction is taking place.

Material shapes which are interacting depend on being connected to one another by a discrete Euclidean shape, that is, a discrete Euclidean shape (of a dimension one-higher than the dimensions of the interacting material shapes) links (at least) a pair of the closest interacting material systems together . . . . , {Note: However, the toral shape which links the interacting material components exists in a Euclidean space which is one-dimension higher than the dimension of the metric-space which contains the interacting material components.} . . . . , to cause a discrete (spatial) displacement for each discrete time-interval, where the discrete time-interval is defined by a (the) period of a spin-rotation between metric-space states (where the metric-space states are defined within the containing metric-space).

But

**

It (the new derivative) can also define a discrete angular transformations, ie Weyl-transformations, between toral components of a stable discrete hyperbolic shape.

Note: This is an operator which can either be identified with the toral components of a geometry in the base space, or it can be related to a finite sequence of group conjugations between maximal tori in the fiber group, where the distinguishing feature of the (a) math construct (eg needed to define a sequence of group conjugations) becomes a finite spectral set, carried by the various discrete shapes which are central to the stability and consistency of the (totality of this) descriptive construct.

That is, the discrete angular transformations, between toral components of a discrete hyperbolic shape may be about "moving the particular toral component" to a "new subspace" of the same dimension, so that the discrete multiplication by a constant (during angular transformations) can remain in a metric-invariant context, where in this viewpoint, it is the adherence to a fixed finite spectral set which defines a coherent descriptive construct.

[In fact, one can hypothesize the quantitative inconsistency associated to the bi-furcations of value within a non-linear context is to be attributed to a (non-linear) system which is (a system which is) without a supporting a maximal torus (ie support, by means of resonances), within the fiber group, (or a set of maximal tori) [or belongs to a particular set of maximal tori] where the set of maximal tori are identified by a finite spectral set.]

These discrete transformation structures which are a part of the differentiating process can also be accompanied by discretely multiplying by a constant. This is an intrinsic property of the interaction structure, where these constant multiplicative factors enter as physical constants, where an example of this process would be the, so called, "radial" measure, r, of the toral interaction-shape, keeps changing (due to the spatial displacements of the interacting materials) with each time interval of the interaction. (now interpreted to be "a new set of constants" associated to an action-at-a-distance Euclidean (toral) geometry defined for each uniform discrete time-interval)

Do the opposite time states diverge (when mixed in unitary rotations)?

It depends on if the system is stable, where stable discrete orbital-spectral systems possess opposite orbital states, since material systems are also metric-spaces which possess within themselves opposite metric-space states.

However, most systems, ie most differential equations of systems, are non-linear differential equations, and thus, they are unstable in their own real-coordinate structure, and thus they have no stable opposite-time state structures associated to themselves.

The derivative is one of many operators which can act on functions

However, in classical physics the derivative is a local, linear, measure which relates function values (of the function's graph) to the function's domain values (or set of domain coordinates), in a quantitatively consistent (local) linear relationship, where physical law is the claim about the equality of force (local vector properties of a force-field based on the geometry of material distribution in space) to a mass's changes in motion, ie physical law defines a differential equation, but the solution function to the system's differential equation only gives accurate useful information if the differential equation can be solved, in which case it gives both global information and a means by which to control the system's properties, but this only happens if the differential equations of classical physics are linear, metric-invariant, and separable (parallelizable and orthogonal local geometric (or coordinate) structure, and consistent with geometric measures), where linear means quantitatively consistent, separable means invertible and commutative, and metric-invariant means the quantities are consistent with geometric measures.

For a quantum system (or a probabilistic-spectral system) sets of operators (including the derivative operator) act on a quantum system's function space (where the function space models the system's random probability properties) in order to diagonalize the function space, or to find the function space's set of spectral-functions for the quantum system, and in doing this "one is also supposed to find the spectral set of a quantum systems observed spectral values," but this has not been true. This process (of diagonalizing a function space by a set of commutative linear operators) is about finding sets of commutative operators which represent the quantum system's physical (observable) properties, but it does not work for probability waves, which (are supposed to) also (simultaneously) carry the system's spectral properties, ie the spectra of the function space are not necessarily the spectra of the observed physical quantum system.

This function space model of (assumed to be random) quantum systems model is based on randomness, and it is used to describe the properties of a system with only a few components, so it is not useful for "system control," yet these systems are stable, definitive, and discrete which implies they emerge from a controlled, linear, solvable context. However, (usually) all the charged components are associated to 1/r singularities, so, for such (general) quantum systems "the function space cannot be diagonalized" by a set of commuting operators, and so the descriptive context does not supply accurate information about general quantum systems. It is claimed to work for one system, the H-atom, but the series solution to the radial equation for the H-atom, in fact, diverges, and is arbitrarily truncated to fit data. Note: Math cannot deal in a consistent manner with, 1/r, singularities (proof by induction).

However, these function space techniques sometimes work if the wave has physical properties and the wave-equation is a hyperbolic equation (wave-equation), and if the differential equation is linear, separable, and metric-invariant.

That is one needs linear, metric-invariance, and "geometric separability" in order to solve differential equations of physical systems, so as to provide information about a system, with physically measurable properties, which is accurate and practically useful. Furthermore, one needs geometry to have a description which is controllable.

Note: The functions which compose a function space need to be simple shapes which allows their local geometric properties to be relatable to the property of commutativity, they need to be shapes related to cubical-simplexes in a metric-space with non-positive constant curvature where the metric-function has constant coefficients and be associated to the simple classical Lie groups SO, SU, spin, and possibly Sp.

Particle-physics, supposedly, adjusts non-existent wave-functions by a linear-geometric-random-non-linear abstract, incomprehensible structure, which, in turn, needs to be renormalized. It is a descriptive structure which only relates to particle-collision information concerning particle-collision experiments and this information seems to be irrelevant in regard to providing useful accurate information about fundamental quantum systems.

The real evidence is (or suggests) that particle-collisions are rare within the nucleus, other-wise all heavy nuclei would be radio-active.

This particle-physics model provides no valid models of the stable spectral properties of: nuclei, general atoms, molecules, or crystals (spectral-orbital properties).

It is a descriptive structure which only provides a description of a handful of general quantum systems wherein the model is still not valid, eg the solution to the radial equation of the H-atom diverges, but is truncated based on no math pattern, but the truncated result does fit-data.

Why should data about particle-collision experiments, which fits into a unitary pattern, but which has no valid interpretation, be the basis for physical description, (or for all of physics)?

There are infinite-extent discrete hyperbolic shapes, which can exist at all dimensional levels, in which discrete hyperbolic shapes are defined.

Discrete hyperbolic shapes can be ten-hyperbolic-dimensions, so the natural containing space is an 11-dimensional hyperbolic metric-space.

Discrete hyperbolic shapes can be bounded shapes up to and including five-hyperbolic-dimensions.

That is, in the context of discrete hyperbolic shapes the infinite extent geometries can be subordinated to the bounded geometries which define a stable, finite spectral set, whose values determine the set of possible patterns for stable shapes and stable properties within an 11-dimensional hyperbolic containing space.

Thus, there can be defined "a finite spectral set" for an 11-dimensional hyperbolic containing space by defining fixed, bounded, discrete hyperbolic shapes for the different subspaces of the same dimension, which are all the various hyperbolic metric-spaces for the 11-dimensional hyperbolic containment set.

Chapter 4

Measurable description of existence

To be useful a description needs to be based on geometry, so that the description is useful for practical creativity.

To describe a physical system which has stable definitive spectral-orbital properties, one also, needs to use a geometric basis which is simple (linear, metric-invariant, "geometrically separable", which allow the vector linear matrix transformation (of the derivative's [or the partial differential equation's] intended purpose) to be diagonal, and it is diagonal continuously throughout the coordinate domain space of the solution function, to the linear partial differential equation).

This also allows for quantitative consistency, logical consistency, clarity, and comprehensibility (understandability), ie so the description is (can be) accurate (to a sufficient level of general precision) and useful.

If descriptions are neither accurate nor practically useful, then certain basic aspects of the descriptive language need to be changed, ie the assumptions, contexts, interpretations, set containment, and organization of language etc,

Note: This is the correct conclusion of Godel's incompleteness theorem. (see Appendix [this chapter])

That is, one needs a description which is:

1. Linear (quantitative consistent),
2. Metric-invariant (geometric measurements and other measured quantities are consistent), and
3. Separable, where this is about non-positive spaces of constant curvature, wherein the metric-functions (symmetric 2-tensor) have constant coefficients, where this is basically

about both the "discrete Euclidean shapes" and the "discrete hyperbolic shapes," though other signatures are (can be) important.

Note: The "discrete Euclidean shapes" and the "discrete hyperbolic shapes," can be called both discrete isometry shapes, as well as space-forms (though the phrases, the "discrete Euclidean shapes" and the "discrete hyperbolic shapes," are more specific).

The contents will be:

I. One has principle fiber bundles with isometry fiber groups, with base-spaces which are metric-spaces.
II. Metric-invariant metric-spaces have physical properties associated to themselves.
III. Real hyperbolic metric-spaces have stable shapes, and material is also a (stable metric-space) shape which is contained in a higher-dimensional metric-space (which also has a shape).
IV. The derivative (or the differential equation) as a discrete operator [material interactions and distinct and independent dimensional levels].
V. Some of the odd-dimensional discrete hyperbolic shapes which have an odd-genus (ie have an odd-number of holes defined on their shapes), which naturally oscillate, and subsequently, generate their own energy, this defines an elementary model of life (see below).

Shapes of both metric-spaces and material components contained in metric-spaces (where the dimension of the metric-space is at least one-dimension greater than the dimension of the material component)

I. ne has principle fiber bundles with isometry fiber groups, with base-spaces which are metric-spaces. This can be interpreted to imply the shapes of the discrete isometry subgroups, and in particular, the stable discrete hyperbolic shapes (or hyperbolic metric-spaces) are determining the base-spaces of the principle fiber bundles, for the different dimensional levels (of a high-dimensional containing space, ie hyperbolic dimension-11) . . . , (but the non-linear spherical geometries are not used in this new math construct) . . . . , where, the discrete Euclidean shapes and the discrete hyperbolic shapes, are related to discrete isometry subgroups, where the fundamental domains of the lattices [of the discrete isometry subgroups' base-space metric-spaces] can be thought of as being composed of "cubical" (or rectangular) simplexes.

(Note: The cube is a direct property of the discrete Euclidean shapes, but the discrete hyperbolic shapes need "the vertices where the cubes meet" [of their fundamental domains] to become separated.)

Yet, in both cases "the local directions of the natural coordinates" of these "discrete Euclidean shapes" and the "discrete hyperbolic shapes" "are always independent (or orthogonal)" [as one continuously traverses the global shapes], a geometric condition referred to as parallelizable shape with local orthogonal coordinate directions, defined in a continuous manner on the entire shape. This defines local matrices which would be commutative, and if linear this allows for inversion and solvability.

(This can be called a "geometrically separable" property, ie in reference to "solvable, linear differential equations, which are called separable.")

The discrete hyperbolic shapes are stable and can carry on their rigid shapes very stable, and discrete, and definitive spectral properties.

Whereas, the sizes and "rectangular" shapes of the fundamental domains of the discrete Euclidean shapes are continuous in regard to their allowable "rectangular" fundamental domains, yet they can carry on their shapes, many values for spectra in a stable manner, where this is done as resonances of an outside stable spectral source, eg resonating with the stable spectra of the discrete hyperbolic shapes. [This can (also) be related to spectra resonating on maximal tori in the fiber group.]

The discrete Euclidean shapes and the discrete hyperbolic shapes are both linear, metric-invariant, and (geometrically) separable.

The local measures of the spectra of these discrete shapes are (can be thought of as) the solution 2-forms of linear, (second-order) metric-invariant, and separable differential equations for shapes, which have their geometric-periods identified by their ("cubical") fundamental domains (L Eiesenhart). Geometrically measurable properties, of a space-form's spectra, surround the "holes" which are a part of the discrete Euclidean shapes and the discrete hyperbolic shapes.

{Note: An n-space-form is contained in an (n+1)-metric-space, and the n-space-form's spectra are the (n-1)-faces of the n-space-form's "cubical" fundamental domain, so (by duality) the local spectral measures are 2-forms.}

[? . . . . such a solution of differential equations, applied to discrete isometry shapes, are sometimes referred to as symmetric-spaces {in a moding-out (or an equivalence topological) math process} . . . . ? They are also called space-forms (already noted above).]

Note: Spherical geometries are spaces of positive constant-curvature, but their metric-functions have variable coefficients, so they are non-linear shapes, and thus {other than for "the one-body system with spherical symmetry"} they are unstable geometric structures.

Note: Spheres work well as bounding geometric shapes.

The sphere is prevalent in Euclidean 3-space, and this is because of the [new] relation of the SO(3) fiber group to a 3-sphere, and the 3-sphere's subsequent relation to the (new) geometry of material interactions which leads to a spherically symmetric geometry for material interactions in Euclidean 3-space {the space of inertial changes} (see below).

(In physical description)

II. Metric-invariant metric-spaces have physical properties associated to themselves. (this may be a necessary math property of metric-spaces which possess within their natural structures stable discrete isometry shapes or discrete isometry subgroups, which compose a dimensional ladder where metric-spaces become models of (stable) material components on the next rung on the ladder (or within the adjacent higher-dimensional metric-spaces within which the (lower-dimension) discrete isometry shapes are contained).)

This is the idea of (behind) E Noether's displacement symmetries.

Spatial displacements are associated to the property of inertia and are contained in Euclidean space. Furthermore, Euclidean space has the property of action-at-a-distance (or non-locality) associated to itself [Aspect's experiment].

Spatial displacements are of two types, translational and rotational, and they correspond to two distinct frames; the frame of the fixed stars, and/or the frame of the rotating stars; and this defines two, opposite, metric-space states (and this also defines matter and anti-matter, the positive and negative energy states of mass, and the advanced and retarded potential wave-functions).

Temporal-displacements identify both charge and energy, and the pair of opposite "positive-time or negative-time" metric-space states are defined on a hyperbolic metric-space, where hyperbolic space is equivalent to space-time.

Thus, metric-spaces have associated to themselves pairs of opposite metric-space states (needed for the material interaction process [see below]) and these opposite metric-space states can be contained in the R and iR subsets of complex-coordinate spaces.

Thus the fiber groups would be unitary, where the unitary operators rotate between the two opposite metric-space states, where at each point in the complex-coordinate space there are always pairs of opposite directions (of either time or mass) associated to a dynamic process.

Thus, there are the R(s,t) real metric-spaces where s is the dimension of the spatial subspace and t is the dimension of the time subspace for (s+t)=n, for n-dimensional real metric-spaces and associated different signature metric-functions and, in turn, their being associated to SO(s,t) fiber groups, or there are complex-coordinate spaces, C(s,t), where each dimension is represented as a complex number, and there are the SU(s,t) fiber groups.

The spin-rotation of metric-space states, expressed in a unitary context (or as a simply connected covering group of the isometry fiber groups), can now be related to either metric-spaces or to discrete material shapes (of either the discrete Euclidean shapes or the discrete hyperbolic shapes). Thus Fermions are spin-rotations of metric-space states defined on material systems, while Bosons are associated to material component properties where the components are a part of a metric-space.

Nonetheless, the spin-rotations of metric-space states on a discrete hyperbolic shape (either material or a metric-space) identify a spin-rotation period (or time interval) which can be related to discrete changes of positions or discrete changes of action-at-a-distance "discrete Euclidean interaction" shapes.

There are no discrete hyperbolic shapes after hyperbolic dimension-10 (ie space-time dimension-11) and all discrete hyperbolic shapes of dimension-6 or higher are unbounded shapes (but with finite volume).

The shapes of hyperbolic metric-spaces allows for hyperbolic metric-spaces to possess on themselves stable spectral properties if the spatial-dimension of the metric-space is five or less, and thus can be associated to a bounded discrete hyperbolic shapes. Assume that the spectra of threw bounded discrete hyperbolic shapes of hyperbolic-dimension of five or less define a finite spectral set for the entire high-dimension containing space (hyperbolic 11-dimensional space). Thus, hyperbolic metric-spaces either as material or as a material-containing metric-space possess stable spectral properties.

Note: There are unbounded discrete hyperbolic shapes which are defined at all dimensional levels in which discrete hyperbolic shapes are defined. Unbounded discrete hyperbolic shapes are good models for light. This can also be related to the idea that the mass of a 2-dimensional discrete hyperbolic shape (contained in hyperbolic 3-space) is proportional to its sectional curvature, ie $m=k/r$.

These are properties of discrete hyperbolic shapes identified by D Coxeter, and summarized by J Humphreys.

III. Real hyperbolic metric-spaces have stable shapes, and material is also a (stable metric-space) shape which is contained in a higher-dimensional metric-space (which also has a shape).

One can assume, that up to (and including) hyperbolic dimension-5, the discrete hyperbolic shapes of the hyperbolic metric-spaces are bounded (though this assumption is not necessary, but it seems to simplify).}

Thus, for the finite number of discrete hyperbolic shapes for all the various dimensions of metric-space which possess shapes, up to hyperbolic-dimension-10, for a containing hyperbolic metric-space of dimension-11, there is a finite spectral set, assuming a bounded, uniform shape for the fundamental domains of each dimensional level, as well as allowing a constant multiplicative constant to exist between dimensional levels, or between different subspaces of the same-dimension.

The discrete hyperbolic shapes imply stable, finite spectra for both material and metric-spaces in hyperbolic space.

This allows for a descriptive math construct to be based on a finite spectral set, which is the spectral set determined by the stable spectra associated to all of the bounded stable discrete hyperbolic shapes which model (some of) the hyperbolic metric-spaces {which form the (bounded) hyperbolic metric-spaces in the first five dimensional levels of the over-all high-dimension hyperbolic metric-spaces}.

This is the math property (III.) in which math people should be quite interested.

The quantitative structure of physical description does not depend on a continuum, but rather on a finite spectral set about which the properties of all "material" can be defined, with the new definition of material, ie material is a (bounded) discrete hyperbolic shape contained in a high-dimensional hyperbolic metric-space, which in turn, divides into many materially-independent dimensional-levels.

IV. The derivative (or the differential equation) as a discrete operator [material interactions and distinct and independent dimensional levels].

Basically, the differential equation is still not understood

Is it:

1. A local (linear) measure defined in a geometric context of existence?
2. An operator on function spaces?
3. An operator which introduces discrete patterns into a geometric context. [In which case, the functions in function spaces should themselves, be simple, discrete, isometric shapes.]

In the new discrete context;

One still uses Newton's law of inertia, F=ma, where F is related to differential 2-forms associated to the geometric distribution of a physical system's material which act on material components* . . . . ,

(ie other discrete hyperbolic shapes which compose either the material system, or the metric-space),

* where the components can either be stable orbits on (within) the discrete hyperbolic shape, or other closed orbits (defined on discrete Euclidean shapes), or "free," closed and bounded, discrete hyperbolic shapes (of lower dimension) contained within the discrete hyperbolic shape.

. . . , (material components) identified in a metric-invariant and linear context where quantitative consistency requires both metric-invariance and separable coordinates for the system.

Thus this is the context of the three second-order, metric-invariant, linear differential equations of:

1. Elliptic

   (the orbits of a stable space-form, where angular Weyl-transformations determine concentric orbital structures, or envelopes of orbital stability, ie the angular changes between toral components determining the shapes of discrete hyperbolic shapes which model material systems [at any dimensional level])
2. Parabolic

   (essentially angular momentum, and it is applied at the dimensional level of the interaction discrete Euclidean shape, this is the about shapes associated to semi-free orbits, which can form a path between stable shapes), and
3. Hyperbolic

   (collisions, physical-waves, and [the new] emerging stable discrete hyperbolic shapes associated to resonances between the newly emerging system and the high-dimensional containing space's finite spectral set, this involves interactions between dimensional levels, and the multiplication by constants [changing the sizes of material space-forms] in adjacent dimensional levels) . . . , types of differential equations.

2-forms of the toral interaction shapes and the geometry of fiber groups is related to the geometry of material interactions

Conical-sections, metric-invariance (for metric-function with constant coefficients), second-order (separable) partial-differential-equations

(1) elliptic orbits seem to be about envelopes of stability for concentric circular orbital structures (where these concentric circular orbital structures are defined by angularly-distorting a discrete hyperbolic shape by fixed, discrete angular-transformations of the shape's toral components.
(2) sets of orthogonal hyperbolic-shapes (hyperbolas which are orthogonal to one another) defined as coordinates on discrete hyperbolic shapes, seem to require that stable "circular" orbits on discrete hyperbolic shapes be the very limited and stable spectral-orbits of the discrete hyperbolic shapes, so that discrete hyperbolic shapes have definitive spectral properties. [Flows on the hyperbolas of discrete hyperbolic shapes, apparently, are not stable orbits.]
(3) planes of (real-numbers) define the normal axis which represent angular momentum, and these planes (given various angular orientations) can identify "discrete-jumps between orbital levels" which the angular-and-energetic structure of a system (or an interaction geometry) allows.

Within this context of orbital-spectral shapes, internal patterns (angular momentum), and emerging stable systems, there are the shapes of orbits, associated to the Weyl-transformations which determine the angles between the toral components of a discrete hyperbolic shape, and this also requires discrete multiplication of that toral component by a constant so as to define concentric envelopes of orbital stability. The Weyl subgroup of a Lie group is related to the conjugation classes of the maximal tori which cover a Lie group (see T. Brocker, T. tomDieck, Representations of compact Lie groups).

The discrete Weyl-transformations affect:

1. Orbit shapes (eg concentric orbital envelopes of stability. That is, the Bohr orbits and Somerfeld elliptic approximations within envelopes of orbital stability, defined for atomic (or other orbital) spectral-energy levels provide a more truthful (geometric) context for atomic spectral structures.)
   [This can be allowed either by lattice structure, or subspace structure (of the same—dimension) contingent on the existence of a fixed finite spectral set which identifies all existence within a particular 11-dimensional hyperbolic metric-space.]
2. Molecular shapes
3. The various different shapes which the discrete hyperbolic shapes can have,
   And
4. The shapes of angular momentum states within an orbital structure.

The property of discrete multiplication by a constant between dimensional levels determines the relative sizes of the stable material components which are contained within a particular dimensional level.

Material interactions are determined by a process which involves action-at-a-distance, discrete, Euclidean shapes which average the mass structure of the interacting stable discrete hyperbolic shapes (of dimension-n) and then (the discrete Euclidean shape) forms an instantaneous toral geometry (of dimension-(n+1)) which is contained in an (n+2)-dimension (Euclidean) metric-space, where the (n+1)-dimensional toral geometry is related to differential 2-forms in the (n+2)-dimension containing metric-space, where the geometry of the 2-form is of the same dimension as the geometry of the fiber group (of the (n+2)-dimension metric-space), so that the 2-form determines the geometry of the spatial displacement transformations (locally, in two-opposite directions) which the fiber group element (which has been determined by its geometric relation to the 2-forms) applies (or is projected-down) to the distinguished points of the positions of the (interacting) hyperbolic-material-shapes (in their (n+1)-dimension containing hyperbolic metric-space).

That is, Gravity is in 2-dimensions but the description is in Euclidean 3-dimensions, and the 3-sphere symmetry of SO(3).

Electromagnetism is in 3-dimensions but the description is in Euclidean 4-dimensions, ie the force-field F of electromagnetism, where the 4th Euclidean-dimension is associated to time [and the 3-Euclidean-dimensions can still be associated to a 3-sphere in SO(4)=SU(2) x SU(2)=(3-sphere) x (3-sphere).

When discrete hyperbolic material shapes are involved in an interaction which is similar to a collision, and if the energy of the collision is within a "correct range," then at some point (of separation) during the collision (the interacting system of discrete hyperbolic shapes and the discrete Euclidean shape) the discrete Euclidean shape can become a toral component of a new discrete hyperbolic shape which has (might have) resonances with the finite spectral set, where this spectral set is defined within the containing space, and thus a new stable discrete hyperbolic shape is brought into existence, where the spectral set (of importance to the newly emerged stable discrete hyperbolic shape) would depend on the dimension of the material interaction, and would be consistent with the spectral set of the high-dimension containing space, and thus, the spectral set "of that particular dimension" would be related to the number of different subspaces of that dimension, which are contained in the 11-dimensional hyperbolic metric-space. The spectra of these metric-subspaces (of the 11-dimensional hyperbolic metric-space), which are modeled as stable, bounded, discrete hyperbolic shapes, would be important to determining the existence of resonances which can bring about the new stable (hyperbolic) material system (through resonance, during a collision interaction which has the "correct" energy).

Thus the derivative acts discretely, and changes by spatial-displacements the positions (of the two hyperbolic space-form's distinguished points) in Euclidean space for each spin-rotation of metric-space state.

Note: Euclidean space is an absolute frame relative to either the fixed or translating distant stars, or to the distant rotating stars.

The derivative operator is also associated to the discrete Weyl-transformations the angles between certain toral components of a (particular) "discrete hyperbolic shape," so that these Weyl-transformations related to orbital shapes of a discrete hyperbolic shape, and it is also associated to discrete multiplication by constant factors of both toral components (whose angles are transformed) and between dimensional levels.

The higher-dimensions are hidden due to the structure of material interactions, and due to the size of the material components which exist within the different dimensional levels due to the multiplicative constants which exist between adjacent dimensional levels. They are also hidden, since light is modeled to be an infinite-extent hyperbolic space-form defined on a particular dimension subspace of the hyperbolic 11-dimensional over-all containing metric-space.

Other notes:

For material interactions of small components, the interaction structure causes Brownian motions which, in turn, result in the apparent quantum randomness, which is observed.

Point-particle events are seen in experience, since the interactions center around the distinguished (vertex) points of the interacting space-forms.

V. Some of the odd-dimensional discrete hyperbolic shapes which have an odd-genus (ie have an odd-number of holes defined on their shapes), when all of its spectral-flows are occupied by "charged" orbital material components, then such a space-form has an unbalanced charge structure on its shape, and thus it starts oscillating, and subsequently generating its own energy (by pushing together opposite metric-space states defined on the discrete hyperbolic shape). This defines an elementary model of life, and when associated to the capacity of maximal tori of the fiber group to hold spectral properties on themselves (and is part of the fiber structure of the containment space, which is modeled as a principle fiber bundle), this would allow an elementary model of a memory, ie a thinking, energy-generating life-form.

   The odd-dimensional hyperbolic space-forms are hyperbolic dimensions: 3, 5, 7, 9 so as to be contained in hyperbolic metric-spaces of dimensions: 4, 6, 8, 10 respectively so that for each such new living-material systems there is a new time dimension associated to the new type of material so that the respective fiber groups would be SO(4,2), SO(6,3), SO(8,4), SO(10,5).

The human being is a center of great creative potential in this high-dimension context of knowledge and existence, which is both perceivable and creatively manipulate-able by human beings.

Appendix

The measurable properties of stable, definitive, spectral-orbital physical-systems cannot be described using the math constructs of: non-linearity, indefinable randomness, and defining (arbitrary) convergences on (within) a continuum, as well as by using the idea of materialism.

An Alternative set of (math): constructs, contexts, interpretations, containment sets, and quantitative structures are needed.

In other words:

Are physical system's only to be based on non-linearity and indefinable randomness, and thus physical systems are simply too complicated to describe?

or

Does the existence of stability imply that "Godel's incompleteness theorem" should be correctly interpreted, and different types of descriptive language structures (including new assumptions, and new contexts) should be considered.

When one looks out to the geometric and spectral patterns of space and material one must ask in all generality:

"What is the shape and dimension of space?"

"What patterns of geometry, size, 'material,' can be detected (perceived), and at what dimension?"

Chapter 5

A new context within which to apply geometry

A new context within which to apply geometry to: math, quantum physics, and the stable solar system

These are amazingly simple ideas, and simple math constructs, which provide a very interesting new math context, within which to describe new math patterns, as well as describe the wide range of stable spectral-orbital properties of fundamental physical systems.

The content of this paper (or talk) is about: How to organize math patterns so as to fit a simple math construct into a more varied and diverse descriptive context, where the main construct is associated to a fixed and finite spectral-orbital set, so as to be able to calculate the underlying observed stable spectral-orbital properties of fundamental physical systems. The more varied structure is the context of a higher-dimensional over-all containment space (11-dimensional hyperbolic metric-space) where the dimensional levels are mostly geometrically independent of one another, and where material can be re-defined.

The motivation for seeking an alternative math structure

One cannot take, seriously, the currently accepted physical-mathematical constructs about physical systems, whose descriptions are based on indefinable randomness and non-linearity . . . , claimed to be true by professional math and physics people.

The modern authorities (2012) attempt to describe the stable spectral properties of quantum systems based on the idea of randomness, so that these properties are described by a global wave-function (representing probabilities of spectral events) whose properties (structures) are to identify a random (local) point-spectral event, wherein the global wave-function, instantaneously, collapses to that local (observed) event.

Whereas classical physics, on the other hand, provides global properties concerning a component of a system (a solution function) from the local properties of a (geometric) system (the system's differential equation).

That is, the quantum description seems to be about "how to lose information about a stable system," rather than "how to provide information about the system which is useful in regard to a reasonably accurate (and sufficiently precise) description which can be used in a practical context."

Nonetheless, the ideas about indefinable randomness and non-linearity, which are believed by the modern authorities, and which are seriously proclaimed to be true by the physics and math professional communities. Thus the public has been led to believe that truth must be represented in a context of extreme complication. This is the main strategy of the propaganda system, namely, that the ruling class is working with ideas which are too complicated for the general public to understand, it is a strategy which implicitly requires the worshipping of inequality.

However,

The currently accepted physical-mathematical descriptive structures do not describe the observed, very stable, properties of fundamental physical systems (of the world) accurately (and) in a general context, ie in regard to a wide array of general quantum systems which are not now (2012) being described to a valid level of precision, for this descriptive structure to be believable, and then there is the issue, that the currently accepted descriptions (of great complication but with very little, useful, insight) are not practically useful.

Furthermore, the properties of particle-collisions are only relatable to nuclear weapons engineering, and are not applicable to the spectral structure of a general nucleus.

One needs to also consider, the proven property of action-at-a-distance for non-local properties of quantum systems, which are defined in Euclidean space (proven by Aspect's experiment). Thus, the constructs of general relativity also cannot be taken seriously.

That is, Euclidean space exists, and it is an absolute frame, which identifies the positions of components within the absolute coordinate frame, in relation to the motions of the distant stars.

If general relativity could describe the (many-bodied) solar system as a stable system, then one could begin to believe in general relativity, but a few data-fits with a handful of mostly irrelevant details (in regard to their relation to practical creativity) is not sufficient to validate the idea of general relativity.

Social issues affect thought in a surprisingly strong way

The US society based on militarization and the monopolistic business domination over society, where business and the military control society by money and communication (propaganda) and a militarized management structure, which also means controlling the thoughts of science and math professionals, where this is done by funding and propaganda (where propaganda is called education). Thus, if these main social forces of monopolistic businesses are being adequately served by science and math, then professional science and math (which serve the dominant social forces) will not be motivated (by these owners of society) to seek new or alternative ideas.

However,

Two well known (and proven) principles concerning math descriptions:

1. Godel's incompleteness theorem states that "precise language is limited in the patterns which it can describe," is an idea which might be best interpreted to mean that "if one cannot build what one wants with an existing (or the traditional, or the authoritative) measurable language of math, then build a new precise language." and
2. The stable (mathematical) basis for shapes are the discrete hyperbolic shapes, might be the best interpretation of the Thurston-Perelman geometrization.

Furthermore, these are issues about equality and the propaganda-education system, which, in turn, serves a monopolistic militarism which dominates both speech (or language) and creativity for all of society.

Should equal free-inquiry, or should an unequal investment social structure, determine "what a (our) culture creates?"

And

A troubling attribute of the attention, actions, and beliefs of humans:

The needed stability and consistency properties of math constructs seem to be ignored as long as there appear to be measures, or distinguishable states, or there appear to be (fleeting) unstable shapes, which vaguely fit into a math structure based on randomness, but furthermore, if the current math language is related to some practical (but unstable) creative context (or process), eg rates of reactions and/or feedback systems, then stability can be made (by the propaganda-education system) to appear unimportant, and thus many of the fundamental properties of stable math constructs, also, are ignored, [even though it is only linear relations between quantitative sets are quantitatively consistent].

Organizing stable geometry so as to form a useful precise language

How to organize the very stable discrete hyperbolic shapes into a many-dimensional context, to be used to describe the fundamental physical systems which possess the properties of both spectral-orbital diversity and great stability?

The math context of a differential equation

Should differential equations "really" be about discrete relations of material components to their surrounding geometry? and Should material components also be related to a new idea about a containment set, which has spectral properties of its own?

Do not believe the authorities (but, within society, this means one must "accept being a failure")

In a nutshell, the causes that might exist for a person's disbelief in the authority of today's (2012) math and physics professionals

The stable, definitive, discrete properties of fundamental quantum systems imply that these systems are to be described within a stable geometric context.

as well as,

How can the stability of the (many-bodied) solar system be related to some stable geometric construct (as opposed to a non-linear geometric context)?

What current (2012) interpretations of math constructs leads the authorities astray, so that the above interpretive contexts are not considered, especially since their methods are not working?

The derivative is an operator which defines a continuum (But . . . . , Is it really?)

Why are containment sets always modeled as continuum's? [Is it, essentially, because of the assumed continuous context of a derivative?]

Quantitative sets, which model continuums, are defined to be extremely large sets.

This seems to allow for too many general structures to have a quantitative context.

Why are such big quantitative sets needed to model quantitative sets?

Consider the point-at-infinity on the (complex) Riemann-sphere, it should be equivalent to any other point on the Riemann-sphere, (say) z. However, even though 2^n approaches the point-at-infinity, and z + ½^n approaches z, but whereas 2^n is always far from the point-at-infinity, z +

½^n gets very near z, yet neither point is ever reached. Is this a valid model of equivalent-points on the Riemann-sphere?

Furthermore, it is not clear that such a set-structure of an infinite-extent space for quantities . . . , within an (actual) existence wherein size might have limits . . . , is valid.

However, infinite-extent (or unbounded) 2-dimensional discrete-hyperbolic-shapes might be placed within a bounded 3-dimensional discrete-hyperbolic-shape, ie infinity is given a bound in regard to the existence of which it is a part.

Note: Such an unbounded, 2-dimensional discrete-hyperbolic-shape which is contained in a (large but) bounded 3-dimensional discrete-hyperbolic-shape might be a valid model a neutrino.

A new goal

One might want to "base mathematics" on finite quantitative sets, where the containment set might possess an "apparent continuum," but it is an almost continuum (based on the rational numbers) whose properties do not need to be explored, if one wants to remain quantitatively consistent, ie mathematically stable, and the described (math) patterns comprehendible, and subsequently, practically useful.

The new math constructs make classical math more discrete, while making quantum description more geometric and stable.

The rational numbers may be "too big," and one may still be able to calculate values . . . , which are measurable, stable, and practically useful . . . , in a quantitative set which is smaller than the rational numbers.

One can say this, because one can base a measurable description of physical systems on a finite set of spectral values, which control the allowable shapes of fundamental stable material systems (as well as their containment sets), wherein the descriptions are measurable, stable, geometric, and practically useful, but now in a many-dimensional containment set where the dimensional levels are composed of closed shapes whose sizes (up to and including 5-dimensional hyperbolic space, yet the over-all containment set is of hyperbolic dimension-11) are controlled by the finite spectral set.

However, the classical geometric and locally measurable interpretations of derivatives and differential-forms are used in this new descriptive construct, but the limit may only be a finitely dependent math process, since the math constructs (for derivative operators) become more discrete (there are clear bounds on the sizes of stable shapes of both material components and metric-spaces).

There are infinite-extent discrete hyperbolic shapes, which can exist at all dimensional levels, in which discrete hyperbolic shapes are defined.

Discrete hyperbolic shapes can be ten-hyperbolic-dimensions, so the natural containing space is an 11-dimensional hyperbolic metric-space.

Discrete hyperbolic shapes can be bounded shapes up to and including five-hyperbolic-dimensions.

That is, in the context of discrete hyperbolic shapes the infinite extent geometries can be subordinated to the bounded geometries which define a stable, finite spectral set, whose values determine the set of possible patterns for stable shapes and stable properties within an 11-dimensional hyperbolic containing space.

Thus, there can be defined "a finite spectral set" for an 11-dimensional hyperbolic containing space by defining fixed, bounded, discrete hyperbolic shapes for the different subspaces of the same dimension, which are all the various hyperbolic metric-spaces for the 11-dimensional hyperbolic containment set.

The obstacle of tradition (and authority, and peer-review)

Consider,

The basic, very complicated, math constructs upon which the professional science and math people dwell

"In general," the current (2012) professional math constructs deal with

1. indefinable randomness [unreliable calculations of either risk or spectra which is placed in a context of randomness] (trying to identify probabilities on elementary-event sets which are not sufficiently limited, eg a proper elementary-event space should be based on a finite set of stable events, this is a necessary attribute of an elementary event space) and
2. non-linearity [unstable (fleeting) shapes]
   (but only linear differential equations can relate a function's values to its domain set's values, in a quantitatively consistent manner, since a (valid) change of scale is based on multiplying a quantitative set by a constant).

"In general," both of these math constructs ((1) and (2)) define non-commutative relations between the operators, which act [on either functions, or on function spaces] so as to identify a (physical) system's (measurable) properties.

But this, non-commutativity, means that these relations cannot be inverted, and subsequently they cannot be solved, in a context which is both stable, and global, and (also) practically useful (where the properties of the physical system are [should be] incorporated directly into the practical usefulness associated to the information gained from a calculation).

Note: These unstable quantitative relations can be used in an undefined context of "relative stability," where the validity of a differential-equation (and its critical points, and associated limit cycles) determine the stability of the context, more so, than do the properties of the patterns which are being measured can be considered to be stable, or quantitatively consistent (leading to chaotic bifurcations of measured properties), but where such measuring would (can) be done for the purpose of feedback, otherwise it is difficult to give these fleeting measurable properties (associated to unstable math, or physical, patterns) a relation to practical development.

This focus on randomness and unstable patterns has the effect of forming a belief structure (in the public) that truth must be very complicated.

Yet . . . . , or That is,

Math constructs need to be stable and controllable so that the set of measurable properties which distinguish a (physical) system are both quantitatively and logically consistent, and the description is not "too abstract" so that it is still comprehendible. That is, if a too-abstract of a description is not-comprehendible then it is of no practical value.

Just as the epicycles of Ptolemy had no practical value except within the limited context of (celestial) calendars, so too, the math constructs of elementary-particle physics have no practical value except within the limited context of nuclear weapons engineering.

Stable, simple, and comprehendible math patterns

A mathematical description (calculation) needs to be measurable, stable, and practically useful.

Both quantities, and measures on shapes (or measures of properties), need to stay consistent with some uniform unit of measuring, ie the changes of scale need to be valid (ie changes of scale caused by multiplication by a constant).

However, this property, of defining reliable measurements, seems to be related to only physical system's whose properties are being described by partial differential equations which are:

1. linear,
2. in a metric-invariant context (for metric-functions which only have constant coefficients) in a coordinate system (which is consistent with the physical system's measurable properties, eg consistent with the metric-function) and

3. separable, ie where for a separable differential equation the local properties of the coordinates determine an independent and orthogonal set of relations with one another, and these properties exist in a continuous manner for the global coordinate structure of the "natural" shape (or physical system).

(note: "to simplify language" this will be called "geometrically separable," where this terminology has no [claimed] relation to current math-use of words.)

Furthermore, the geometrically separable coordinates need to be consistent with the coordinates which make the metric-function diagonal (where a metric-function, or symmetric-metric-2-tensor, can be represented as a matrix, which always has a coordinate system in which the metric-function's matrix is diagonal, ie the matrix of the local vector properties of the coordinates).

Measuring and quantitative consistency

The simple geometries of circles and lines are the shapes which are consistent with quantitative sets to be used for measuring, so as to be characterized by their being consistent with a uniformly defined unit of measuring.

Furthermore, solutions to polynomial equations, where polynomials have the structure of numbers . . . , are complex numbers whose geometry is based on lines and circles, and complex analytic geometry is about circles and discs (where the non-linear shape of the sphere will always be avoided in the discussion of this paper).

Thus, one should try to piece together these shapes, ie lines and circles, so as to achieve shapes which are quantitatively consistent, when placed within a set of math techniques, which are used to identify shape, and other measurable properties (in a quantitatively consistent manner).

Included in this, fairly limited, set of allowable shapes are the "cubical (or right rectangular) simplexes" where a cubical-simplex, is a basis for these shapes, along with an equivalence relation (or a moding-out process), allows for both "geometrically separable shapes" and for the commutativity of local (matrix) operators defined on a physical system's natural coordinates, so that these (matrix) operators are continuously commutative, in a global manner, along the coordinates, ie along the circles (or curves) "parallel" with the (simplicial) cube's "edges." Note: Upon (2-dimension, but also in general) the discrete-hyperbolic-shapes, it is "hyperbolas" which are "parallel" with the "edges," but where the "edges" are circles, where the other "edges" of the fundamental domain (on the lattice) and their (other) parallel hyperbolas are perpendicular to both the original edges and the original hyperbolas (which are parallel to the original edges).

The resulting shapes (wherein the metric-functions only have constant coefficients), are of great interest to the descriptions (or models upon which calculations are to be based) of physical systems, and these shapes are both:

1. the "continuous" discrete Euclidean geometries ("continuous," in that their right-rectangular-shapes can adapt to (or fit into) any size [or into any sized-space defined by material-separation in a material interaction]), and
2. the very stable discrete hyperbolic shapes which possess stable discrete spectral properties [the natural models of stable (energetic) material systems], where hyperbolic space is equivalent to space-time, eg a (projective) velocity space, but the discrete hyperbolic shapes (the set of hyperbolic lattices) are separated from one another by discrete jumps, ie all values of velocity are not allowed (if the velocity is to be represented by a discrete hyperbolic shape), yet all values for positions are allowed by discrete Euclidean shapes.

These discrete shapes (both Euclidean and hyperbolic) were made famous by the Thurston-Perelman geometrization theorem.

Definite physical properties associated to metric-spaces

Furthermore, other properties of "the relation of space to geometric and the other measurable properties of physical systems," such as action-at-a-distance, need to be a part of the descriptive properties, since action-at-a-distance was shown to be an observed property of quantum systems, ie quantum systems have non-local properties (Proven by Aspect's experiment).

That is, physical properties need to be attached to metric-spaces of different types R(s,t) {where R stands for real-numbers, s is the dimension of the spatial subspace, and t is the dimension of the temporal subspace of this metric-space}. The R(s,t) metric-spaces, of various dimensions, are associated to isometry fiber groups, SO(s,t), eg space-time is R(3,1) associated to the fiber group SO(3,1), and Euclidean 3-space is R(3,0), associated to the fiber group SO(3).

(It would be fairly accurate to say that this is what E Noether did, namely, attach physical properties to metric-spaces, eg wherein the time-part of space-time was associated with energy (by Noether), where this led to E=mc^2.)

Physical properties associated to particular "dimension and signature" metric-spaces, eg a material type and a symmetry relation with the coordinates, leads to the idea (a math construct of) metric-space states, eg the symmetry and its opposite.

For example, Euclidean space is associated to: inertia, and to spatial displacements (ie associated to an absolute frame, which also allows for action-at-a-distance), and its two opposite frames (or two opposite metric-space states) are the fixed stars and the rotating stars.

Spin-rotation of metric-space states, can be defined on discrete isometry shapes (for metric-functions with constant coefficients)

Thus, one can also consider C(s,t) {where C represents the complex-numbers, ie each dimension (either, s or, t) is represented by a complex-number set}, so that C(s,t) represent complex coordinates of complex dimension-(s+t), wherein "the two metric-space states" can be fit into C(s,t), ie the two opposite metric-space states associated to any well defined physical property of a metric-space space, where these two opposite states fit into the real and pure-imaginary subsets of the (Hermitian) complex-coordinate system.

For C(s,t), the fiber group of the principle fiber bundle is unitary, and spin-rotation can be defined as the spin-rotation between the two opposite metric-space states associated to any well defined physical property of space (which fit into the real and pure-imaginary subsets) of the (Hermitian) complex-coordinate system.

The properties of the very stable discrete hyperbolic shapes, ie hyperbolic space-forms, which possess definitive spectral properties, and which are very stable spectral properties

The properties of the very stable discrete hyperbolic shapes, which have been determined as the dimension of the shape changes, and they have been determined by D Coxeter, using discrete hyperbolic reflection groups.

The possible discrete hyperbolic shapes end-in hyperbolic-dimension-11 (there are no 11-dimensional discrete hyperbolic shapes), and all discrete hyperbolic shapes of 6-dimensions or above (up to hyperbolic dimension-10) are infinite extent shapes (perhaps with finite volume).

There are both bounded and unbounded discrete hyperbolic shapes for (hyperbolic) dimensions-one through dimension-five.

Note that discrete Euclidean shapes are always bounded, but discreet hyperbolic shapes can be unbounded, and, when placed into a physical interpretation, this pattern may be about (or have a relation to the Euclidean property of) action-at-a-distance.

Many-dimensions

Then one can construct . . . , from this context of shape . . . , "a finite-, but many,—dimensional" over-all containment set, where all the lower-dimensional (or lesser-dimension), discrete-hyperbolic-shapes, can be contained-in an 11-dimensional hyperbolic metric-space, which can define an over-all, high-dimension containing space, for all the lower dimension subspaces (or lower-dimensional metric-spaces) it contains, and where the different (distinct) dimensional levels all possess the property of being discrete hyperbolic shapes.

These discrete hyperbolic shapes are very stable, and thus such a set, composed of discrete hyperbolic shapes of different dimensions, then the bounded, stable shapes (contained-in an 11-dimensional hyperbolic metric-space) define a finite "spectral set" for the over-all containing space.

A metric-space which has a bounded shape, can (may) be contained in an adjacent higher-dimension metric-space (which itself may also be a bounded shape). Within this math construct of adjacent dimensional levels it is the lower-dimension shape which is the correct construct of a material component, and its distinguished point (or vertex, or vertices of its fundamental domain) is what "has determined the appearance of the point-like properties" of (small) material components (the basis for the point-like spectral-events, in quantum theory).

Finite spectra

The stable bounded shapes of the different dimensional levels of hyperbolic spaces, up to and including hyperbolic dimension-5 discrete-shapes (which are bounded shapes), allows the (over-all) containing-space to possess the properties of definitive and finite spectral set.

There are infinite-extent discrete hyperbolic shapes, which can exist at all dimensional levels, in which discrete hyperbolic shapes are defined.

Discrete hyperbolic shapes can be ten-hyperbolic-dimensions, so the natural containing space is an 11-dimensional hyperbolic metric-space.

Discrete hyperbolic shapes can be bounded shapes up to and including five-hyperbolic-dimensions.

That is, in the context of discrete hyperbolic shapes the infinite extent geometries can be subordinated to the bounded geometries which define a stable, finite spectral set, whose values determine the set of possible patterns for stable shapes and stable properties within an 11-dimensional hyperbolic containing space.

Thus, there can be defined "a finite spectral set" for an 11-dimensional hyperbolic containing space by defining fixed, bounded, discrete hyperbolic shapes for the different subspaces of the same dimension, which are all the various hyperbolic metric-spaces for the 11-dimensional hyperbolic containment set.

Because the different dimensional metric-spaces possess shapes, they also define a spectra, and the spectra of the over-all high dimension containing space can define a finite spectra set which would (could) determine the shapes of the discrete shapes which are allowed within the containment set (through the property of "a necessary resonance" between an existing metric-space [which are discrete hyperbolic shapes of various dimensions] and (with) the finite spectral set, that the high-dimension over-all space contains).

The resonance can be modeled in relation to the maximal tori of the (unitary) fiber groups of the principle fiber bundle math structure of the spaces. This resonance on a maximal tori of a fiber group could be a model of memory for both the material existence as well as for living material systems (see below).

Material interactions

Material interaction can be defined as 2-forms on the discrete Euclidean shapes which mediate (or connect) between [pairs of] stable and separated discrete-hyperbolic-shapes by an action-at-a-distance averaging-structure associated to a (separated interacting materials) connecting discrete Euclidean shape, where, in turn, the 2-form (defined on this connecting torus) is related to the fiber group whose base-space contains the discrete Euclidean shape (ie one-dimension more than the dimension of the connecting discrete Euclidean shape) within which a group element is selected by means of a geometric relation it has to the 2-form (note: the 2-forms of a metric-space and its fiber group always have the same dimension) so as to transform the distinguished points by the group element selected [of the pair of interacting discrete hyperbolic shapes] as spatial displacements (in Euclidean space), the magnitude of the discrete spatial displacements are related to the geometric properties of the material-connecting Euclidean torus, and this spatial displacement is defined for each time-instant, or for each time-interval, where the time-interval is defined by (the period of) a complete spin-rotation through metric-space states, where for each time-interval a new discrete spatial displacement is defined for the interaction.

This is possible due to action-at-a-distance.

That is, this is a restatement of Newton's law, wherein (or however) the derivative is similar to a connection-derivative, unless the interacting material components are already a part of a stable orbital structure.

New interpretations of well-known math constructs

This is a second-order differential equation which depends on defining 2-forms for the geometric context of the equation.

Thus, this second-order differential equation (of Newton) is either

1. elliptic (orbital), or
2. parabolic (bound, but aimed at freedom, eg angular-momentum, and perhaps the spherical harmonic of spherically symmetric probability-waves, where spherical symmetry

in Euclidean 3-space is a result of the geometric relationship which exists between the interaction 2-form and the SO(3) fiber group), or

3. hyperbolic (free, eg collisions, or waves which also possess [intrinsic] physical properties).

When a system is defined within an orbital shape (elliptic), the discrete shapes can be identified as a stable orbit-spectral structure defined on a stable geometry, while for collision type interactions, ie the hyperbolic case, wherein if the interaction system is within a "correct" energy-bound, and if the interacting shapes have the properties needed for resonance with the entire (over-all) containment set's spectra, then the collision interaction might . . . , instead of identifying a collision (system) . . . , result in resonance and the, subsequent, formation of a new stable spectral-orbital system.

Consider, "how to interpret the differential equation?"

Consider a new interpretation,

The differential equation is interpreted to be; defining the "local" discrete changes of the system's properties being described (or calculated)

Thus, there is a new model of a differential equation as a discrete change of relations which are defined on a finite spectral set in turn defined on many dimensional levels.

That is, are differential equations based on the ideas of:

1. Local measures, which try to be linear within a separable differential equation (geometric, classical physics)? or
2. Sets of operators defined on function-spaces (random, also waves which have physical attributes)? Or, (rather, they are)
3. Discrete "changes of geometric relations" in Euclidean space, where these changes are defined on a finite spectral set, in turn, defined on many dimensional levels? (.)

a continuum vs. discreteness

(applied discretely by derivative operators (or by other operators) at particular points of the system, which are part of the descriptive context),

(eg applied . . . [during an apparent collision interaction, when instead of the dynamics of a collision there is a new system (which is) formed] . . . to (what was) the interacting (or connecting) discrete Euclidean shape)

{wherein a new resonance (of the new system resonating with the containing space's spectra) might be (or is) defined, to form a new stable material system}.

(1) and (2) are the traditional interpretation of differential equations and they are assumed to be constructs, associated to limit processes, and the associated idea of nearness, defined on a continuum.

This has led to unnecessary, general contexts, concerning general-shapes and spectral sets.

Is a continuum necessary?

However, in the interpretations of differential equations (in regard to the paradox concerning large quantitative sets), the descriptive construct (of a differential equation and its solutions and contexts) has come to be associated with very little quantitative consistency (within non-linear descriptive structures), and/or little quantitative meaning, ie it has also led to abstractions which have lost meaning (in regard to the practical creative development of new ideas associated to differential equations).

There are also the questions, in regard to indefinable randomness:

"Do function space techniques (applied to wave-functions which possess no physical properties) identify probability density functions? or
"Do they identify spectral values?"

. . . , this is not clear . . .

(It is assumed that differential equations identify spectral values {though these spectral values are seldom found [for general quantum systems] or determined from calculations based on physical law}, but "what would be the basis?" for this assumption, eg for physical systems (or oscillations with physical properties) there are many harmonic patterns associated to differential equations).

. . . , and the difficult to interpret properties of differential equations defined on a continuum has also led to abstractions which have lost meaning (in regard to the practical creative development of these ideas).

Defining the discreteness of operators, in a new context of discrete geometries

On the other hand, the context of (3) . . . . , ie wherein differential equations are interpreted to be discrete "changes of geometric relations" in Euclidean space (where these changes are defined on a finite spectral set, in turn) defined on many dimensional levels, as well as being defined in hyperbolic space, wherein changes to interacting systems can occur, ... , the derivative as a discrete operator can be interpreted to mean that:

1. New systems can form due to resonances (of a new system) with the (high-dimension) containing set's spectra. It should be noted that when a new discrete hyperbolic shape (or hyperbolic space-form) is formed then the "interaction toral-component" of that interaction can be multiplied by a constant so as to identify the new shape for hyperbolic space-form (as it is formed).
2. Conformal factors can be introduced between dimensional levels, thus affecting the relative sizes of the system-interaction when one goes from one dimensional-level to the next (causing less interference due to material interactions between dimensional levels. This is a natural context for small material systems to be contained in large metric-spaces, where the size of the containing metric-space's shape identifies [in a natural way] the sizes of interacting material systems, when described in the dimensional level adjacent to the (original) metric-space (size))
3. The shapes of discrete hyperbolic shapes can be determined by discrete angular changes, which are defined between toral-components of the discrete hyperbolic shapes, where these discrete angular changes are modeled as Weyl-transformations which typically act on maximal tori, eg in fiber groups. Conformal factors can be introduced during this discrete transformation (of angles between toral components of the discrete hyperbolic shapes).

   This means that, when Weyl-transformations occur, a metric-space [represented as a discrete hyperbolic shape {then the metric-space}], can have a metric-function which changes conformally in a discrete manner on the toral components which have been Weyl-transformed.

   Must the spectral properties of the (particular) toral component be related to the spectra of some discrete hyperbolic shape (or metric-space) which has not been Weyl-transformed? [The simplest math structure would require an answer of yes.]

   That is, for a discrete hyperbolic shape any toral component can be a different size than the other toral components of the (hyperbolic) shape, but so as to still define a valid lattice. This could be defined either as a discretely applied conformal factor, or as a lattice structure in hyperbolic space which allows for such a structure, for any particular toral component of a discrete hyperbolic shape.
4. Discrete spatial displacements of the interacting components, where again conformal factors can be introduced during this discrete transformation, indeed this is related to the 1/r^2 force-field formula for material interactions in 3-spatial dimensions (see below).

What is a differential equation?

A differential equation is about identifying discrete constant factors (which allow changes of shapes within a metric-space but the metric-function is not changed) . . . . , applied to:

1. toral components, so as to change the shape of a discrete hyperbolic shape (Weyl-transformations, also accompanying a [simultaneous] conformal transformation, but the metric-function is not changed, ie the space remains metric-invariant), {note: the shaping of a system's orbital constructs are related to parabolic differential equations which are associated to angular momentum (identifying a relation between free and bound components defined between toral components of a discrete hyperbolic shape), most often defined on a system whose discrete hyperbolic shape is that of concentric circular tubes, where these "concentric circular tubes" also define envelopes of orbital stability.} or to
2. the toral shapes of material interactions which change, eg caused by discrete spatial changes due to a simultaneous metric-invariant transformation as well as a conformal (or perhaps a discrete) change applied to the Euclidean torus, but the discrete hyperbolic shapes are not changed, ie the metric-function is not changed, or to
3. entire discrete hyperbolic shapes, which are modeling a metric-space (as a shape), applied during transitions between adjacent dimensional levels (the metric-function is changed, but in this case one is transitioning between different metric-spaces), . . . , multiplied during the application of discrete operators acting on both shapes and dimensional changes, so as to form shapes, so as to affect their size and shapes, and so as to (effectively) form (create) independent dimensional levels (when applied between adjacent dimensional levels).

Answering obvious questions

Apparent randomness (is due to discrete structure of interactions, which, in turn, results in Brownian motions for "small" material components)

The (properties of) higher dimensions are, apparently, hidden (due to both the structure of material interactions and the relative sizes of material systems which exist between adjacent dimensional levels)

Spherical symmetry of material interactions in Euclidean 3-space (is due to the geometric relation between 2-forms defined on a 2-dimensional discrete Euclidean shape of interaction (which is contained in 3-space) and the SO(3) fiber group).

The descriptive construct allows for many different types of base-space and associated fiber group constructs to, simultaneously, be a part of this description, though this is mostly about the simplest of the classical Lie groups: SO, SU, spin, Sp, (but Sp Symplectic has not been examined in this paper), and their associated R(s,t) and C(s,t) base-spaces.

New Mysteries

Find a complete, finite, spectral set for an 11-dimensional hyperbolic containment set, ie find the bounded discrete hyperbolic shapes up to dimension five, which define such a finite spectral set.

Can the infinite extent discrete hyperbolic shapes which can oscillate (see below), in turn, define their own spectral sets?

The dimensional structure of material, ie the R(s,t), which can be meaningfully associated to material properties,

The organization of the time-subspace in the metric-spaces of new material types [the natural energy-generating oscillations of certain, odd-dimensional and odd-genus discrete hyperbolic shapes (which would possess a charge imbalance), which define the new types of materials, and possibly a general definition for life]

How to organize a life-form based on a new definition of life as "an energy generating discrete hyperbolic shapes which has a memory based on the life-forms relation to the maximal tori of the fiber group," where maximal tori can resonate to the finite spectral set of the high-dimension containing space.

Determine the finite spectral set of some over-all high-dimension containing spaces

Are different finite spectral sets associated to (1) different maximal tori or (2) different principle fiber bundles?

Are these high-dimension spectral sets associated to?

Planets
Stars
Galaxies
Universes
?

Universe expansion could result from conjugation of maximal tori

And

Most importantly, Determine how to build on this structure, and "how to build on it" so as to change its properties, so as to allow for greater possibilities, but the building process must be consistent with "what already exists."

Chapter 6

Geometry and abstraction

Do descriptions (of existence) deal with stable geometric components, which possess stable spectral-orbital properties, which are defined by either fixed geometric measures (of stable shapes) or envelopes of orbital stability, associated to the very stable "cubical" simplexes (or fundamental domains) of discrete hyperbolic shapes, and so that the context is many-dimensional, where these stable discrete shapes define both stable material components and stable metric-spaces, where the metric-spaces (and material components) are associated to physical properties which define metric-space states, which subsequently define pairs of opposite metric-space states, so as to move from a real context to a complex-coordinate context, so that these shapes (or spaces) and their pairs of metric-space states are related to: SO, spin, and SU fiber (Lie) groups (in a structure of containment which fits into a principle fiber bundle).

On the other hand, twists and rotations (other than an angular-progression around a spectral-orbital construct) of a shape's bounding geometry are not (so much) about the spectral orbital geometric properties themselves, but rather they are related to either rotations of metric-space states, or it is a property which belongs to a free component in either rotational frames or translational frames, or (related to) partially-free components, which moves between orbital envelopes (or between spectral-orbital measures), but this unified movement has associated to itself either conserved energy or conserved angular momentum (or both) where this rotational-orbital movement allows a flow (of a component, which is fit into an orbit) between orbits (or between spectrally equivalent constructs).

Note: Such "angular momentum constructs" can (may) define steps between stable orbital-spectral properties (or systems).

Twists and rotations within an occupied stable orbital-spectral shape would create a context of shape-instability, where the twists could store energy, which, if released by the occupying component, could destabilize the spectral-orbital system (or shape).

Groups of translations, reflections, and rotations

The metric-invariant properties which are locally linear and solvable seem to be about local coordinate transformations which are the: translations, and reflections, and rotations, wherein these local coordinate transformations are only used (or defined) in a locally independent and "geometrically separable" context.

That is, local rotational transformations of coordinates must stay in a commutative context in regard to local vector structures and/or local changes of coordinates, ie shapes which are globally, continuously (linear) commutative in regard to local rotational changes of their shapes (or in regard to local changes of one's position in the coordinates), eg cylinders, tori, and discrete hyperbolic shapes.

Math ideas of professionals seems to be about their identifying an abstract context, which essentially has no meaning. They tend to describe things in an undefined context, in regard to a practical use for a math construct, where a valid context in which a definition of purpose is (would be) given, and it would be about either a practically useful informational context (associated to coupling and control), or about a relatively (or a sufficiently) accurate description of observed patterns [and a description which has wide ranging generality associated to a wide class of observed patterns].

Arbitrary contexts are developed (by the math professionals) but a stable useful (meaningful) relation to quantity, . . . as well as a distinguishable geometric pattern which is measurable, couple-able and useable . . . , is (usually) not made.

Instead the focus (by math professionals) is on irrelevant Platonic truths such as considerations about: convergences (in a continuum), function—spaces, and algebras of operators, where the context is non-linear and/or vaguely random (ie the elementary event space is [almost always] not well defined), where boundaries are unstable and arbitrary, and the spectra have an arbitrary [most often a non-commutative] context, so that the math procedures used to find spectra, result in leaving the spectra unidentified, yet arbitrary spectral cut-offs are often used (especially in a vaguely random context, eg quantum physics and particle-physics and then they try to find a context in which convergences exist in an unbounded containing space).

Professional mathematicians make-up their own language.

Where, in fact, anyone can (and should) also make-up their own math language.

But the language of the professionals describes a mathematical land of illusion.

This is because they do not stay: quantitatively consistent, logically consistent, and their constructs are not stable, ie they describe fleeting patterns, which are essentially of no practical use, and they make their constructs within sets which are far "too big (within a continuum)," so that they describe incomprehensible patterns. A continuum does not necessarily allow for quantitative consistency (or logical consistency). That is, incompatible math patterns, eg one based on randomness and one based on geometry, can be related to the same quantitative sets by means of different definitions of convergence (and different standards for defining a convergence, eg two divergences equal a convergence) so as to appear to be contained in the same set of quantities, eg the same system-containing coordinate system, and thus incompatible math patterns can appear to be measurably related to one another, but they are not compatible patterns.

They are paid to place interpretations onto their own incomprehensible patterns, and this has mostly been a failing endeavor.

Furthermore, the professionals claim that only the professionals, "who have been tested for their dogmatic purity," can have a say in "peer reviewed" journals about math patterns, and only the professional mathematician has a say about the relation that these math patterns have to practical creativity.

However, the relation of a "truth" to practical creativity, or to a sufficiently general and sufficiently precise of a description of observed patterns, is the best measure of a precise description's (or math construct's) truth.

That is, the "professional duty" of professional mathematicians is to make math into a very narrow (an over-emphasis on describing complicated, and unstable, fleeting patterns, which are irrelevant to useful [practical] creativity), and they provide in their professional journals mostly useless patterns, and they define within society and unequal endeavor of what are defined "by the meaningless communications of the society's propaganda system" to be the society's technical "masters." This is ignoring the wisdom of Socrates who championed equal free-inquiry related to either useful practical creativity or to accurate descriptive measurable patterns which are sufficiently close to the observed measured values of "material" systems.

One can model greater variety of mathematical patterns associated to more valid and relevant descriptions with simple geometries (eg the simpler translational and/or reflection symmetries, rather than the more complicated rotational symmetries [though rotations can be built from reflections]).

However, the simple, stable geometries can be placed in a many-dimensional context (without the inter-dimensional relationships decaying), rather than putting the idea of shape into a relationship with abstract geometric properties (more often) related to abstract algebraic patterns, eg rotations, or deformations of circles, which are placed within a continuum.

In regard to the deformations of the circle, it is the more "quantitatively consistent" pattern that the circle (with its fixed shape) is a linear measure, and that it remain this way. The stable circle might be best used to define holes and their related spectra (where holes associated to circles are stable in the discrete hyperbolic context of shape).

However the circle is most often introduced (and used) by the math professionals into the "math literature (which provides a map into a land of illusion)," in unnecessarily complicated geometric patterns, and quantitatively inconsistent patterns (see below).

In fact, the rotating stars, or the fixed stars (the pair of opposite inertial frames) is a big enough of a complication (or is a context which is complicated enough).

"How a circle can be deformed" is unrelated to quantitative consistency, and thus is unrelated to stability.

A deformed circle is related to an SU(2) connection (an expression of non-linearity and non-commutative relations which are not solvable), where the deformed circle is considered a real-curve.

However, the circle stays rigid in the complex-number-plane, in order to provide (or maintain) the property of quantitative consistency.

That is, knots (deformed [real] circles, whose curve is expanded to become tubes) represent an unstable relation to quantities, where a non-linear "connection term" is needed to describe their local twisting, curved geometric vector properties in 3-space . . . . , {where apparently their braids (or twists) can be related to branched relations on sheets of the complex-plane, ie to try to create a single-valued function on layers of number-sets, defined above a "multiple-valued domain point" in the complex-plane} . . . . , but these geometric shapes are non-linear (the twists and turns are non-linear), quantitatively inconsistent, and unstable, resulting in descriptions of unstable fleeting quantitative patterns.

One needs a stable component to be related to other stable components.

Chapter 7

Circle-spaces and holes in space

Part I

A new context in which to apply geometry to: math, quantum physics, and the solar system, etc

Is the circle to be deformed or stay fixed?
Do holes determine shapes?
or
Do circles determine holes within shapes?

This new context is that circles stay fixed, and it is circles which determine holes within shapes, and higher-dimensions are what cause the existence of either the lower dimensional potential functions, ie so differential forms are exact, or the higher-dimensional stable spectral-orbital properties, that are observed for the fundamental physical systems within existence (which possess stable definitive spectral-orbital properties), or within math descriptions. or the higher-dimensional stable spectral-orbital properties to be separated from the lower-dimensional context in which potential functions (and formula for potential functions) are defined. Furthermore, the properties of potential functions and their relationship to 2-forms depends on the geometry of the (metric-space base-spaces') fiber groups for their valuable physical properties.

This many-dimensional descriptive context of potential functions (exact forms) and spectral properties can exist for any dimensional level. It is especially defined for the two types of discrete shapes. Namely, the discrete Euclidean shapes and the very stable spectral properties of discrete hyperbolic shapes.

Because of their stability (and rigidity) of both shape and spectra, the discrete hyperbolic shapes are the main shapes to consider for the descriptions of stable properties of stable physical systems.

It should be noted that D Coxeter determined the natural dimensional limits for the very stable discrete hyperbolic shapes, where the last discrete hyperbolic shape is 10-dimensions, where these 10-dimensional discrete hyperbolic shapes are unbounded, and the last compact (or bounded) discrete hyperbolic shape exists as a 5-dimensional discrete hyperbolic shape. Thus the containment of all such stable discrete hyperbolic shapes can be contained in an 11-dimensional hyperbolic metric-space (which has no shape associated to itself) and because the last hyperbolic dimension which possesses bounded discrete hyperbolic shapes is five, thus if the hyperbolic metric-spaces are all given discrete hyperbolic shapes, which are bounded shapes, then the total number of different subspaces of any particular (hyperbolic) dimension which is five or less can also determine the total number of spectra which are needed to define all the stable discrete hyperbolic shapes which are resonant with the given set of metric-spaces which possess bounded discrete hyperbolic shapes, so that this number is finite.

One cannot deform the circle, nor twist it, nor relate it to branch-cuts spiraling between circular discs, nor relate rigid circles to non-orthogonal juxtapositions amongst themselves, because these patterns are quantitatively inconsistent and non-linear, and non-commutative, and result in unstable and thus imprecise patterns which have little relation to the observed stable properties of physical systems.

It is only lines (line segments) and rigid circles which can be used (so as to be pieced together) to compose a parallelizable, orthogonal set of coordinates, which can contain systems and their measurable properties, in a quantitatively consistent manner, so as to be a "descriptive construct" which can provide quantitatively consistent, and solvable, and controllable descriptive context. That is, within a principle fiber bundle with a many-dimensional base-space, where the base space is a metric-space of various dimensions and various signatures, eg related to R(s,t) where s is the dimension of the spatial subspace and t is the dimension of the temporal subspace of an s+t=n, n-dimensional metric-space, and an associated fiber group. Descriptions within such a math construct can be in a practical creative relationship to the (existing) stable, measurable, definitive spectral-orbital structures of physical systems.

Such a base space within a principle fiber bundle provides a descriptive context in which at some lowest-dimensional level the space has a potential function, and at higher-dimensions the base space possesses stable spectral-orbital constructs which can be solved and controlled

The controllable aspect of the description must be related to the 2-forms, since the 2-forms have the same dimension as the dimension of the fiber group, thus allowing local geometric vectors to affect the geometry of material interactions, where the geometry of the fiber group can always be related to the geometry of a maximal tori.

For a material containing hyperbolic n-space, the highest-dimension material shapes that an n-dimensional hyperbolic metric-space contains is an (n-1)-dimensional discrete hyperbolic shape, which is also a hyperbolic metric-space (or an (n-1)-dimension spatial subspace) which, in turn, has on itself (n-2)-dimensional spectral-flows, which can be occupied by charged material, and the geometric measure of the (n-2)-spectral-flows, ie the (n-2)-differential forms, are isomorphic to the 2-forms of the containing n-dimensional metric-space.

The (n-1)-discrete hyperbolic shapes (which are determining a material interaction), in turn, determine an n-dimensional Euclidean "rectangular" simplex, whose moded-out shape is contained in an (n+1)-dimensional Euclidean metric-space. The 2-forms of a material interaction are defined in this (n+1)-dimensional Euclidean space with an SO(n+1) fiber group.

The 2-forms of a material interaction have the same dimension as either the SO or spin fiber group, so that local geometric relations can be determined between 2-forms and the fiber Lie group's geometry.

The Euclidean torus, ie the discrete Euclidean shapes, can take any rectangular size (so as to be continuous), and they fit as toral components of the discrete hyperbolic shapes, where a torus has a rectangular fundamental domain, whose faces can be determined from both averages over the (pairs of) discrete hyperbolic shape, eg the center-of-mass coordinates, and the (actual) spatial separations of the interacting discrete hyperbolic shapes, and the property of action-at-a-distance provides a rigidity to the material interactions, in regard to a geometric-complex of both discrete hyperbolic shapes (representing the pair of interacting materials) and a discrete Euclidean (interaction) shape, which fits spatially in-between the pair of interacting discrete hyperbolic shapes. This geometric-complex is an n-dimensional structure whose shape is contained in an (n+1)-dimensional hyperbolic metric-space.

It is not clear if potential functions exist, though they appear to exist, due to an open-closed topology of metric-spaces which are discrete shapes, as zero-forms, as well as 1-forms, but 1-forms would exist because it also appears to an observer, in an open-closed metric-space which is also a discrete hyperbolic shape, that material of dimension-(n-1) are what determine the 1-dimensional material-currents.

Part II

There is a belief that the physical world is too complicated to describe, even though the observed patterns (or properties) of many fundamental physical systems: nuclei, general atoms etc, are stable, definitive, and repeatable, which is a set of properties which really suggests a stable, controllable, geometric structure, and this belief in complication and indescribability, which are, apparently, the patterns of material systems, is brought about by a self fulfilling belief structure within math and science, wherein the descriptive patterns used to describe these systems' properties are: non-linear, non-commutative, and random, and not simply random but rather a form of indefinably random construct where the elementary event spaces are undefined and always changeable, furthermore, the point of math has been to find the most general context and find the most complicated and abstract patterns in such a context with which one is to deal (or consider). Circles are deformed a twisted out of their simple commutative patterns so as to describe a wide range of unstable fleeting (math) patterns.

But classical physics is about relatively simple laws, though somewhat abstract due to its wide range of applicability, yet classical physics is applied in very useful ways over a wide range, but the useful part is the simplest; the linear, metric-invariant, separable, commutative, and solvable differential equations, which model system's inertial and force field properties for relatively simple geometries.

The new ideas are about how to describe the stable definitive observed properties of physical systems, and the strategy is to extend the simple math context of "solvable classical math" but the context is even more simplified with constraints of both stable "material" systems and "the shapes of metric-spaces of various dimensions and signatures," to the linear, metric-invariant, separable, commutative, solvable shapes of the discrete Euclidean shapes and the discrete hyperbolic shapes. However, the possibilities for the organization of the properties of such physical systems and metric-spaces (for such structures) are expanded due to the description being contained in a many-dimensional context.

In this new containment context, of which the material world is a proper subset, the spectral properties of systems are related to a finite spectral set associated to the properties of shape, and also associated to the containment set, and there is a new found set of structures through which complicated systems, eg living systems, can now be organized at a relatively simple construct of dimensional organization and hierarchical control over lower-dimension subsystems.

The descriptions of all changes, ie material interactions, are contained in a high-dimensional space dominated by stable discrete shapes, ie discrete hyperbolic shapes. All systems contained within the dimensional bound of a 5-dimensional hyperbolic metric-space may be assumed to have bounds to both their shapes and spectra, ie metric-spaces have stable discrete hyperbolic shapes, while some shapes contained in a 6-dimensional hyperbolic metric-space can be assumed to be bounded but the 6-dimensional hyperbolic metric-spaces themselves have discrete hyperbolic shapes which are unbounded.

Chapter 8

How to model higher-dimensions, where metric-spaces possess physical properties

Claim: In order to describe stable patterns one needs mathematical stability (or quantitative stability) based on stable geometric patterns.

That is, one needs quantitative consistency, and one also needs to be talking about a definitive, identifiable (math) pattern, where math is about quantity (measured properties of a system [coherent set of related components] contained in a coordinate space) and shape (a geometry with an associated subset of fixed and measurable sub-structures {or constructs}). Or [math can also be about random events contained in a {stable, well-defined} set {whose elementary-events can be counted in a consistent manner}; and there would also be a further context within which the random events can identify a "useful" function, eg calculate probabilities in regard to a bet or a decision.]

It will be assumed that it is the stable definitive patterns (not the random properties) which form the fundamental structure whose measurable patterns one wants to describe.

To do this one needs:

(1) stable sets of geometric components which are distinguished by:

shape,
distinct-component-ness,
stability,
bounded or unbounded properties, and

dimension;
pieced together as "inter-related components" so as to form (or build) a system (set of components and their frame of reference, or containment set), and

(2) a quantitative construct based on a finite set of stable, spectral values {or set of discrete properties which are sufficient to identify either systems or components}
(or spectral values can also be thought of as a set of lower-dimensional (than the metric-space frame of system containment) geometric measures defined on a "cubical" simplex).

The context of containment is:

(I) many dimensional (the [dimensional frame of the] viewpoint of the observer may be varied) and the context of an open-closed metric-space topologies, where
(II) the metric-spaces of the different dimensional levels have stable shapes associated to themselves . . . . , (and where within a "frame" one is isolated from other dimensional levels [because a metric-space is open-closed and a stable shape], yet the descriptive patterns depend on a set of inter-dimensional relationships, ie lower-dimension free components and higher-dimension constructs of material-interactions in (adjacent) higher-dimensional metric-spaces, where all, but the highest metric-space level, possess stable shapes) . . . . , and whereas
(III) these same metric-spaces also have "physical" properties (or properties of existence, or types of quantities) associated to themselves.

Physical properties of metric-spaces and a need for a complex containment set

For example, (in regard to (III)) Euclidean space is related to spatial position and spatial displacements, and these spatial displacements of system components are associated to the inertia of a "material" component whose spatial position is given in relation to the distant stars, while hyperbolic space is associated to time-displacements, energy, charge, and velocity, where time is locally given both a positive and negative direction, ie retarded and advanced potentials respectively.

These pairs of opposite metric-space states require that two opposite (metric-space) states, which are (in the case of time) locally oppositely related to one another, so as to define separate, distinct, opposite, real metric-space states for a system, which in turn, can be contained in the real and pure imaginary subsets of complex-coordinates, so these states can be unitarily rotated, within a set of complex-number containing coordinates, so that there always be locally opposite states, but distinctly "real" opposite-states (or opposite real-number processes).

On a space-form there are identifiable opposite-state spectral-flows which can be spin-rotated so as to periodically identify a pair of real but opposite set of material processes. The period of the spin-rotations . . . , between the real, but opposite, metric-space states . . . . , define the time-intervals associated to discrete dynamic processes associated to material interactions.

The natural restrictions on the very stable shapes of the discrete hyperbolic shapes

The containment set, which has a stable shape, is a hyperbolic metric-space (which is equivalent to generalized space-time metric-space), which can be of at most 11-dimensions (there are no 11-dimensional discrete hyperbolic shapes which are known, ie theorem of D Coxeter). Furthermore, the last bounded stable discrete hyperbolic shapes are 5-dimensional (these are properties of discrete hyperbolic shapes identified by Coxeter [apparently, determined by the properties of reflection groups]).

The definition of a finite spectral set upon which all the other quantitative sets depend

Thus, the bounded shapes of the hyperbolic metric-subspaces (ie subspaces of the hyperbolic 11-dimensional containing metric-space), up to and including 5-dimensional hyperbolic metric-spaces, determine a finite spectral set, and it is this finite set "to (or about) which all the bounded components, "of hyperbolic dimension five or less" must be in resonance."

(where components are confined to dimensional levels by:

(1) their containment within metric-space shapes as free components,
(2) the process of material interaction, which is dimensionally dependent, and
(3) the relative sizes of material for each dimensional level, where relative size associated to interacting material systems within any particular dimensional level is determined by multiplication by constants between dimensional levels {these constant factors are related to physical constants, eg h})

But of the several (or many) components (or hyperbolic metric-spaces which are bounded shapes) . . . . , [which can (do) exist in higher dimensions] . . . . , one might ask, "which ones are the distinguished sizes and shapes for the various subspaces of the same dimension for all the different dimensional levels within the containing 11-dimensional hyperbolic metric-space?"

This may be answered as "simply the ones which define a universe for a particular finite spectral set which identifies an existence (ie a universe)."

Low-dimension free-components contained within hyperbolic metric-spaces are confined to a particular subspace of the 11-dimensional containing space, but their shapes must resonate with the set of stable spectra which are defined either by particular shapes associated to this set of particular subspaces of a given dimension, or by the particular dimension spectra.

"Measuring and functions," "function values and the shape of a space," "potential functions as aids to measuring (or aids to calculating [and a pre-conceived model of containment and quantitative inter-relationships])"

The point of a function is to relate a measured property of a system to the coordinate values of "the containing space of the system" upon which the measured property is defined.

For example, in 3-space a derivative should relate a function's properties to both coordinates and to material surfaces which can also be contained in 3-space.

However, it is assumed that metric-spaces do not have holes which are a part of their structure.

This would (also) be true for metric-spaces which are of a particular dimension, and they are both open-closed and possess shapes, where these shapes are built from "cubical" simplexes (or fundamental domains) of a metric-space (which possesses non-positive constant curvature), that is, if they are defined on such a lattice then there is no evidence of holes being within the metric-space, since one is an observer on the lattice (within a fundamental domain whose topology is open-closed {one sees the indefinable fundamental domains which "fill space" with no apparent property of holes in that space}).

However, "for spectra" there is naturally a hole structure (or there should be a natural hole structure) associated to the metric-space (or the domain space of a function, whose values are defined on a system contained in the domain space) upon which the spectral properties are defined. That is, for a space within which there are observed stable spectral properties, one should look for (or assume) a space, which has a stable hole-structure (upon itself) about which a stable spectral set can be defined . . . . , as (having) stable properties.

or

One should look for system components which are stable shapes which also possess a hole structure.

But this is usually assumed to "not be true."

Furthermore, material interactions are always assumed to be spherically symmetric for material which is in a frame where the material is not moving, and it is assumed that spectra result from a well defined potential function which (for relatively motionless material) has the form, $1/r$.

But the spectral properties of systems composed of (many) material components which possess $1/r$ potential-functions cannot be determined.

If there are no holes (or no detectable holes) in an open-closed metric-space, then there are always potential functions "for any order differential form," ie without holes in a metric-space then a function (a differential form) can always be assumed to be single valued, ie a function's values are all path independent.

But this is the wrong context for stable spectral properties, stable spectral systems naturally are defined in (on) hyperbolic metric-spaces which possess holes in their shapes.

In 3-space the 2-surfaces of material effectively define the holes (or material currents) which define an assumed single-valued 1-form which is supposed to be a potential-function for an electromagnetic 2-form (force field).

However, such a condition, of spaces which posses (stable) holes in their shapes, can be realized in regard to it being related to the sub-components (with spectral properties) which are contained in a metric-space of any particular dimension.

That is, stable spectral properties are best described in regard to holes in the system-containing metric-spaces, or the system's model as an open-closed metric-space shape, where the metric-spaces are discrete hyperbolic shapes.

The fiction of electromagnetism is that space-time defines a 4-dimensional metric-space, wherein 1-forms are defined by both (relatively motionless) point-charges (with spherically symmetric force-fields) and material currents (which are defined by the 2-surfaces of moving material charge, and) which define a magnetic field.

This is a fiction because stable currents can (also) be defined as 1-flows on 2-dimensional discrete hyperbolic shapes defined in hyperbolic 3-space, but the force-field of electromagnetism is defined in hyperbolic 4-space, where the geometry of the SO(4) fiber Lie group of the inertial Euclidean space (associated to material's spatial displacements) is equivalent to SO(3) x SO(3).

On SO(3) x SO(3) both the same 3-space and the same 2-space along with the 4th direction of R(4,0) is defined, but

(1) the material of R(4,1) is the wrong size (needs to be the size of the solar system)
(2) the same 2-space and the 4th direction can be associated to a new material as well as to a new time dimension, ie in the newly identified space of R(4,2).

Furthermore, the same 2-space (which is also associated to the 4th independent direction of R(4,0)) may also define the plane of the solar system, and where Newton's gravity might be best described as a force in 2-space (or 2-plane).

In regard to spectra one needs holes in space.

If partial differential equations are to be related to stable spectra through solution functions, (and not through a system's containing metric-space's properties),
then the functions need to be the shapes of hyperbolic space-forms, which are solutions to linear metric-invariant and separable partial differential equations.

Note: Hyperbolic 4-space should have holes which are characterized as being 3-dimensional-holes (with an attendant set of lower dimensional bordering, and thus marginally important, hole structures), in the metric-space, but this would imply that 1-forms and 2-forms should be exact, but the potential function might have an implied set of holes (or currents built from 3-dimensional material systems, which can be formed into "what appear to be approximate" either 1-dimensional and 2-dimensional set of closed material "currents") . . . , (just as 2-dimensional material components in 3-space can be formed into "what appears to be an approximate" 1-dimensional set of material currents, as in electromagnetism theory).

The derivative as a discrete operator

In the context where there exists a rigid containment set . . . , (ie made rigid by stable rigid shapes of which it is composed) . . . , then the derivative becomes (can become) a discrete operator which is related to either a finite spectral set (stable shapes), or to discrete time intervals (dynamics), or to discrete Weyl-transformations (orbital envelopes of stability).

This transcends both the idea of a derivative being a model of local measuring associated to geometry (the model which the new discrete-derivative construct most resembles) as well as the idea that spectral properties of physical systems are most associated to the eigenfunctions of function-spaces, where it has been assumed that the measured properties of spectral sets are more closely associated to a set of eigenfunctions, which identify a function-space (and their associated set of operators) than to the coordinate space which contains the spectral system, but The eigenfunctions of function-spaces are to (or should) become either stable metric-spaces or a combination of both discrete Euclidean shapes and stable discrete hyperbolic shapes, where the two sets of shapes together define a rigid spatial context within which the material interactions of physical systems are both described and contained.

This is necessary since algebras of local linear operators defined on function spaces (as if defined on coordinates of a geometric pattern) . . . . , for functions which have no apparent physical properties, and only possess random manifestations associated to local measured properties which when randomly identified (ie when identified, by an observation, in a context of local randomness of point-particle-spectral random events in space and time) destroy the system's global wave-function . . . . , where this should be contrasted with classical description, wherein functions are used to describe the system's properties so that the functions identify a set of global measuring properties . . . . , but nonetheless these functions which represent abstract (non-physical) properties of a quantum system cannot be used to identify spectral values of these, general, quantum systems.

The "indefinably random" model of description (of quantum systems) is not logically constructed to identify stable spectral values of systems whose spectral properties are stable definitive and discrete.

Instead, or to remedy this dysfunctional-ness, one must be confined to a containment set whose properties are both more constrained, and more rigid, ie the Euclidean discrete shapes can be used to fit into the very stable and very rigid discrete hyperbolic shapes so as to do this in a very constrained and rigid, yet continuous, manner (so as to be able to describe the properties of a stable spectral system (physical system), as well as the properties of macroscopic material dynamics).

Arbitrary quantitative patterns, eg algebras of operators defined on function-spaces (whose functions represent random point-like-spectral events in space and time of particular spectral values), have no implied context through which a stable pattern of spectra can be identified.

Measuring in space and time requires stable geometry in regard to: context, pattern interpretation, and assumptions.

An implied spectral pattern, which is assumed to be in a context of randomness, has no clear pattern of reference, so any construct which fits with data can be used, so as to use sets of inconsistent constructs [particle-collision geometry and randomness and non-linearity] which are forced together by defining (arbitrary) "convergences" on a very large set, ie the continuum.

Thus description (today, 2013) comes to be about a correlation identified in regard to a process of data-fitting (which also happen to be consistent with using information about probabilities of particle-collisions, which fit into models of (nuclear) reactions used in weapons engineering).

but

The observed spectral patterns of material at all size scales are stable a definite, and only have an appearance of being random (point-particle events in space and time).

Limitations of general relativity

Is there a context in which general relativity (inertia is about following geodesics) can describe observed patterns?

It is only within a linear, solvable shape (which represents a metric-space), that general relativity can describe an observed pattern, ie general relativity make any sense, eg

(1) the 1-body system with spherical symmetry, and
(2) stable spectra on hyperbolic space-forms the sets of pairs of "orthogonal hyperbolas" as the natural coordinates on discrete hyperbolic shapes require that the geodesic structures be very limited on discrete hyperbolic shapes, and these highly confined geodesics also define the very limited properties of spectral measures on discrete hyperbolic shapes.

Space-forms allow fundamental questions to be re-framed in a context of a geometric map into higher-dimensions, rather than a context of unrelated abstractions about symbols representing things of different types, so that vague correlations (but [nonetheless] consistent with big business [or military] interests) become the basis for physical descriptions.

Consider:

$m=k/r$ (defining gravitational mass for a gravitating body, as the curvature of a 1-loop on the plane of the solar system) and
$k'/r=v^2$ (defines the kinetic energy of one of the spectral values of a 2-dimensional discrete hyperbolic shape as being related to its sectional curvature).

This implies $m = (k/k''c^2)v^2$, which, in turn, implies $mc^2 = E = (k/k'')v^2$
[Is k" about 10^{-11} or 10^{-15}m/kg? ie related to the mass of electron or proton?].

Chapter 9

Varied discussion

How to usher-in a new cultural context for all of mathematics as well as physics?

This is a new math context within which the problem which has not yet been solved . . . , The problem concerning "how to account for the stability of fundamental systems" such as: a nucleus, or a general atom, or a molecule (including its shape), or a crystal, or the solar system . . . , can now be solved, and the math context for the solution is very simple.

Namely, use stable discrete hyperbolic shapes to model both material and metric-spaces within a many-dimensional containing space, ie an 11-dimensional hyperbolic metric-space, where the discrete hyperbolic shapes may exist at any size scale for any of the many different dimensional levels, yet certain size scales are selected to be (or do tend to be) the predominant size scale used to model material systems for each particular dimensional level, and then adjust the way in which operators associated to local linear measuring and differential equations are constructed and used all in the context of the very stable discrete hyperbolic shapes . . . . (or other non-positive constant curvature metric-spaces, and their associated discrete subgroup geometric relations to lattices, and to the various "cubical" simplex fundamental domains for metric-spaces of various dimensions and various metric-function signatures).

This gives the answer, just as Copernicus and Kepler gave the answer as to the properties of the solar system, and Newton gave a re-adjustment of these answers within a precise global context. That is, Newton gave, sort of, an adjusted answer in the (complicated) form of a global solution to a partial differential equation, for the 2-body problem, in center-of-mass coordinates, though the stability of the solar system still had no valid description, . . . , until now!

That is, the current way in which to try to describe the observed properties of fundamental, stable physical systems is a miserable failure . . . . , ie quantum physics, particle-physics, general relativity, and all the other theories derived from this, so called, "basis for physical law," eg string-theory etc, are all great failures, which verify their descriptions (or calculations) by pointing-out irrelevant marginal aspects of data (either very far away "electromagnetic properties" of "stars," and the very minute (and useless) patterns of particle-collisions [which are incorporated within such a complicated descriptive structure, that they are totally useless descriptive structures] {though they remain consistent to the idea that reaction rates are related to probabilities of random particle-collisions used in nuclear weapons engineering}) which support the contexts and interpretations of these failed theories, ... , so it is strange that an alternative context is so readily marginalized and without interest by the professional technical communities. But both social domination and social authority are more about propaganda than about a valid, practically useable truth.

Though this new descriptive context seems overly simple, where the math and the geometries are the very simplest . . . . , yet it is only in the linear, solvable context where the differential equations are related directly to geometry, that science has actually provided descriptions which result in practically useful development, and wherein the information provided is global in nature (and generally accurate to a sufficient level of precision) . . . . , since the new context essentially says that the discrete hyperbolic shapes are the solutions, they define the stable spectral-orbital properties of the above mentioned "fundamental, stable physical systems," but "How can these shapes be fit together to describe the formation of new stable discrete hyperbolic systems?" . . . . , where this is done in the context of a finite spectral set, upon which an entire galaxy (or universe) of existence depends.

and

"Can this new description account for the jumble in which the stable components of material structures mostly exist?"

It can!

And it can do it without depending on a containing space which has the set-size of a continuum.

but

Now the new quest is to find such the finite spectral set for a finite number of subspaces in an 11-dimensional containment set, but the spectral set only needs to be defined up to and including the 5-dimensional subspaces of discrete hyperbolic shapes.

This new descriptive construct is a great simplification over the current descriptions which are based on indefinable randomness and non-linearity, where these types of (current) descriptive languages (the set of assumptions upon which these unstable descriptive languages are based) cannot describe a stable pattern, and at best only fit data, just as Ptolemy fit data, and mostly these current descriptive languages cannot do this (that is, fit the observed data) . . . . , because it is "stable patterns," which are "what is observed" for the fundamental underlying material components, as well as for large scale material organization, while these components (most) often mix together in, what appears to be, chaotic patterns.

Does one give bare definitions and then technical discussion, such as can be done for a new context, or does one try to introduce the new culture in a "culturally-talkative" presentation?

This is a talkative presentation.

The main idea (in regard to the new context) is about many-dimensions, and that there are shapes associated to the different dimensional levels (as well as shapes associated to "material" components, which are contained in the metric-spaces, and it is these shapes, along with the discrete Euclidean shapes, which determine the properties of component interactions) up to hyperbolic-dimension-11.

However, in real experience, the stable shapes (which form a scaffold [or foundation] of stability for existence) are most often in a jumble, and this is also true within the new ways in which to organize math patterns, so as to describe the basic fundamental and stable properties observed for physical systems which are contained in space.

The basic shapes of both the fundamental stable material components, and the basic shapes upon which the stable scaffolding are built, are linearly consistent in regard to, local linear models of measuring physical properties (or properties of stable patterns).

Namely, these are the shapes built from both line segments and circles, and these shapes are placed in a many-dimensional context of containment. In particular, "cubical" (or rectangular) simplexes and their associated "circle spaces" of both tori (discrete Euclidean shapes) and discrete hyperbolic shapes, which are, themselves, composed of toral components.

"In space" these stable shapes form the stable borders of the possible shapes which can exist, a border (or a rigid stable scaffolding) for stability, so that a descriptive language is actually trying to describe a stable set of patterns (which are now {ie what is being described in this paper} assumed to be fundamental to existence), which are consistent with a set of partial differential equations which are: linear, separable, metric-invariant, and solvable quantitative descriptions of global properties of systems (or shapes) contained in some coordinate space which is also associated to stable patterns.

Note: physical systems and their components are, in this paper, assumed to be the stable discrete hyperbolic shapes, and this is also true for the component-containing (hyperbolic) metric-spaces.

Many-dimensions are organized so that dimensional levels seem to be hidden from one another.

The, apparent property of hidden dimensions, is partly about the model of material interactions, and partly about the size of material systems which interact in any particular general dimensional level, and the size of material components which interact is related to constant multiplicative factors which are defined between dimensional levels, where these constant factors manifest as physical constants, eg h etc.

Other hidden structures, whose existence it is best to assume, deal with [or are associated to] the hidden properties of "holes in the shape of space," which (because they are, presently (2013), assumed to "not exist" within the space, which we experience and measure) causes a belief that material spectra, eg energy properties of quantum systems, emerge, not from the existence of holes in space (or from the holes in the discrete hyperbolic shapes which must be used to model stable material-spectral systems), but rather from, 1/r, models of potential-energy functions (for each charge component of a system), where these, 1/r, potential functions are used to describe material interactions. When, in fact, the properties of stable, discrete, definitive spectra of quantum systems are a result of the existence of holes in the stable shapes which must be used to model these systems. The stable discrete hyperbolic shapes must be used to model quantum systems, since they also must be used to model the stable properties of metric-spaces (which also possess stable spectral properties), where in the new descriptive context, a material system is simply a (hyperbolic) metric-space contained within an adjacent higher-dimensional metric-space of the same type of metric-space, but one-dimension higher.

Thus, the question about "How shapes with holes manifest within existence?" is of central importance, and it is implied in the assumptions about "how to construct stable, and quantitatively consistent, math patterns, which can be used to describe, in a general manner, the stable discrete spectral-orbital properties of the most fundamental physical systems, which exist at all size scales, from nuclei to planetary orbits, and whose stable properties go without valid description."

Stable holes are associated to stable spectra, which, in turn, are associated to stable shapes of circle spaces, most significantly the discrete hyperbolic shapes.

Stable spectra come from stable shapes which are circle-spaces, where the stable spectra are defined about the holes which are defined on the stable shapes of the circle-spaces.

In part, the assumption of "needing to base descriptions . . . , of existing systems which are stable . . . , on stable shapes," depends on a need to associate properties (or a need to associate a quantitative type) to each different type of metric-space (dimension and signature).

A stable existence has particular types of properties, which are both mathematical and physical.

Thus a 3rd fundamental attribute [besides (1) many-dimensions and (2) both metric-spaces and material components are associated to stable shapes] which is needed to describe stable shapes, as well as stable physical systems . . . , [within a general idea about shapes whose existence and properties can be measured in a reliable, stable manner, and yet many of a metric-space's "properties of shape" remain hidden, when viewed in a particular dimensional context] . . . . , is that metric-spaces, which are distinguished by dimension and metric-function signature [a signature of a metric-function is related to R(s,t), eg R(3,0)=Euclidean 3-space and R(3,1) is space-time, which, in turn, is equivalent to hyperbolic 3-space, etc] . . . , are (also) associated to "physical" properties . . . . , [but also mathematical properties about measuring and stable (geometric) patterns] . . . . , of definite types, which are associated to shapes of particular metric-invariant spaces, whose discrete isometry subgroups, possess stable shapes.

These metric-spaces, with stable shapes associated to themselves, are the (discrete shapes of) metric-spaces of non-positive constant curvature.

There are also a set of interaction processes (which are dimensionally dependent, and) which can be defined between the stable shapes.

What are the key set of properties which must be associated to the different dimension and different metric-function signature metric-spaces? [(1) spatial position, and (2) stability, or conservation, properties.]

(1) There is a property fundamental to "space", the property of a position, eg the position of some existing component contained within the space, is a fundamental property, which is needed to describe measurable patterns. There is a metric-space (Euclidean space) which has a spatial type which is associated to the property of position and subsequently to the determination of spatial displacements.

This distinguishing feature of the components' (or systems') spatial positions, when contained within a coordinate space, must exist in order to define a stable and quantitatively consistent mathematical context.

This property of spatial position and spatial displacement has two states in regard to the distant stars [How is infinity and how is bounded-ness to be realized in a mathematical context?], namely, (1) a fixed absolute space and (2) a rotating space, where both states are defined relative to the distant stars.

Absolute space has an absolute position determinable in an inertial frame (or constantly moving frame) and this allows action-at-a-distance . . . , if the space contains a stable set of rigid shapes within itself . . . , as demonstrated by Aspect's experiment.

Note: General relativity only makes sense within a context of existing stable shapes, where "material" components possess stable measurable properties of position, as well as a stable property of existing.

(2) Then there is the property of existence, ie the property of stable "material" components. This is about:

energy conservation, and
conservation (or time continuity) of mass and charge.

Energy is conserved when a system, which is defined by it energy, is invariant to time displacements (E Noether).

Time displacement is associated to hyperbolic space, ie the "projection space" of space-time into velocity-energy space.

While the conservation of momentum is a property of spatial displacement invariance.

The discrete hyperbolic shapes imply a space of energy, or the "space of an existence" which is conserved.

This is (also) about defining stability mathematically, where types of measurable properties are types of mathematical patterns (or mathematical properties), energy and conservation properties are associated to the mathematical shapes of the discrete hyperbolic shapes, and where inertia (or mass) is about the relation of discrete hyperbolic shapes to discrete Euclidean shapes, and the relation that Euclidean space has to identifying a stable component's position in space, as well as its displacement in space.

Thus, discrete hyperbolic shapes determine stable components, which possess stable energy, and these components can be given position in Euclidean space, and these positions can be related to spatial displacements by means of discrete Euclidean shapes and their relation to discrete hyperbolic shapes, where discrete Euclidean shapes can be related to discrete hyperbolic shapes because of the fact that discrete hyperbolic shapes are composed of toral components, and because discrete Euclidean shapes "can be of any rectangular size" so as to be able to fit in-between the separated (and stable) discrete hyperbolic shapes in a rather rigid manner (spatially).

Properties of discrete hyperbolic shapes

The discrete hyperbolic shapes can be bounded (in their shapes), up-to and including five-hyperbolic-dimensions, and the last infinite-extent discrete hyperbolic shapes is at ten-hyperbolic-dimensions (these are properties of discrete hyperbolic shapes identified by D Coxeter). This means that the discrete hyperbolic shapes of the subspaces of an 11-dimensional hyperbolic metric-space, up to 5-dimension, have a stable "hole" structure, associated to a finite number of these shapes which, subsequently, define . . . . , [stable holes in stable components of systems (patterns) which conserve energy (ie an existing pattern), and] . . . . , a finite set of stable spectral values.

Thus, metric-invariant, linear, and separable (in regard to a system's description [or global solution function] by means of a partial differential equation which describes patterns which have enough stability for the information and the control associated to a globally measurable description to be used in a practical way) and this implies metric-spaces which are of non-positive constant curvature, where the metric-functions of various signatures and dimensions (or dimensional levels) only have constant coefficients, and this, in turn, implies the discrete Euclidean and discrete hyperbolic shapes . . . . , where the discrete hyperbolic shapes are associated to stable existence of (material) components, and the Euclidean spaces define spatial positions and the discrete Euclidean shapes . . . . , allow for a rigid "filling of space" by the relation that the discrete Euclidean shapes have as an approximate toral components of a discrete hyperbolic shapes, as well as the inter-linking of discrete Euclidean shapes, with the properties of discrete hyperbolic shapes, where hyperbolic metric-spaces have discrete hyperbolic shapes associated to themselves so that this results in the capability of descriptions based on stable shapes (or the descriptions of patterns which exist and are stable and measurable), in turn, to be based on a finite spectral set, where the dimension of the highest-dimension spectral value is (hyperbolic) 5-dimensional.

The wide range of descriptive possibilities for the new geometric basis for physical descriptions

Though this seems to be an extremely limited, and overly constrained geometric construct, to be used to describe physical phenomenon, nonetheless, it fits into the usual 2-order (metric-invariant) partial differential equations related to classical physics, but now it has some new interpretations and new contexts in regard to orbital systems (elliptic systems), while the parabolic (or within an orbital structure) context becomes the language of angular momentum, which is usually defined within an orbital context, and, finally, the hyperbolic context is that of collisions, as well as electromagnetic and mechanical (or sound) waves.

The hyperbolic 2nd order partial differential equation descriptive context is either about an emerging well defined stable orbital structure and its associated angular momentum properties, or it is about collisions, which "on the microscopic realm" identifies Brownian motions in relation to the interactions of ever changing charge properties of small charged systems, and thus it defines either the context of randomness of quantum theory (as was shown by E Nelson), or it defines a jumble of stable components or it defines the context of spectral-orbital stability.

Note: Theses discrete hyperbolic shapes have a distinguished point associated to their shapes related to the vertices of the "cubical" simplex from which such discrete hyperbolic shapes are formed (or described), and this distinguished point is the cause for the model of small material components to be interpreted not as geometries, but rather they have been modeled (or interpreted) as point-particles.

That is, if a collision is defined in a "correct range of energy," and if the collision gets to the size level where the hyperbolic and Euclidean toral-component complex (associated to the interaction of, relatively stable, discrete hyperbolic shapes) can begin to resonate with the (finite) spectral-set of the over-all high-dimension hyperbolic containing metric-space, then such a collision can result in the formation of a new stable discrete hyperbolic shape.

This is the very type of material interaction process, where a new stable system emerges from a material interaction, which the current descriptive constructs (2013) simply has not been capable of describing.

The new descriptive structure is based on stable discrete hyperbolic shapes in a many-dimensional context where the higher-dimensions and the shapes which determine the higher dimensions are macroscopic shapes.

This is sort of equivalent to the way in which Copernicus and Kepler together gave the answer as to the properties which the solar system possesses, and then the calculus, and a partial differential equation modeling inertia, was associated to a force-field which is based on an inverse square law which also possessed spherical symmetry, so that solution functions of Newton satisfied the already given answers.

That is, the statement is that one can understand and use the stable spectral-orbital properties of physical systems at all size scales if one bases one's description on the very stable discrete hyperbolic shapes placed in a many-dimensional context associated to the macroscopic stable geometries of the discrete hyperbolic shapes, where multiplication by a constant factor between dimensional levels is allowed.

This is the answer, and now "how can (or find) a specific set of spectral quantities (which can) be put together to form 'what is observed to exist?'"

The new context re-invents the derivative as a discrete operator in regard to:

1. material interactions,
2. angular relations between the toral components of a discrete hyperbolic shapes which are needed to identify envelopes of orbital stability, and
3. the existence of constant multiplicative defined between dimensional levels

That is, the new question is:
What set of discrete hyperbolic shapes determine what exists?
Or
What set of discrete hyperbolic shapes determine a particular existence?

That is, what is the finite set of spectral values which represent geometric measures of different dimensional representations of the circle spaces of the discrete hyperbolic shapes (up to and including hyperbolic dimension-5) which make-up our existence?

What is the range of spectra from which a spectral set can be chosen, so as to form a finite spectral set, upon which an existence is (can be) defined (or determined)?

What is the (or an) entire set of both spectral values as well as patterns (or processes) which "one should think-of" as an existence?

What is the spectral basis for the types of discrete hyperbolic shapes, and what are an (what is the) associated set of interaction processes which (together) determine an existence?

Chapter 10

Master-plan, a guide to a new context for individual creativity

(A primer on how "very deep practical creativity" is related to math at an elementary level)

A new description of existence, far superior to the descriptions of the authorities, and the new description is based on simple shapes and elementary calculus, it is a new context which describes the properties of stable systems in a general and precise manner.

It is a new context which is expansive, and geometric.

Using simple math patterns to describe the properties of various fundamental material systems, so as to do this in a very general context, and in a relatively (or adequately) precise manner, and to do this in a geometric context, so that it is a very practically useful set of descriptions, and it is a description which transcends the idea of materialism.

[The intellectuals of society are trained to believe that truth is only expressed by the propaganda system, and the set of experts and authorities which are the pillars of the narrow set of values and narrowly defined knowledge (and associated creativity which the owners of society depend and maintain) which the propaganda system expresses, yet everyone is left wondering "why society becomes so destructive?" Society has become destructive to itself, because it is too: narrow and domineering, thus leading to wide ranging fear.]

For example, apparently N Chomsky believes the dogmas of the math-science experts at MIT, even though these experts are wage-slaves who serve the propaganda system.

Society is very destructive because people believe in inequality, where inequality is the main idea expressed by the propaganda system, a propaganda system which is controlled by so few.

Instead, one must trust oneself and one must trust nature, and not trust the propaganda system, . . . , people are equal (and they are not naturally selfish, and they are not naturally self important).

These new ideas about math and physical description are revolutionary, ie they oppose orthodoxy.

Many of the assumptions, contexts, and interpretations upon which these new ideas are based will "in your minds" (which have been bent to conform to orthodoxy, current-authority, and high social-value) be readily dismissed as being clearly wrong, except for the fact that, the orthodoxy of today's authorities "does not work!" in regard to describing fundamental, stable quantum and orbital systems, so that the description's possess the properties of: generality, accuracy, and practical usefulness! [The descriptive properties which are possessed by the linear, solvable classical descriptions.]

That is, there exist many fundamental physical systems which possess stable definitive properties, but today's authorities claim that these systems are "too complicated to describe," in regard to using the descriptive language based on what are now (2013) called "the laws of physics."

Yet, the fact remains, that they are stable systems with definitive properties, and thus they must emerge from some controllable process in which they acquire these unique and stable properties.

If the descriptive context for what now passes for physical law . . . , ie randomness, non-linearity, quantitative inconsistency, and logical inconsistency eg descriptive structures based on both randomness and geometry, where such a contradictory structure for language has been allowed by properties of convergence which are defined on a continuum, a continuum is a set which is "too big," so as to allow such logical inconsistencies to easily exist, . . . , was, in fact, true, then the properties of these systems could not be stable and definitive, to the point of their possessing unique identifications, based on their stable measurable properties.

In fact, if (after over 100-years) stable definitiveness is not both generally applicable and calculable then the, so called, "laws of physics," in current usage cannot possibly be true. Trying to fit observed definitive and stable patterns of fundamental physical systems to patterns of randomness has not been practically useful, nor accurate to sufficient level of generality, to be at all convincing.

How can consistent descriptions result from patterns which are based on fleeting, unstable, random events?

The possibility that such a descriptive structure, based on randomness, could possibly be true, if such a basis for physical law can be used to describe the observed stable properties of fundamental systems, such as nuclei, atoms whose atomic numbers are greater than five, etc, but these laws cannot be used to do this.

There is no reason for today's orthodoxies to have any authority, yet it is demanded of the, so called, professionals that they demonstrate "dogmatic purity" in regard to believing and using these laws.

How can anyone be serious about calling this basis for physical description, "the verified laws of physics" (or even valid math patterns)?

Apparently this has become orthodoxy, since it fits into the business-military interests, so as to result in its far over-blown high social-value.

The propaganda system of the US expresses only the ideas of the dominant business interests, so as to exclude all other ideas, so that the public will not take seriously any ideas which oppose the desires of oil-military-banking interests.

By the militarization of the management of US institutions . . . , which has taken place since WW II . . . . , the education system has become part of the propaganda system, in a similar manner as the Food and Drug Administration (FDA) now serves corporate interests, in opposition to the FDA's main (or intended) function of protecting the public.

That is, education best serves the interests of monopolistic business interests, than it serves the development of valid knowledge, to be used by the entire public for their creative purposes.

All the ideas expressed by the media are failing and causing destruction, while all the actions taken by the owners of society express extreme violence and are taken to regiment the society. That is, Truman's militarization policies have led to a military which serves the narrow selfish interests of the domineering forces of the oil-military-banking monopolies.

Math should be related to creativity [ie proof of a math pattern should be about the pattern's relation to a wide range of practical creativity]

Consider a new context for both mathematical and physical description, ie a new context for creativity.

Classical physics assumes the local linear measuring of physical properties are to be associated to geometry, so as to obtain (when the differential equitation is linear and separable) a global, controllable description. This descriptive structure has wide applicability, it has sufficient accuracy, and it is very useful in a practical sense because it is geometric and measurable.

On the other hand randomness assumes a global wave-function structure, and a set of operators, which identify its set of (simultaneously) measurable properties, so as to, in principle, uniquely define the quantum system's properties, but this descriptive structure is used to descend down to a local, random, and discontinuous (the wave-function is collapsed) context of particle-spectral events, ie it gives-up information. It is not general, it usually lacks sufficient precision to be a valid description, and it is without any valid (or only a most limited) relation to practical usefulness.

How to describe the stable, definitive, discrete spectral-orbital properties over a wide range of very stable fundamental material systems, which compose, "what we consider to be" the material world?

A. Stability, quantitative-consistency, and geometry are considered to be the set of math properties which can be used to describe stable properties of systems in a practically useful context.

Consider a descriptive context in which stability is a central pattern to be used to describe the existing, stable, definitive properties of material systems, and to allow for measurable consistency.

This is a demand for rigid geometric structures, which are associated to the very stable discrete hyperbolic shapes. These shapes are circle-spaces built out of moded-out cubical (or right rectangular) simplexes. Circles and line-segments can be quantitatively consistent and cubical simplexes resulting in circle-spaces lead to commutative relations between the local linear properties of function values and their domain space coordinates, represented as diagonal matrices defined locally on the vector properties of the coordinate system's local directions.

Rigidity of shape is related to the existence holes in stable shapes, since the holes, within the shape of a discrete hyperbolic shape, are an important property which characterizes these shapes as circle-spaces, in turn, the discrete hyperbolic shapes are to be used to model material components, ie circle-spaces or space-forms or "moded-out cubes" are used to model the material shapes of stable material systems.

Very stable and definitive spectral values are (can be) defined around the set of holes which exist in very rigid and stable circle-spaces (or discrete hyperbolic shapes) which can be used to model material components (as well as modeling metric-spaces).

But these same stable shapes would allow for very stable properties . . . , to be related to a local descriptive context (of partial differential equations) which is (are): linear, metric-invariant, and parallelizable and orthogonal (shaped) coordinate system context, for metric-spaces of non-positive constant curvature (for metric-functions with constant coefficients, so that along with R(n,0) and R(n,1), there is R(s,t) which can also be considered [where s-space, t-time dimensions]), . . . , such as a cause for stable orbital systems, as well as allowing the containing space, itself, to possess stable spectral properties, so that on a finite dimensional, but high-dimension, containing metric-space quantitative measurable properties can all be derived from a finite spectral set, thus identifying a clear challenge to quantitative containment being based on the properties of a continuum.

The observed property of non-locality, or equivalently action-at-a-distance, can be interpreted to be an expression of great rigidity for the existence of stable (geometric) patterns.

How can math patterns be organized so that stability and quantitative consistency are properties of the world which manifest as stable systems with definitive properties?

Re-iterating

Using stability, and quantitative consistency, as the basis for mathematical models of fundamentally stable physical systems, instead . . . , of basing the descriptions of stable systems on indefinable randomness, and non-linearity which are defined on a continuum . . . , [consider stability, and quantitative consistency, which have a geometric basis, as the proper basis for physical description].

To do this consider new ways in which to interpret Thurston's geometrization, and consider the very stable "discrete hyperbolic shapes" as the stable basis which underlies the geometry of physical systems and their descriptive (or mathematical) structures as well, where the stable, rigid "discrete hyperbolic shapes" are also to be the reason that measuring is also stable (measuring as well as the observed measurable properties of physical systems are stable).

Reject the math patterns of both randomness and non-linearity as being fundamental to physical description (randomness and non-linearity are not fundamental, either physical or mathematical, properties needed to describe the stable properties of the very stable physical systems).

Rather, randomness and non-linearity are (can be shown to be) either derived properties or are the properties of "free" material systems, (respectively), which are not a part of either a (strongly) stable shape or a stable orbit.

Discrete hyperbolic shapes defined within a finite dimensional hyperbolic metric-space (11-dimensions), which models the description's containment set, allows for the definition of a finite spectral set, where such a set is defined so that all quantitative descriptions will depend on such a finite set of quantities, ie a model of a "math pattern" which is very far from being dependent on a continuum (where in a continuum, convergence has been used to define "set containment," and thus it has also been used to determine the set of operations where are assumed to be applicable to the descriptive structure (or construct), this can lead to logical inconsistencies [eg geometry and randomness co-existing because of the ability to define convergence properties in the two logically incompatible properties, eg "Are there smooth shapes or does space have a randomly chaotic structure at its very small scale (or very small size-scale)?"]).

The stability of discrete hyperbolic shapes is related to rigid geometric patterns.

It is a need for rigid geometric structures and the relationship which holes in these geometric structures have to the property of rigidness, and spectral stability, in regard to both bounded-ness and quantitative consistency, thus leading to the idea of "circle-spaces" as a stable, "quantitatively consistent" basis for descriptions of (observable) measurable patterns. The very stable, and geometrically rigid, circle-spaces allow for the existence of: stable spectra, the discrete hyperbolic shapes, themselves, or relatively stable force-field properties.

The force-fields of the electromagnetic materials are based on the hyperbolic wave-equation which has both advanced and retarded potentials, or wave solution-functions, so that these wave-functions are dependent on either holes in stable shapes or holes determined by rigid material structures, which are similar to holes in space.

Thus, there is a dependence on a set of harmonic wave-functions associated to the property of metric-invariance on space-time, {where space-time is one-to-one and onto (or isomorphically) equivalent to hyperbolic 3-space}.

The currents associated to an electromagnetic force-field have the properties that they surround "materially made holes," so that "what is described" by these currents is fundamentally spectral (or harmonic, but these spectral properties (of currents) may not be resonant with the finite spectral set, which defines all of what can exist in some given 11-dimensional hyperbolic metric-space).

The rigid nature of material, made-up of components which are discrete hyperbolic shapes, allows for this material to also define "holes in space," such as 1-dimensional lines (or bounded curves) to be identified in space by the material components, eg electrically conducting wires, so that the currents (equivalent to rigid spectral flows which surround spectral sources) can flow about these closed material curves, which are composed on "free" material components, which are "condensed" so as to be close together.

The idea of "condensation" is based on resonances which many (gathered together) "free" material components have with either lower dimension spectra, or related to resonances with subspaces of the same dimension, but different subspaces than the containing space of the rigid material.

The condensation is not a strong bond. This is because the material components are not of the right ("correct") size for strong material interactions within the dimensional level, and/or the particular subspace, within which the material is contained, where the dimension of the new "condensed system of material components" is the same dimension as the material containing metric-space, ie a dimension above the dimension of the stable material shapes which the given metric-space contains.

The main issue comes to be about rigidness built from (or upon): (either) the "holes" (and related spectra), upon which the rigid structure depends, are built of material, composed of "free" components of (condensed) material (where bounded, closed, 1-dimensional curves of material current are analogous to the 1-flows of the stable hyperbolic 2-dimensional material which are contained in hyperbolic 3-space) (or) resonances which can be used to define a very rigid, and very stable, sets of discrete hyperbolic shapes (where each such shape is composed of stable holes, and has a property of resonance with the spectra of the high-dimension containing space), eg the stable hyperbolic 2-dimensional material components which are contained in hyperbolic 3-space, and have spectra which resonate with the finite spectral set, defined by the shapes of the hyperbolic metric-spaces of the different dimensional levels of the 11-dimensional hyperbolic containing metric-space.

Note: The 2-dimensional hyperbolic material components also define a confining orbital structure (or "free" material-component metric-space structure) for 1-dimensional hyperbolic material components (where, in turn, the 1-dimensional free material could define 0-dimensional holes in the 2-dimensional hyperbolic metric-space).

That is, the stable hyperbolic material components can also be considered to be hyperbolic metric-spaces with open-closed topology, resonating to the spectral values defined by a finite spectral set contained within an over-all high-dimension containing hyperbolic space (an 11-dimensional hyperbolic space).

The condensation of "free" material components is a result of resonances with either lower-dimensional spectra (ie the faces of the condensed material component resonate with the finite spectral set), or resonances with different subspaces of the same dimension (as the dimension of the hyperbolic material-containing metric-space).

B. Within this new context, this new set of interpretations, and new sets of assumptions, there are many new questions:

Are the existing states of material mainly about the distinctions . . . , between the hyperbolic material in a containing hyperbolic metric-space . . . , being related to

I. orbital properties of the metric-space or

II. related to being "free" components within the metric-space, but the "free" components fit into a very rigid structure [of both the containing metric-space and the set of other very rigid "free" material components contained within the metric-space].
This can be:

1. free material components by themselves and interacting with one another, or
2. the spectra of a space-form (resonance, size-scale, and energy are all consistent, so that new stable space-forms can emerge), or
3. components conforming, by a structure of material interactions, to orbits of the material component's containing metric-space (eg a material component of the correct dimension) occupying a metric-space's stable face (or stable spectral-flows) structure), or
4. condensed material components (with only weak resonances where: dimension, resonance, size-scale, and/or energy, are all not consistent to the condensed shape of the set of material components), but some of these key-factors . . . , related to (the existence of) material/metric-space structures . . . , are present, so as to allow material components to condense together)?

What is a derivative?

Is a derivative a discrete operator used to distinguish the relation between the "free-ness" or "orbital-ness" of the discrete components of the descriptive context, eg material components or the component which identifies the metric-space (which, in turn, contains the material components).

What is space?
What is material?
What is its dimension?

or

How are space and material related to one another by a dimensional property?

Different (or adjacent) dimensional values distinguish material shapes from metric-space shapes.

What shapes can space have?

The same types of "discrete hyperbolic shapes" that material components can have (but of a different dimension).

Is a discrete hyperbolic shape . . . , of either a material, or a metric-space, component . . . bounded or unbounded?

They can be either bounded or unbounded up-to hyperbolic dimension-6, where for six-dimensions and above, all discrete hyperbolic shapes are unbounded.

How can: containment, interaction, shape, material, metric-space, and dimension be organized?

How can: containment, interaction, material, and metric-space be organized, in regard to both dimension and shape and size?

What are functions?

In a containment space which is "very rigid," are functions primarily to be modeled as, or related to, stable shapes?

Yes.

Are functions and operators primarily used in regard to distinguishing "free" material components from orbital material components? Essentially, yes.

Operators (derivatives, multiplication by constants, local transformations, etc) and processes (measuring, interactions, discrete angular transformations, etc) have the same types of structures for each dimensional level, but they are associated to different properties for each particular dimension, in regard to both metric-spaces and material components contained in metric-spaces, eg dimensions are different and the set of "normal sizes of material components" of a particular dimensional level, or of a particular subspace, can be quite different.

C. The main issue of being human is creativity.

Creativity is related to "knowledge," which can either be directly related to perception (we have within ourselves a deeper mystery than we expect, and it is a deeper mystery that can not be unraveled by an assumption of materialism), or it can be an abstract model within which creative actions are realized, ie this would be a geometric abstraction since actions are in a context of both space and stability (eg conservation of energy).

These are the two main (mathematical) properties (position in space, and stability of a pattern) which distinguish Euclidean space from hyperbolic space, (where they are mathematical properties, since it is a set of stable math patterns which also needs to be preserved within math constructs).

Re-iterating

It must be noted that the primary failure of today's math-science descriptive context is that it is failing to describe the observed stable properties of material systems, so as to consistently make such descriptions of stability, to a sufficiently precise level, for general (but fundamental and stable) systems.

The claim is that these systems . . . , such as the relatively stable nuclei, or atoms with atomic numbers greater than five, or (ironically) the apparent stability of the proton . . . , are too complicated to be able to describe, based on what is currently (2013) considered to be the basis for physical law, where one of the assumptions associated to current physical law is an assumption of materialism, ie material defines existence, and whereas now (2013) physical law is based on randomness and non-linearity (and materialism).

Language can be re-made

However, if one wants to create something, and the language which others are using is not adequate to realizing what one wants to create, then one should try to form a new language, and do it at the level of "elementary ideas" about precise language, ie quantity and shape (this is the correct interpretation of Godel's incompleteness theorem).

What to consider when re-making a precise language

Consider the elementary level of; axiom (or assumption), context, interpretation, and definition, upon which precise language can be built.

An example of a list of such elementary properties is:

1. assumption
2. Context (and definition)
3. Interpretation (eg in particle-collision experiments, which patterns are the more important:

    "The particles and their types ?" [this may really be about the particle-type's relation to bounded or unbounded shapes],

   or

"Are the properties of unitary-ness and higher-dimensions more important properties, to be fit by a (one's) descriptive language?"

4. Measuring (quantity)
5. Containment (sets)
6. Dimension (measuring modeled as a set of lines)
6. Independence (orthogonality or commutativity)
7. Geometric measures (differential-forms)
8. Metric-invariance (if a metric-function is non-linear, then it is not quantitatively consistent. Because there are many fundamental systems whose properties are observed to be stable, then this is an inductive proof that non-linear description has a limited range of (valid) applicability.)
9. Does space have a shape?
10. Continuum (too big of a set) Or,
11. A Finite basis for quantitative sets
12. Space (position in relation to the distant stars)
13. Time (stable energy in time, defined in relation to space-time, eg m=E)
14. Stability, continuity considered first as being related to a stable, existing set of patterns, which exist in time.
15. "What are the stable shapes?" The easiest answer (might only be a partial answer) which is a working hypothesis, is that they are the set of "discrete hyperbolic shapes."
16. Material, is it a shape, or a scalar multiplicative factor, or a point, or an illusion? It is a metric-space shape (ie the shape of a metric-space) associated to certain (physical, or mathematical) properties.
17. "Do different materials, which are stable, demand different temporal dimensions associated to their existence?" That is, material is identified as that which can be continuous in time, ie conservation of energy, where in turn stable material can be related to position in space, but a spatial position is determined in regard to an absolute space, ie Euclidean space and its associated property action-at-a-distance between material at different spatial positions, but "What if there was a new material types, different from inertia and charge?"
18. Forces, ie time's relation to the changing properties of material in space, ie the relation of material to its position in space and the way in which a material component changes (or is caused to change) its position in space.
19. Complex numbers, circles are observed to be consistent with linear measures
20. holes in space, (the relation of potential energy to either a system's conserved energy, or to stable spectral values).

That is, Which set of values is more fundamental, "energy [and whether energy is single valued] or spectra?"

21. potential energy [Is single-valued only if the space within which it is defined has no holes in its structure.]
22. Functions (see below)
23. spectral properties (stable spectra can be related to stable holes in shapes)
24. Discrete Euclidean shapes
25. Discrete hyperbolic shapes
    {These are both circle-spaces, or shapes obtained from moding-out rectangular (or cubical) shapes}
26. The shapes of spheres are non-linear, and when perturbed the results cannot be described, ie the spherical shape is an unstable pattern.
27. A uniform stable unit of measurement. (Why are measurements so apparently stable when they are used to build things? Why do some fundamental systems possess such stable properties?)
28. Is stability dependent on stability of shape? Apparently, yes.

Any of these concepts can be considered at an elementary level, and these considerations can be used in regard to creating a new measurable language with which new creative contexts can be developed, and these creative contexts are not confined to a (necessarily) material world, yet they can still be about practical creativity within a real existence.

Interpretations of math patterns

Comparison (or measuring)

(as in the case of a measuring rod, or in relation to metric-invariance, or where a stable unit of measuring can only be related to a quantitatively consistent linear partial differential equation and metric-invariant containment spaces [or domain spaces for measured values, ie functions, which are modeled as (solution) functions (to linear, separable partial differential equations)]. Note: When the partial differential equations are separable the directions for measuring (with measuring rods) are locally independent everywhere on the global shape, a shape which is defined by the shape of the coordinates which "separate" the partial differential equation)

Is measurement about comparisons, eg lines compared with rods, rods compared with material extension in space?

Do material shapes represent rigid (discrete hyperbolic) structures, so as to be able to define a rigid geometry associated to either stable uniform measuring capability or Is measuring mostly about spectral values associated to holes in space, since holes in space are mostly (or in their simplest form) about the structure of circle-spaces, which can be used to identify both metric-spaces with stable measuring properties, as well as stable material components, which possess stable spectral properties, and this is (can be) modeled in a quantitatively consistent manner by the shapes of circle-spaces, which are linear, metric-invariant, and the tangent directions (thus, re-introducing the idea of measuring directions or rigid rods) of the circles (upon which uniform measuring is consistently defined [or depends]) are independent, so as to fill the dimensions of the containment space everywhere, so that each point possesses such an independent pattern of local measuring directions, (ie commutative local numerical relations, or the matrices [used for the local linear approximations of functions' values, in regard to a local model of the inter-relationships of a system's measured values] are diagonal, everywhere) etc,

Or
Is measurement about,
Functions?

"Is measurement fundamentally about a function?" since measurements would be related to (measurable) properties of the system, where the system is contained in space, where, in turn, the containment space represents the domain space for a function whose values represent the measured values of the system's properties, so the function should be represented by the domain variables, since the system is contained by the domain space, where the function's values are measured in a local construct of a linear approximation to the tangents of the function's graphs, in relation to the domain space's measuring directions, ie tangent lines represented as linear graphs (or linear functions) defined in the variables of the domain space.

Or

Stable spectra (are the correct basis for a descriptive measurable language, since they can be used to describe the set of discrete hyperbolic shapes which together compose an entire many-dimensional existence)

Spectra may be considered as representing "the fundamental" set of measured values associated to stable sets of circles defined on stable shapes, which are always assumed to be bounded.

[or more substantially, one can model a quantitative structure as being built upon a finite set of spectral values, so this means that holes in space, around which the stable spectral values are defined, where space is most often a stable (discrete hyperbolic) shape, ie a circle-space.]

Spectra may also be considered to be related to sets of differential equations whose solution functions define a function space. This is the usual math model of a spectral structure.

Spectra may be associated to sets of harmonic functions, eg sine and cosine functions associated to a fundamental wave-length and the set of all of its harmonics (ie integer multiples of the fundamental wave-length), but such a set of functions can only represent periodic functions. That is, the spectral-functions in function spaces are placed in an infinite series (or sum) so as to converge to any arbitrary, but periodic, function.

Comparing solution functions associated to force-fields to solution functions associated to spectral values

But this comparison is done in a context which is related to circle-spaces

Forces are related to potential energy (related to a conserved, stable, unified structure in time) Potential energy can only be defined in a space which does not have any holes in its shape. However, cubical fundamental domains of relatively stable circle-spaces would qualify as space without holes, if they are considered at the "correct" dimensional level.
vs.
Spectra, where a finite stable spectra can be related spaces which possess a set of geometrically stable holes.

Potential energy is defined within spaces which have no holes, so that a potential function will be single valued (this allows energy to be defined and a physical system can be associated to conservation of energy, ie the system identifies a stable pattern).
vs.
Spectral values are stable (in the discrete hyperbolic case) when they are related to stable holes, in a stable shape.

Thus, there is a "natural" structure of space and its stable shapes,

1. forces appear to be related to stable energy values since it appears at the dimension at which space is defined that no holes exist, since one's viewpoint requires that one see the

(apparent) system containing space's fundamental domain (or right rectangular simplex), while

2. spectra appear, due to the circle-space structures of lower dimensional material components (or open-closed, bounded metric-spaces) which are the material systems contained in the metric-space of the given (initial, or apparent) dimension.

Forces are related to relatively stable material geometry which surround the point in space where a material component's motion is (or spatial displacements are) going to be measured in a local manner. The local measurement of spatial displacements define a second order differential equation related to geometric properties (of material distribution) defined in a metric-invariant, second order structure of differential-forms, in turn, related to a stable potential energy value for the system. That is, spatial displacements are related to a stable system geometric pattern associated to conservation of energy, so that this is done by means of a metric-invariant second order set of partial differential equations, which are solvable only if they are linear and separable.

Note: This is the context of circle-spaces. Outside the context of circle-spaces, forces define unstable fleeting patterns, either collision interactions, or condensation of a set (usually a large set) of material components.

vs.

Function spaces defined by metric-invariant, spectral (or harmonic) equations (heat equation)

[note: spectral values are also related to wave-equations, where in this case the wave has identifiable physical properties.]

However,

Function spaces do not work in regard to finding a general random system's spectral structure, because functions are defined in a context which is too general, especially in regard to the rigid structures of both material components and metric-spaces (where both structures are being modeled as discrete hyperbolic shapes), upon which existence is being built.

Interpretations of stability and containing sets

There are models where a containing metric-space has a (stable) shape.

There are models where a material system has a stable shape.

There are also models where a (solution) function (to a linear separable metric-invariant partial differential equation) has a shape.

This is the context of stable circle-spaces, which seems to be the main mathematical pattern through which stable properties of physical systems can be consistently developed and precisely described.

Higher-dimensions are placed into a containment set, which is dominated by a need for both stability and quantitative consistency, so that this stability is related to stable shapes, ie the very stable discrete hyperbolic shapes which also possess very stable spectral properties. The property of quantitative consistency is (can be) related to a quantitative basis which is a finite set, a finite spectral set, which determines the sizes and spectral properties of the discrete hyperbolic shapes upon which the forms of both material and space depend.

There are models where higher-dimensions have shapes (the shapes of stable circle-spaces) associated to themselves.

Such a context requires new models for:

1. the processes of material interactions
2. Scale of a stable geometry, then size-scale of material, or of a system, or of a metric-space, also is a contributing influence on how higher-dimensions can influence the various dimensional levels of existence.
3. This allows for a definition of a finite spectral set upon which an existence can be based.

A physical system's differential equation is metric-invariant so it is either a generalized Laplacian, eg a wave-equation in space-time,

Or

It is similar to an SO(n) connection-form

This is about material either being "free," but requires that the free-component be identified with a position in space.

Or

Material shapes identifying a spectral-orbital geometry, or a free material component being related to the orbital properties of the discrete hyperbolic shape, within which the material component is contained.

Holes in space can also be related to homotopy, or to obstructions to the continuous deformations of circle-spaces to single points . . . . , (where the continuous deformation of a shape . . . , [a shape which in its "lowest dimension containing space" can be used to define an inside and an outside (of the shape, which is contained in a coordinate space {of lowest dimension})] . . . , is obstructed by the holes in the space), ... , is sort of a definition of a "hole in space."

This idea seems a bit too general, since holes in space can be defined by circle-spaces, (though the properties of a hole-in-space might be somewhat abstract in higher-dimensional spaces).

or

"Holes in space" can be defined by material structures which identify a closed-loop, or a surrounded geometric construct, eg material is capable of identifying a rigid discrete hyperbolic shape which is a chain (or closed loop) which possesses the property of great stability equivalent to the circles (or circle-shapes) in circle-spaces, ie the material chain is defined by resonances (or by "condensation" of "free" material components into a relatively stable discrete hyperbolic shape, where a circle-shape surrounding one of these discrete hyperbolic shapes also cannot be deformed to a point, where a discrete hyperbolic shape is defined by the geometry of its circle-shapes).

Appendix:

The context of culture, in regard to the basic nature of being human, ie the desire to seek knowledge in order to be creative in a practical context (as well as in a selfless context), can be summed-up in the over-all containment space of society where there is either equality or inequality.

The society which supports people as equal creators will be a creative society, while a society which organizes society in the context of social inequality will be an extremely violent society.

The propaganda system as well as the system of law, eg based on property rights and minority rule, is now being used to define and support inequality and extreme violence.

The nature of authorities who possess dogmatic purity (so as to be published in the authoritative [peer reviewed] propaganda system) is that they are performing their expressions of dogmatic authority so as to "get the attention" of the owners of society, so as to "qualify" so as to get to work, as wage-slaves, on "what the owners of society want created" (for the narrow selfish interests of the owners of society).

This essentially degenerates into the question of being able to service instruments of great complexity vs. realizing new scenario's of existence and new processes in which to develop other complex instruments which are (would be) new, or more to the point realizing "how life is a complex instrument" associated to fundamental creativity. That is, the complex instruments, controlled by the owners of society, are simple compared to life.

Thus, it is natural to consider new contexts within which to view existence and the position of life within these new contexts.

Whereas dogmatic purity and fixed ways in which to interpret observed patterns is in opposition to the core of what it means to be human.

The instrument which is organized and used to most oppose new ideas and to oppose great variety in regard to human creativity is the media, which is a set of instruments which are owned and controlled by a few, where the laws organized so that "only the ideas of the owners of society" are given expression, and the tone of that expression is over-bearing authority and domination of the many by the few. This is the same model as the Holy-Roman-Empire, where then (after the Emperor Constantine), the only expression given authority was the church, which was then the media of communication as well as of education, wherein only those ideas which possessed dogmatic purity were allowed to be expressed.

This all boils down to law being based on property rights and minority rule (a republic), but the US society was born from a society whose law was "to be based on" equality, yet the violent selfish elements of the settlers who came from Europe so that a selfish, violent society won-out, so as to become the Puritan-US-Empire, where Protestantism was originally about reading scripture, and interpreting that body of literature for oneself, ie it was about equal free-inquiry. Furthermore, there were fine examples of the US society of around 1776 (or within the 18th century) as being equal. Most strongly, the Quaker settlements (settlers), around Philadelphia, who negotiated with the native peoples as equals.

Today dogmatic purity of the science and math community is about the owners of society wanting physics and math to be about building weapons and organizing instruments so as to place ever more control with the owners of society . . . (so as to have a society based on central-planning for selfish interests, whereas the propaganda of the similarly oligarchical communists is "central planning for society's needs" but inequality only sees illusion), . . . and not having the people of the society being equal and creative [where not being equal, and not being equal-creators, also results in a highly controlled market structure, ie the market is currently (2013) an expression of social domination, a hierarchy of false value within which the wage-slaves toil].

Chapter 11

Ways in which to simplify broad contexts

It is a good idea to always review the basic constructs and contexts and assumptions and interpretations of a subject.

What is the subject matter trying to describe, ie what motivates the subject, and how is it trying to do this, eg what (math) patterns are used and how are they organized?

However, the intent of a professional math or scientist is to "correctly" apply their dogmas; so as to continue to get a "big salary" from their paymasters, to whom the dogma might be useful, in regard to the narrow "productive" interests of the owners of society.

While only a few professional scientists and mathematicians are allowed (or even try) to consider new ideas, but they are straight-jacketed by the narrow dogma which defines their value to the owners of society.

Math is about:

Math is about quantities (or measurable properties) and shapes.

Formation of quantitative sets, which seems to have included developing the ideas of the operations of addition and multiplication

Quantities are used to measure, so as to build things and use the measurable properties of particular types

The stability of quantitative sets and measurements depend on a stable uniform unit of measuring

Defining equations, two different representations of the same quantity (or the same property), using numbers (variables), functions, and operators, where math constructs number properties around the ideas of algebra, which focuses on solving algebraic equations, so algebra organizes quantitative constructs around the idea of an inverse operator, or inverse value.

Solving equations

Algebra

Order of operations and inverses, where the distributive property allows for the structure of a number system, and the commutative property for a many-variable coordinate space (or containment space) [commutative or diagonal matrix construct] encodes one-to-one and onto (square matrices), but where the property of onto demands either a linear or hyperbolic (in a correct geometric context, eg one-to-one increasing functions, etc) function-value to domain-value inter-relationship (defined continuously, in a global manner) so that an inverse can be found, and thus geometric measures can then remain quantitatively consistent, in regard to solutions (eg solution functions to differential equations).

Functions

Function values represent measurable properties, and a function's domain space (or coordinate space) is the space within which it is assumed the system (and its properties which are being modeled, or measured) is contained, ie set containment.

The idea of nearness in regard to function-values and its domain-values can be represented as the properties of sets in regard to the image set of a function and the inverse-image set

Nearness is easiest to consider as converging sequences of numbers

The operators on functions

Derivatives and its (almost) inverse integral operator

The derivative is also a function which represents the slope of original function's graph, and which identifies a linear quantitative relation between the original function's values and the coordinate values of the original function's domain space. It provides a local linear quantitatively consistent measure of the function's values in terms of the domain set's values.

The derivative allows a model of a quantitative set (the measured values which a function's value represents, when defined on the domain space) defined within a continuum, where these local linear approximations are used so as to become a local set of quantities which can be used within algebraic constructs matrix operators acting on vector-component values.

It also allows for a discrete operator to act on a system whose underlying stable structure is a simple geometry.

Derivatives allow differential equations to be defined and (possibly) solved

However, there are very many other operators which can act on function spaces, eg the multiplication by a function operator.

It is assumed that these operators on functions are required to be defined in a continuum

Thus, quantitative sets, often put-together as coordinates, where each independent measuring direction, it is assumed, must have the properties of a continuum

But continuums are very big sets, perhaps "too big"

There are metric-spaces which have metric-functions (symmetric 2-tensors) which can be represented as matrices, where there can always be found coordinates where the metric-function's matrix is diagonal (when this is continuously true for the entire coordinate system then this is a model of geometrically separable shape), and there are other geometric measures of: areas, volume, etc

Shapes

There are the (usually) unstable non-linear shapes and the stable (linear and geometrically separable) shapes of the non-positive constant curvature metric-spaces whose metric-functions have only constant coefficients and (for example) these are the discrete Euclidean shapes and the discrete hyperbolic shapes

Function spaces

These are applied often successfully for physical waves in the hyperbolic wave-equation

And

They are applied to probability waves in either parabolic or relativistic wave equations (there is very little general success with this method of quantitative description)

Operators are applied to function spaces so as to diagonalize the function space, or to find a spectral-set of functions for the function space.

Physics is about:

Review of the math of modern science (2012)

Classical physics:

Based on materialism, works on an assumed continuum, based on geometry (or material), the differential equations is based on the idea that it is a local measure of a function's value, where if the function is velocity then the local geometric relation is associated to the properties of a 2-form related to the material geometry of the component's bounding region of containment. The solution of the differential equation which is defined locally extends information from the local to the global, where the size of the geometry is from the nano-level on up.

Quantum physics:

Based on both materialism (probabilities of point-material events in space and time) and (indefinable) randomness, the system is modeled as a function space, and the operators (such as derivatives) which act on the function space represent physical properties, where the fundamental operators is the energy-operator (or wave-equation) which is a parabolic second-order differential equation which is plagued with 1/r singularities, the point is to identify the stable spectral set of a general quantum system by diagonalizing the function space with the set of operators. The process is essentially starting with a global view of the very small system and reducing the description to a (practically un-useable) random local property, which when observed causes the system to dissipate, ie it is a (math) process of losing information concerning a system whose properties are very stable and definitive.

The systems with few components based on probability, where the components (often a [or set of] nuclei and a set of electrons) are often charged so that all components possess a 1/r singularities associated to their position in space so that no valid solvable wave-equation can be found for such systems. That is it is a useless descriptive structure.

Function space techniques work for waves with physical properties, and usually in regard to the second-order hyperbolic differential equations.

Particle-physics:

Model material interactions as random point-particle collision events (though it is based on randomness so the geometry of a particle-collision in a continuum is not valid) where the interaction is supposed to correct the quantum wave-function (which basically does not exist) by perturbing it, but it does this by introducing internal particle-states both as part of the particles and as part of the wave-function, similar to the spin-states of both particles and the wave-function, yet it is also a set of local vector properties of an imaginary (internal) space where the interaction is modeled similar to a classical potential-energy, but also a connection term, whose (coupling constant), eg charge, acts in the context of materialism while the geometry of the connection acts on the local particle-state vectors, ie the most fundamental assumption of materialism (the geometry of material containment) is maintained, but the wave-operator (which permutes the wave-function) does so in a non-linear manner (a connection term implies non-linearity) but it is not clear that these internal particle-states associated to a wave-function have any meaning, especially, since neither general sets of atomic properties nor any nuclear spectral properties have been described in any convincing manner by this descriptive structure. Such an empty relation to accurate descriptions of a wide range of general systems proves that this descriptive structure is incomprehensible.

Where internal particle-states are vaguely related to different dimensional representations of the Lorentz group (or Poincare group [Lorentz x translations]) apparently which are acted upon by the different, A, potentials (or connection terms) in regard to the perturbing structure applied to the artificial (internal) vector properties (arbitrarily) attached to the wave-function.

The vast dysfunctional realm of quantum and particle-physics, and all of its ambiguities

Note: Nearby points, related to different sites of particle-collisions, are points in space which are completely independent of one another within the laws of particle-physics, but the "chaotic structure of space," as presented for renormalization purposes, claims that near-by points are not necessarily distinguishable from one another, thus independence is (to be) questionable.

One of many ambiguities in the parade of ambiguities associated to quantum physics and particle physics

The idea of materialism, defined on a continuum, with a wide range of allowable possibilities, to be built into a basic math structure of an indefinably random context, has not led to any wide ranging sets of valid descriptions, which in turn are (also not) related to a wide range of practical creative developments.

These ideas about particle-collisions are only related to nuclear weapons engineering, otherwise they are almost completely useless ideas about precisely describing existence (or even the more narrowly considered, material world).

Parade of ambiguities of quantum thinking

There are too many issues in the math construct of quantum physics, so as to make it too complicated and without meaning

The issue as to the scalar field (associated to Higg's particle) of symmetry breaking so as to allow for mass to exist within the particle-physics set of symmetries, which are claimed to determine all physical law,

That is, the collision patterns of elementary particles in high-speed collisions (one is required to believe) somehow determine stable spectral structures.

[yet there is no descriptions which would lead one to believe such an outlandish claim], but it is not clear that components occupying the spectral structures of quantum systems, eg atoms, nuclei, etc, actually do collide, ie Is a random particle-collision common and fundamental?

Or

"Is it a fairly uncommon occurrence?

But the assumed results of a random particle-collision, ie being associated to spectral properties of physical systems, are properties of the containing space, rather than related to the intrinsic properties of these random collisions, (though the energy range of the collision will be important)"

(are these high-energy collisions really probing higher-dimensions of macroscopic shapes

ie the shapes are the same types, ie discrete hyperbolic shapes, and they are unitary mixtures of metric-space states, which are states which are related to the dynamics of material interactions)

That is, symmetry breaking is just another ruse (red-herring), since there is nothing in the description which actually follows the rules which are claimed to exist:

For example:

Quantum properties are low-energy properties, but particle-collision patterns are determined for high-energy collisions,

Quantum properties are related to the size of the system, but no distinguishing size is provided, vaguely somewhere around the nano-size there is a transition from classical to quantum, yet large crystals have quantum properties

Mostly general quantum systems the stable spectral properties of, eg quantum systems with several charged components of nuclei and electrons, cannot be determined through the wave-equations since there are too many 1/r singularities existing for such a system, this would include most atoms, and nearly all molecules.

The description is aimed at local properties of point-spectral values recognized as the spectral properties of a stable quantum system, yet the global wave-function collapses when this information is determined.

That is, the direction which information is flowing in quantum descriptive math structures is from global to local to dissipation.

It is a random description of a system with few components, ie the information has no practical relation to creative development.

It is a descriptive structure which cannot describe the stable properties of the most fundamental quantum systems (as mentioned above).

Furthermore, what structure of a point-particle allows it to carry spectral information so that the spectral information is dis-associated from the motion of the particle.

Particle-collisions are modeled on a spatial continuum but the particle's spectral properties are discrete.

Angular momentum as a system which is semi-free, the parabolic second-order differential equation

The only part of the description of quantum systems within the context in which most of the fundamental quantum systems exist (atoms, nuclei, and molecules) is the description of angular momentum, in 3-space, which seems to still have some validity, even when there is no spherical symmetry, ie when there are many charged components which compose the system, and this is primarily for atoms, where the nucleus still provides a point for spherical symmetry for a positive charged component.

{should this be interpreted to mean that the parabolic case is the better case within which to consider angular momentum where the parabolic differential equation is placed into a context of discrete hyperbolic geometry}

That is, "what geometry is [the information of] the heat-equation (parabolic equation) a part of?"

That is, thermal physics is related to the geometry of its containment vessel.

Thus, if angular momentum is stable, then why is it stable?

Answer: It must be bounded by some containing (geometric) structure.

Back to the ambiguities of quantum thinking

Spin properties are modeled as representations of rotation groups, but for the property of spin "what is being rotated?" (a point-particle?)

Perhaps there is a need to spin-rotate a new property of a metric-space state where metric-spaces are attributed with specific properties (either math properties or physical properties)

The series solution to the radial wave-equation of the H-atom diverges, yet it is truncated without a mathematical context for the truncation, other than it fits data.

The inner-states of elementary point-particles when the data of particle-collision experiments is interpreted follow the rotation structure of a unitary group, but this would be related to the circle and disc geometries which are so prevalent in the geometry of complex coordinates, these patterns are related to the collision geometry of point-particle collisions in a continuum, yet the renormalization context requires that space not be a continuum.

The biggest gaffe (blunder) in particle-physics is that it is about modeling material interaction as particle-collisions which perturb the wave-function of a general quantum system, but there are no wave-functions known for general quantum systems.

Furthermore this is about a linear metric-invariant global wave-function which particle-physics places in a non-linear geometric context associated to the connection term which identifies the internal particle-state structures of elementary particles which occur (or change, or rotate states) during particle-collisions

The rules, laws, or assumptions of the entire math construct of quantum physics are broken when ever it is convenient to do so.

Thus broken symmetry (so as to allow for mass to be defined) is a ruse (red-herring) concerning a math construct whose rules are arbitrary.

Is the description random, within a continuum, and with geometric properties?

or

Is the description random, and without geometry, and not within a continuum?

Is mass a geometric construct, as is claimed in general relativity?

or

Is mass an intrinsic property of elementary particle whose collision patterns fit into a unitary rotation of particle-states, and is contained within a global wave-function, in turn, contained within a metric-invariant context?

To clarify word usage in particle-physics:

Symmetry breaking is done in the context of a continuum, and in turn, is used to describe discrete properties which are random, and thus outside the context of a continuum, as if the entire construct does not discretely break the indefinable rules of the construct.

If a function space (representing a general quantum system) is to be diagonalized [and a spectral set (for the function space) found] then the set of Hermitian operators, which identify the quantum systems set of measurable properties, must commute, but in general, such a set of commuting operators cannot be found for general quantum systems.

General relativity is also disfunctional

General relativity:

General relativity is a non-linear theory, so it has effectively tried to turn a workable (when linear and separable) classical theory into a non-workable, unsolvable, and unstable non-linear theory. This only provides vague information about a 1-body system with spherical symmetry.

New ideas about measurable description where the measured quantities are stable reliable and remain quantitatively consistent

New ideas:

View the differential equation as a discrete operator in a many-dimensional context, based on the stable discrete hyperbolic shapes, which model both material and metric-spaces but at different dimensional levels.

Newton's law still is considered to be true, so that one can still consider second-order differential equations of:

elliptic (orbital),
parabolic (angular-momentum etc), and
hyperbolic (collisions and waves)
types of differential equations.

material interactions are modeled either as orbits and associated angular momentum properties or they can still (or often) be modeled as hyperbolic (or collision) type systems, but during the collision they can be discretely related to an outside spectral set so as to resonate and to (or if they are resonant then they) become a new stable system, during a collision.

However, all the actions are related to discrete structures and the set can be determined by a finite set of spectra.

The idea of materialism is lost when many-dimensions are used.

Discrete multiplication by constants can occur during discrete processes, if the new shapes created by the conformal factors are allowed by hyperbolic lattices.

Chapter 12

General Relativity

General relativity vs. A new basis for (physical) description, based on stable math patterns

The new ideas about math description are based on stability and quantitative consistency (as well as a many-dimensional containment set) are quite at odds with general relativity.

Descriptions, which are measurable, should be structured so that there are assumptions and interpretations, etc, and any such set-up for a precise descriptive language is to be directly related to useful patterns for practical creativity and placed into a context of reliable models for measuring.

However, to obtain careful word agreement based on a set of assumptions, can (often) lead to a precise language which is dysfunctional. Such is now (2012) the case of the current precise descriptive languages used (now) to interpret the observed data patterns.

The basis for measuring in the new ideas

The core agreements, or assumptions, of the new basis for description are that there are stable objects, and to identify the changes in their positions in space (spatial displacements of an object's position) these object's positions in space must be measurable in relation to the distant stars.

This idea of measurable positions in space implies an absolute Euclidean space defined in relation to the distant fixed (or rotating) stars. [This is upheld by the findings of A Aspect's experiments on non-locality. (or the data of that experiment can be interpreted to support the idea of an absolute Euclidean space)]

The objects that exist are stable patterns, and this means that they "continue" their stable properties, in relation to changes in time (ie objects which remain the same during temporal displacements). These are the properties of hyperbolic space, or of the discrete hyperbolic shapes, or equivalently, the stable (energy) properties of material systems in the metric-invariant structure of space-time. Furthermore, in a (the usual) metric-invariant model of space-time, temporal invariance implies mass is equal to energy, and this same pattern also relates space-time to a hyperbolic metric-space (or a velocity space, or an energy space), and subsequently to the very stable discrete hyperbolic shapes, where a hyperbolic metric-space is a space characterized by energy and time.

The formulation of general relativity

In the formulation of general relativity, there is the core idea that no measures in a physical system's containment set, which is locally space-time, are valid (since an object's frame, associated to [an object's] motion, is indeterminable), but it is assumed that it is only possible for an object's relative changes [within the containment frame] to be contained within a pattern of an equation (representing a physical law) where that equation's "form" is invariant to arbitrary coordinate changes (of a frame) . . . , (an equations invariant-form is supposed to imply that physical laws be tensor equations), . . . , so as to identify a valid description of the object's measurable properties of change.

or

An, equivalently, object's properties can only be [measurably] described if the equations which describe the object's properties are invariant to any coordinate transformation (or being generally covariant, ie consistent with a tensor's invariance to arbitrary local coordinate transformation properties applied to the tensor), then the solution function to such an equation can be used to identify a valid description of change, in the local coordinate space, for that object).

However, this descriptive structure does have an implied relation to a "conserved systemic-entity," {or a property of conservation of energy is assumed to be related to the system's changes} (or an object's motions can only be related to the covariant structure of its equations, which identify the relation that locally measurable properties of the system (or object) have to material (or energy) distribution in a spatial region [so as to be only identified within a local pattern]).

That is, general relativity is about the laws of local measuring in regard to motions of stable objects, where:

1. it assumes that the object does exists in a continuum,
2. it assumes "no" coordinate structure associated to space,
3. it assumes "no" structure within which to measure time, but

4 (a). there is (also) an implied assumption that there are "no" holes in the continuum, metric-space, containment set,.

4 (b). Thus, by this last assumption, the inertial system (of mass and/or energy) has a well defined relation to energy.

5. Furthermore, it seems to be assumed that there is a property of conservation of energy, so that the solution function can be found before the key components of the inertial construct can dissipate, ie the equation, representing the physical law, stays valid for some identifiable (macroscopic) interval of time.

However, if one cannot measure, then there are no stable math patterns, which are valid.

That is, fleeting unstable local relations cannot be a basis upon which a consistent quantitative set (a model for measuring) can be built.

The existence of stable measured (sets of) properties of systems is the property upon which the laws of change depend for their usefulness, where, in the classical context, this means laws (and their solution functions) can be associated to both accuracy and control of a system's properties. (1) (over)

The new foundations for measurable descriptions

A containment space and stable forms (or stable patterns) are inter-related and are the domineering concepts of existence and their inter-relationships with descriptions of change. The new model of inertial interactions does happen to be covariant in regard to Euclidean space. In the new model for describing inertial (material) interactions (or inertial change) objects are determined to have absolute positions relative to the distant, fixed stars in Euclidean space. The describable pattern of a material (inertial) interaction is related to a stable object's relation of spatial position in regard to the distant stars. The existence of such absolute positions in Euclidean space is a property which is upheld by A Aspect's experiment on non-locality (space is non-local, action-at-a-distance is real).

That metric-spaces of different dimensions, and "different signature for their metric-functions," have (physical, or mathematical) properties associated to themselves; those properties being:

(1) spatial position in a Euclidean metric-space, and
(2) the property of stability (or temporal continuity) of form, or stability of a material object, (or conservation of energy) is a property which belongs in hyperbolic metric-spaces.

These properties of metric-spaces (position, stable form, etc) identify opposite states, and the spin-rotation of metric-space states can be defined between these opposite states.

These opposite metric-space states are a part of the dynamic or interaction process.

The opposite states:

(1) in Euclidean space are: "the fixed stars vs. the rotating stars," and
(2) in hyperbolic space it is: (+t) vs. (-t).

Note: It is only within a discrete hyperbolic shape that one can define a many-body system whose many stable orbits are related to a dynamic adjustment of the inertial properties of an object (within one of the relatively stable orbits) due to the shape of the space within which the object is contained.

Inertial properties are contained in Euclidean space, while stable energetic properties are contained in hyperbolic space, the needed Euclidean properties of inertia of a stable material system comes into Euclidean space from a property of resonance, which a inertial Euclidean shape (a discrete Euclidean shape) has with the very stable discrete hyperbolic shape which is the stable basis for a material interaction.

How is stability of a math-form to be expressed? either by an invariant form for an equation for any local coordinate transformation (or covariant tensor form for an equation which represents how an object's motions change), but which (in general) is a non-linear equation, and thus it is a quantitatively inconsistent expression of quantitative inter-relationships, which exist between the solution function and its domain space, or by the stability of a mathematical pattern of existence, ie by the stability of shape in the space of energy, [where only the frames of differing velocities are not detectable {and thus need to be accounted for by local coordinate transformations} within a hyperbolic metric-space].

In regard to an invariant form of an equation, it is the geometric context of the domain space (or the context of solvability) which would allow a solution function to be found, (so as to describe a stable measurable pattern by means of the existence of a solution function).

Whereas an equation which describes properties in a quantitatively inconsistent manner (ie a [numerical] solution to a non-linear equation) has no ability to identify a stable pattern of existence.

If there are no ways in which to identify coordinate shapes and measuring (comparison) structures in the domain space, then an equation defined in such a context cannot be solved, but if numerically solved it has no quantitative consistency.

It is the stable discrete hyperbolic shapes which are contained in a hyperbolic metric-space which identify stable shapes, and it is only in the context of these stable, globally commutative shapes, that stable, controllable, solutions can be found to exist, and subsequently controlled.

Surprisingly, hyperbolic space is associated to the physical property of energy, and where "invariance to time displacements" implies conservation of energy, which in hyperbolic space means that the discrete hyperbolic shapes are invariant to time displacements.

Chapter 13

Mystery

Consider the prospect of creativity in regard to expressing ideas with words, and also consider the true realm of the truly broad expanse of existence, in regard to perception and interpretation of observed patterns and the unknown, in regard to identifying which interpretations and which key patterns, and what method of organizing language which can be used to (correctly) describe the observed patterns, then in the context of language this identifies the wide ranging realm of mystery.

It is this vast realm of mystery to which the science and math communities are to be related in regard to creating a measurable descriptive language which describes (stable) properties and allows one to use these properties through measuring coupling and control of the described measurable patterns of existence.

But instead these communities are: dogmatic, authoritative, narrow, elitist, and they oppose those who do challenge their authority, and so does everyone else whose minds are inhabited by the authority of the propaganda system, so as to convey the message that only they, the superior intellects of the world, can discern truth.

Consider looking to the stars.

Are the distant stars similar copies of the sun?

or

Does space have hidden shapes?

and

Is physical law, really (at best) a regional law?

as now seems to be the correct interpretation of the (apparent) fact that distant stars (galaxies) are related to data whose conventional interpretation is to be that the distant galaxies do {are interpreted to} possess the property that many of these whole galaxies are, in fact, traveling faster than the speed of light.

Just as A Aspect's experiment showed that Einstein was wrong about non-locality, which can be interpreted to mean that action-at-a-distance is true, where it should also be noted that guiding small (man-made) satellites through the solar system it is best to assume Newton's action-at-a-distance in feedback systems of dynamic control of these satellites, so that Einstein may also be wrong about special relativity and the ability of material to exceed the speed of light, but perhaps it is only distant material which can possess this property, but maybe not.

It is difficult to justify all the expression of dogmatic authority in regard to science and math, since physics and math, since about 1910, has contributed so very little to the development of (new) technical contexts, rather quantum properties have been attached to classical systems in regard to quantum's relation to technical development, apparently only the laser was developed based only on quantum ideas.

Apparently, in the past, the solar system has gone through great upheavals in regard to its orbital structure {one does not really know if these claims are valid}

Nonetheless, it also is believed (by authorities) that the solar system has been stable for over 1-billion years, whereas life formed almost immediately after the earth's crust cooled (if in fact that is a valid model) 3.5 billion years ago.

Furthermore, the magnetic field of the earth apparently has changed its structure (eg polar orientations) very quickly in the past punctuating periods of (long time) stable structures.

If they are true, . . . , then how can these large-scale fairly-rapid upheavals be explained?

Consider the properties of the solar system, so as to consider a new form by which space is shaped, if the 3-dimensional 'discrete hyperbolic shapes" . . . . , in "our" dimensional subspace within a particular subspace set, . . . , which characterize the size of such shapes which form, at this level of 4-dimensional hyperbolic containing space, so that these material sizes are the size of the solar system in order to manifest in this particular subspace (both dimension and spatial set) as stable shapes (or stable material objects) then the interaction of our solar system (with its orbits occupied by condensed material objects) could interact with a (1) (over) similar sized shape but perhaps the other structure is materially occupied by material which has the shapes of 2-dimensional flows contained within the 3-dimensional orbital shape which our solar system can interact in the 4-dimensional hyperbolic metric-space which models the containment space in a higher dimensional construct of existence. Then this would mean that we would not see this other shape nor would we feel it, due to the rigid-rod model of the discrete hyperbolic shape in (by) which our solar system is modeled, yet during a "close" interaction of these two shapes there could be sudden large scale upheavals which take place, ie orbits shift etc.

So that big changes happen relatively quickly, then the system resumes a stable pattern, though most often it would be a new pattern which has been changed from the old pattern during the (relatively) quick transformation process.

We are part of "what is best considered to be an unfathomable mystery" which speculation based on authority should not ruin, as is now the standard practice, where equal free-inquiry is resisted with great physical violence and intellectual violence and intolerance.

Consider the indefinable random model of evolution, It cannot explain the development of the complex instruments possessed by life-forms, eg eyes etc.

Yet on both sides of this mystery are extremely violent factions who want to destroy the other side as to their respective belief's and thought processes, so as to institute a monolith of authority.

But on the one side (the dogmatically religious side) there will never be a victory, similar to the right-to-life community never winning its victory, even though they are given all the advantages which the propaganda system can give to them. This is because their role is to express the moral basis of the entire culture (a culture which is based on inequality and violence), and (these pawns are) the community fighting for their (dogmatic and oppressive) religious beliefs, is to represent the (free) social struggle for that moral base (even though they are all about coercively imposing their dogmas on the public, ie they are the people in the society who are most opposed to equality and freedom). That is, these religious zealots are pawns in the game, just as the original European settlers (used as pawns) to help slaughter the Indians, have been pawns to monopolistic social power, ie social power about which it is claimed is based on "free" markets, but which is really about violence of the justice system against the public to protect the value of the owners investments, and a political system which is based on the propaganda system which is only (effectively "only") is used to "sell" the ideas which support the monopolistic interests of the owners of society.

When on looks at the stars,

What is one seeing? What shapes are really there? How can they be hidden? What hidden ways can we be connected to unknown structure? How many interpretations of the "mystery of stars" can there be?

What is knowledge?

A. It is based on perception
B. it is based on a measurable language

One cannot view knowledge exclusively as being based on the formalism of a precise language because this is far too limited of a viewpoint.

Yet, western culture is all about imposing authority and other restrictions upon (authoritative) language structures. Western culture wants to squelch the idea of the real mysteries to which the human-life-form can attach itself and subsequently create in a wide variety of ways. The failure of western culture is related to the narrowness of its authority and its narrowly defined set of allowable creativity. Nonetheless, even with failure, the human culture is pushed ever more away from the deep knowledge of existence, which makes human-life magical, referred to as magical, since it transcends that which western culture allows to be expressed as ideas, in regard to the relation of life to creativity.

Perception is (could be) based on existence (thus the need for a stable (or repeatable) context for verification).

Language can be related to building (or inventing) . . . , but the building could be (only) the limited context of building; measuring, fitting (coupling), controlling . . . , material instruments,

Or

Language could be related to art, which, in turn, is also related to experience, it is related to a cultural context and an interpretation of the dominant set of ideas of the culture, that is it is experience which is pushed upon and transformed by violence and authority

Chapter 14

General essay

How science and math get immersed into a social context, due to wage-slavery and the rise of the professional scientist and mathematician, and the subsequent loss of the validity of science and math

When one writes about science and math and one has new ideas, or when one writes about society, and one expresses ideas different from the "ideas" expressed by the media, and one is not responding directly to the media, then one needs to understand the organization of power within society so as to understand what is allowed as (valid) expression of ideas within society. This is about how the US justice system has used extreme violence and vast resources in order to turn the public into wage-slaves to serve the owners of society. Science and math are wage-slaves who are used to support the owners of society.

The value of money is empty, its value is upheld by extreme violence, and accumulating it comes from (1) exploiting narrow traditions, which are violently protected, that is, accumulating the riches of money depends on fixing and stabilizing knowledge within society, ie controlling and maintaining particular markets and controlling how information is distributed and used within society, and (2) lying and stealing land, material, and the knowledge of the culture, and (3) controlling how information is provided to the society and controlling what language (eg particular words etc) means and how language is used. The politician and the justice system are institutions which support such destructive and violently formed social traditions, which have been violently formed and maintained by particular people within the society.

Money fashions a society which supports those who invest and the value of money is upheld by violence and corrupted law and corrupted justice.

According to the Declaration of Independence US law is to be based on equality, which gives one freedom to gain and possess knowledge and to use that knowledge to create, within a society which supports all of its members with both material and the knowledge of the culture.

The society is highly manipulated by a few owners of society (individuals and institutions) and the proof of this is that the (US) justice system's main function in society is to enforce wage-slavery on the US population. Furthermore, politics is a branch of the propaganda system, where the politicians "sell law to the owners of society," and then the politicians promote that law, ie politicians (as well as an army of wealthy vassals, who fit into the way in which the very rich have organized and managed economies and markets) are also wage-slaves who are good at brown-nosing the very wealthy.

In such a society the math and science is first and foremost about expressing the idea of inequality in regard to the dogmas which support the narrow business interests of the owners of society, and their servants, where the servants are other big businesses, such as the military, and justice systems, which the political structure supports.

In this context the education system is a filter used to find people who quickly adhere to the dominant culture, ie intelligence is the measure of the arbitrary property of "quickness of cultural acquisition" for the purpose of further developing the narrow dogmatic aspects of a culture which serves the narrow interests of big business, ie the few owners of society (maybe 10 main people). In doing this, the culture is greatly manipulated, and the experts are about memorizing disconnected sets of ideas which pass for dogmas and this is best done by autistic types who still possess language capabilities though their language skills do not have to be "all that well developed," it is better if they express ideas as if the ideas represent a mania associated to an absolute truth (dogma).

The main point of the experts is that people are not equal, and only the experts are allowed to talk about the technical issues of fundamental importance to a society, such as creativity, and "complex" social arrangements, eg fraudulent contracts.

The technical idea (from the math-science experts) that suits this best, is the idea of "quantization of the world" by means of indefinable randomness and non-linearity, which are the types of patterns most often observed, but they are patterns which (these quantizing methods are not capable of describing) and which ignores the fact that the fundamental properties of the world are (all) about:

1. the stable, definitive observed spectral-orbital properties of fundamental material systems, as well as
2. the stable properties of living systems, and

3. about the stable properties of numbers and shapes, such as stable uniform units of measuring, which can be used in a reliable manner in order to build things, as well as
4. very stable shapes, which possess stable spectral-orbital properties, and which model the stable and controlled context by which stable systems emerge with well defined (definitive) stable measurable properties, and
5. They are the stable shapes which are needed to maintain a mathematical language which is quantitatively and geometrically and logically consistent.

One has to actually be talking about some stable property (or pattern) to which the language is constructed to be consistent. If one can describe a pattern but that pattern is not stable, then such a description cannot be used in a practically creative manner in a stable context (except for quickly responding feedback systems). That is, in the descriptions of patterns which are fleeting (and unstable) the accuracy of such a description is not verifiable, and its practical relation to a stable controllable form of creativity does not exist.

In the context of math:

Should a measuring set be a continuum, or based on a finite set?

Should the derivative be defined on a continuum or does it only fit into math in a consistent manner when the math structures have great limitations on their set structures, so that the derivative identifies a discrete structure that may or may not exist within a continuum (though there may exist an apparent continuum)?

The current fraudulent structure of the math language:
Within a manufactured context of randomness and instability (and quantitative and logical inconsistency), it may be assumed that the truth "must be very complex," and thus (it is claimed, that) only a few are capable of mastering this complexity.

Thus, (it is proved that) people are unequal and no one is qualified, but a few, who are rigorously filtered and tested, to discuss the main issues about truth, so that this educational-filter identifies the truth which most supports the interests of the owners of society.

The propaganda system becomes the voice of the one-truth which the society's experts support.

Thus, if one says that physical systems with stable properties need to have the descriptions based on stable and simple geometries, then one is not believed.

However, it is an idea supported by many patterns of intellectual thought: Gödel's incompleteness, the properties of complex numbers and their geometry, the simple nature of the most useful mathematical descriptive patterns, eg linearity etc, the inductive knowledge of "what works" and "what has not worked," and perhaps most strongly by the geometrization theorem of Thurston-Perelman.

Furthermore, if one says that the stable simple geometries can be related in a strong way to a many-dimensional model of material, which also happens to be (or is) greatly attached to the spirit, then this cannot be believed by those who believe the (Empire's) experts, eg the religious leaders who uphold the society's (arbitrary and subjective) spiritual values.
or

If one says that equality is central to creativity and knowledge, this is not believed since there are superior intellects found by society who can "discern truth" for society.

But the truth found by the experts (who are confined to a competition defined by narrow dogmas) is only "the types of truths" which support the interests of the owners of society.

In such a society "authority has become truth," and it is a truth upheld by extreme violence and militarism (where justice is an expression of militarism), as was the case for the Holy Roman Empire.

In other words, if US law is not based on equality, and the subsequent freedoms to be protected by the Bill of Rights, namely the first amendment (where the second amendment is about not having standing armies) so that the focus of society is equal free-inquiry in regard to developing knowledge which can be used in a practical manner, so that each person is an equal creator within society, then there cannot be anything close to the existence of "free markets," and the (current type of) markets are based on a fixed way of doing things within society, which leads to the destruction of resources and a destruction of the people within society, as is now the case within the US society of oligarchy, the new holy roman emperors.

In the descriptions of quantum systems the wave-equation is linear and the model for a quantum system's components is random, where in quantum physics a function-space and its dual-space imply an uncertainty principle (see Powell and Crasemann "Quantum Mechanics").

Yet the adjustment (or correction) to a "solution wave-function" is based on the explicit geometry of point-particle collisions, and on a non-linear term in the operator structure, where this non-linear term is associated to elementary-particle collisions (used for the wave-function adjustment) and [which is the quantum model for material interactions] and is central to the wave-function adjustments (or perturbations), so this math construct (which model material interactions) is neither quantitatively consistent (ie it is non-linear) but it is also logically inconsistent allowing both randomness (requiring an uncertainty principle) and geometry (which depends on a metric-space whose values at-least require the rational numbers to determine limits).

These are opposite properties:

linear vs. non-linear; and
random vs. geometric;

yet convergences, or rather a pair of divergences are considered equal to finding a convergence, and this has been done two or three times, so that these three (or so) "convergences" (which are claimed to be a valid calculation) are interpreted to mean that these, just mentioned, opposing sets of ideas are valid, and these opposing sets of ideas are claimed to be properly related to one another within the quantum description of material systems and their material interactions.

These are the types of inconsistencies which "sets which are 'too big,' allow," where the continuum is a set, which is "too big."

So math and science have become an endless sequence of speculations concerning both

(1) math patterns (not knowing how to organize them so as to consistently identify the observed physical patterns) and
(2) interpreting observed physical patterns,

but this speculation is often accompanied by careful and sometimes (or apparently) rigorous appearing discussions, and this provides the speculations with a sense of authority . . . , eg we will simply trap collision processes which result in nuclear fusion and determine how to extract the generated energy, . . . , and even though the speculations never succeed at their descriptive goal, they can lead to other speculations, which are again endowed with an apparent authority . . . , and on and on it goes . . . eg perhaps we will understand the strong-force and control how energy flows etc. . . . , so absurd math models are devised to describe forces, eg the strong force is modeled as non-linear changes of particle-states during particle-collisions, etc, where such non-linear models of particle-collisions, it is claimed, can describe "why the un-seeable quarks can bind-together to form protons" (where protons are claimed to be unstable) etc, (This is more like religion, and a scientist's absolute authority has become the model "of truth for society.")

However, this process is always consistent with the wants of big business, eg nuclear weapons engineering etc, eg collision probabilities are (remain) related to rates of reactions etc.

The oil and coal companies never get any meaningful business competition from a business supplying a clean-cheap energy source, since such a source never gets developed, and never will by using the way knowledge about quantum systems is expressed. It is in the oil and coal business interests to have the mathematicians and scientists confused about quantum properties and quantum-material interactions.

This type of "science" can only be done within a society which mistakes authority for truth, and it is clear that it is the media (the propaganda-education system) which is modeled as "a single-voice of authority," for an entire society, ie a single authoritative voice which has taught the people of society to consider truth to be authority.

Namely, the authority of the media (or propaganda-education system) which expresses the interests of the owners of society, but it is a society (and propaganda system) which, essentially, only serves the narrow needs of society's owners and is related to a process of "central planning for selfish purposes" which is done by a few very rich people (may be 5 or 10 of these barbarian [selfish, ignorant, and psychopathic] people) who own (possess the controlling shares) and operate a narrowly defined market. The so called market is a narrowly defined and highly fixed way in which to use material resources (though it is masked as "always developing," but it only develops material uses, eg electric circuits, based on 19th century science, eg it also depends on reaction-rates being related to probabilities of component collisions).

It is the owners of society who define both creativity and high-value within society, and it is defined to serve their selfish needs and wants.

It is these few owners of society whom the political-justice system serve, and the ideas and desires of these few owners of society are all expressed through the media, and these wants are backed and supported by a militarized state, a society which has been militarized since 1940.

Science and math is part of such a set of militarized management, and it is used to support the (mostly militarized) interests of the owners of society.

Perhaps, the militarization is the basis for the value of money, and a violent social-system which forces people into wage-slavery, as opposed to the natural inclination of people to know and to create things which are to be given to others as gifts (creativity and developing knowledge for selfless purposes).

Is value knowledge and creativity, or is social value all about controlling money, ie arbitrary value, and controlling the voice of authority, and (with the aid of law and propaganda [ie politics]) violently forcing people into a social structure of wage-slavery. Where the basis in law, for such a society, is property rights and minority rule, but the Constitution also had a Bill of Rights, but the government has never upheld these laws, ie the Constitution is invalid since the government has never upheld the laws which the Constitution proclaims, eg free speech and no state involvement in religion. Today the US is essentially equivalent to the Roman Empire but the emperors are the owners of society and the government serves these de facto emperors, ie the pay-masters of the wage-slaves. It is the interests and the security of the emperors to which the full power of the state is now applied.

A new Continental Congress needs to re-instate the basis of US law to be "equality," and the freedoms to know, ie equal free inquiry, and to use what one knows to create, in a society where everyone is an equal creator.

Only in an equal society, can there be, truly, free markets.

A society must be equal since there is not a measure by which to judge all the various creative efforts, but with equality there must be self-less-ness as the motivation for creativity.

Chapter 15

SO(4) and SO(3) and the geometric structure of space

How does one re-organize and re-interpret both math constructs as well as the observed physical patterns of stability, respectively, to be able to describe; to a sufficient level of: generality, accurate-precision, and useful relation to practical creativity (or technical development); the stable, definitive, discrete spectral-orbital properties, which can be (or are) used to identify physical systems, and where these stable spectral-orbital properties exist at all size scales (from nuclei to solar systems etc)?

Answer:

Use the stable patterns of the "discrete hyperbolic shapes" placed into a many-dimensional containment space, ie an 11-dimensional hyperbolic metric-space, where each dimensional level is a hyperbolic metric-space (of that dimension) modeled to be a "discrete hyperbolic shape." But where the entire range of the set of metric-spaces which possess non-positive constant curvature can also be used within the (this new) context, defined by the hyperbolic metric-spaces modeled as the very stable "discrete hyperbolic shapes" defined in a many-dimensional context.

The fact that "discrete hyperbolic shapes" are "circle-spaces" and are composed of toral components, where each toral component models each one of the holes-in-the-space, where the number of these holes is counted so as to define the genus of the "discrete hyperbolic shapes," and this close association (between discrete hyperbolic shapes and tori) leads to the "discrete Euclidean shapes" (or tori) also being a part of the interaction process, and in doing this, it allows that the material interactions, in the new model of existence, be essentially similar to classical material interactions.

This new model for material interactions also allows for the new descriptive context to be such, that the apparent fundamental randomness of quantum description has its (the randomness's) source identified, these types of interactions between atomic and molecular systems so as to be determined between a continual sequence of very small, discrete time intervals (~10^-18 sec), where such an identification of the cause of a random pattern (based on discrete, geometric interactions applied to small material components) also depends on the fact that "discrete hyperbolic shapes" have associated to themselves a highly distinguished "vertex" point on their shapes. This distinguished "vertex" point causes their interactions to be centered around this "vertex" point, and this causes the appearance of point-like material interactions, which exist in a random context.

The goal of this paper (chapter) is to consider the SO(3) and SO(4) link between a geometric structure of material interactions in regard to both 3-space and 4-space, and the various forms of material interactions which exist (and are already used) in these spaces (in regard to force-fields and material interactions),

But on-the-other-hand, one wonders, if we are always a part of a circle-space then,
Why does our "material containing space" not (appear to) have any holes?

[We are within an open-closed and bounded metric-space, which defines a lattice, but we are only aware of the blocks of the lattice {of the discrete hyperbolic shapes} which extend away from the lattice's fundamental domain, so we see an unbounded metric-space, which, we believe, contains no holes, but the fundamental domain, within which our metric-space is (truly) contained, is related to a circle-space.]

How does one describe the patterns of "spectral-orbital stability" for "the so very many," very fundamental many-bodied systems, which exist at all size-scales (from nuclei to solar systems)?

These spectrally and orbitally stable material systems are systems which are identified, by the professional math-science community, as being associated to patterns which are (to be) based on non-linear and/or random (indefinably random) properties, so as to form a set of chaotic and random systems, whose descriptions, according to these experts, are subsequently, irrelevant to practical development, since it is claimed that these simple, fundamental systems are "too complicated to describe."

How can math be a subject about "quantity and shape," if there is no "math property which focuses on stability," or no concern about, the math properties which deal with, or are concerned about, the math property of "the stability of a mathematical pattern?"

(Or concern about determining the math properties needed for the stability of the patterns which math describes?) This is most often related by the professionals to defining set containment, but the set which "does the containing" is modeled to be "too big of a set," namely, a set which is a continuum.

Answer: One should be concerned about this issue, and one should also try to base mathematical descriptions (of systems) on finite quantitative sets.

Assume the containing space is primarily R(n,1), which is contained within a many-dimensional context {Note: R(3,1) is space-time}, ie R(n,1) is a hyperbolic metric-spaces of dimension, n, but each dimensional level is a "discrete hyperbolic shape," and the construct possesses the further property, that for each subspace (of each dimensional level) there is a "discrete hyperbolic shape" . . . , (representing either a metric-space or a material component, depending on the dimension of the frame from which the "discrete hyperbolic shapes" are observed [see below]) . . . , which has a maximal size, where it should be noted that for hyperbolic-dimensions "two to five," inclusive, there exist open-closed and bounded "discrete hyperbolic shapes," which can be used to model metric-spaces. Furthermore, these (bounded) hyperbolic metric-spaces (or discrete hyperbolic shapes) can be either metric-spaces which contain lower-dimension discrete hyperbolic shapes or they can be models of material components which are contained in open-closed, hyperbolic metric-spaces.

D. Coxeter: "Discrete hyperbolic shapes" which are 6-dimensions and higher are all unbounded, and the last defined "discrete hyperbolic shape" has a hyperbolic-dimension of ten (so the proper containing space for stable geometric shapes is a hyperbolic metric-space of dimension-11).

The assumption of the existence of a maximal-size for subspaces of hyperbolic-dimensions five and less (inclusive), for such (possibly bounded) shapes, which are contained within an over-all containing space of 11-hyperbolic dimensions, implies that such an over-all high-dimension containing space has its spectral-orbital structure . . . , {associated to the (mostly bounded, low dimension) discrete hyperbolic shapes which identify its stable geometric-spectral structure} . . . . , determined by a finite set of "discrete hyperbolic shapes" which, in turn, determine a finite spectral-orbital set. Thus, the "discrete hyperbolic" shapes which exist within such a containment set must be in resonance with this finite spectral set, ie both the hyperbolic metric-spaces and the material components, where both of these constructs are modeled as very stable "discrete hyperbolic shapes."

It is this finite spectral set upon which the quantitative structure . . . , used for the descriptions of all the stable material components which compose (or are contained within) this set . . . , depend. {for their precise descriptions and math constructs}.

This construct is both the quantitative and geometric basis for the stable patterns which are observed physically, and which exist as (stable) math patterns within the set.

That is, all stable, bounded "discrete hyperbolic shapes," which are contained in this (or any such) particular 11-dimensional containing hyperbolic metric-space, are in resonance with the finite spectral set, which is defined by the different dimensional levels (each dimensional level modeled as [with] discrete hyperbolic shapes).

Note: There can also be multiplicative-constants which are defined between dimensional levels, as well as being defined between different subspaces of the same dimension. [ie subspaces which have the same dimensional value]

Math properties, or physical properties, associated to metric-spaces and material-component interactions

Both the properties of "position in space" and concerning material interactions of stable material components (or stable math patterns), where the interactions are determined by spatial relationships, are defined within Euclidean and by hyperbolic metric-spaces, respectively, where spatial positions of the stable shapes {which are defined in the hyperbolic context as the stable "discrete hyperbolic shapes"} exist within a spatial set, which depends on R(n,0), or Euclidean n-space.

These properties of spatial position and stable existence of a pattern, are used in the (new) model of "interacting (n-1)-discrete hyperbolic shapes" (ie interacting material components) which are contained in an n-dimensional hyperbolic metric-space, where the interaction is determined by the existence of an n-dimensional discrete Euclidean shape (torus) which spatially links the spatial positions (ie the distinguished vertex-points of the discrete hyperbolic shapes) of the interacting material components, and this interaction torus exists in a (an) Euclidean metric-space, where the linking Euclidean n-torus is contained in an (n+1)-dimensional Euclidean metric-space.

The math of the interaction is about a differential 2-form defined on the n-torus which is contained in (n+1)-Euclidean space, where this 2-form, defined in (n+1)-Euclidean space, is from a local (fiber) space of differential 2-forms which have the same dimension as the (tangent to the) SO(n+1) fiber group. Thus, the local transformation of the distinguished points, which identify spatial positions of the material components, is carried out in relation to the toral geometry of the Euclidean n-torus (where the interaction torus forms for each small time interval, defined by the period of the spin-rotation of metric-space states (see below), so as to form with the property of action-at-a-distance for each time interval), thus the spatial displacements of the distinguished points [of the discrete hyperbolic shapes which model the stable material components which are interacting] can be placed in the Euclidean lattice defined in n-space and locally transformed by

SO(n+1) in relation to the n-torus geometry, upon which both the 2-form is defined, and which is contained in R(n+1,0).

Re-iteration concerning quantum randomness

This model of material interaction defined on very small material components, which can be either neutral or charged (and changing very rapidly in regard to their charged properties), would define a Brownian motion, on the set of discrete time intervals [identified by the period of the spin-rotation of opposite metric-space states]. Thus, this description is equivalent to the properties of quantum randomness applied to the distinguished points of the discrete hyperbolic shapes, ie the interaction would be both random and appear to be point-like. [E Nelson showed such Brownian motion {defined for a set of distinguished points} is equivalent to quantum randomness, 1967, 1957 (?)]

Physical properties of metric-spaces

The (math) properties of quantity (in regard to spatial measures of position) are contained in Euclidean space, while the properties of stable math patterns, whose positions of their distinguished points are measured in space, are the very stable "discrete hyperbolic shapes."

That is, position and inertia are contained in Euclidean space while charge and energy are contained in hyperbolic space.

This construct of properties associated to metric-spaces (of non-positive constant curvature) implies the existence of pairs of opposite metric-space states. For Euclidean space these are the positions, which are defined in regard to either the fixed distant stars or the rotating distant stars. For hyperbolic shapes this is (+t) or (-t), where energy conservation is associated to invariance to time displacements (E Noether), so energy is also associated to time.

Containment within complex coordinate systems, ie finite dimensional Hermitian spaces, and thus related to SU(n), or SU(s,t)

The pairs of opposite metric-space states can be placed in the real and pure imaginary subsets of complex-coordinates. Thus, the new description naturally has a context associated to unitary fiber groups as well as a spin rotation fiber groups. The period of the spin-rotation of a metric-space's opposite-time states, defines a time interval (~10^-18 sec) within which distinguished points have their positions locally transformed, and the transformation can be in relation to the opposite metric-space states, which are defined between (spin-rotation) the set of time intervals, which define a sequence of time intervals.

"Condensation" of material components, and resonance with the finite spectral set of the 11-dimensional containment (hyperbolic) metric-space

The n-dimensional discrete hyperbolic shapes model stable material components (which possess distinguished points), which are contained in both (n+1)-dimension hyperbolic space, and their distinguished points are identifiable within an (n+1)-dimensional Euclidean space.

However, it might be noted that, the n-dimensional hyperbolic material component can be defined in any dimension metric-space which has higher dimension than (n+1). However, these stable material components would tend to "condense" on (or within, by penetration (within) by means of [relative] high-energy collisions) the stable, high-dimensional "discrete hyperbolic shapes" which it encounters, due to either interactions or collisions.

Crystals are about such condensation, but it seems that atoms are also about such condensation, where both crystals and atoms are held together by either resonance with the finite spectral set (of the entire 11-dimensional containing space), or (if) by the condensed material being statistically of lower energy (than its thermal environment) so a "discrete hyperbolic shape" can form, but it is not (may not be) in resonance with the (over-all stability defining) finite spectral set.

Thus, the energy of a condensed set of material is defined to be as much about either its (unstable) "condensed" discrete hyperbolic shape, or about its inertial-interaction, and its relation to a discrete Euclidean shape.

[Note: The discrete hyperbolic shape exists in a sea of thermal energy, which apparently is relatively low thermal-energy, which allows the condensed state to be slightly stable and to define a "discrete hyperbolic shape" which may not be in resonance with the finite spectral set which defines stability. However, atoms clearly have the property of being in resonance with the finite spectral set, which defines stability within the containment set.

How come atoms are strongly resonant, and thus relatively stable, but many crystals are not "as strongly" stable?

Consider hyperbolic 3-space (ie space-time):

In an 11-dimensional hyperbolic metric-space there are, [11C3]=165, such 3-dimensional subspaces, where each such 3-dimensional subspace has a discrete hyperbolic shape associated to itself (assume they are all bounded shapes, but this assumption is not necessary).

This defines a spectral set for "discrete hyperbolic 3-shapes," but the 3-flows, which are defined on the "discrete hyperbolic 4-shapes," also define part of the hyperbolic 3-dimensional spectra for the 11-dimension hyperbolic containment set, etc.

However, because we (believe that we) exist within a 3-space; it seems that it is most natural that we exist in a spectral-orbital structure, wherein, when we transition from 3-space to 4-space, the discrete hyperbolic shapes get bigger so that both the "discrete hyperbolic 3-shapes" are contained within the "discrete hyperbolic 4-shapes" as free material components, eg atoms, and the 3-spectral-flows defined on the "discrete hyperbolic 4-shapes" determine the orbital structure of our planets.

Thus, one asks "What allows atoms to have a resonance with 3-shapes?"

In regard to the 3-space, within which we (believe we) are contained, there are, 11-3 = 8, separate dimensions, which are separate from the 3-space within which we are (or appear to be) contained. Thus, there can exist, [8C3]=56, 3-spaces which are separate from the 3-space within which we are contained.

Thus, each of these 56 separate 3-spaces (in the 8-dimensions which are separate from the "given 3-dimension subspace" of 11-space model of a containment set) can be a part of a "spectral-dimensional-size" sequence of discrete hyperbolic shapes which model the metric-space structure of these 3-spaces, and from which a sequence of spectral values is defined.

One can assume that for the set of the 56 different sequences of spectral values either the spectral values descend in value (decrease in size) as the dimension increases or the spectral values increase in value (increase in size) as the dimension increases.

In the increasing sequence, of spectral values, there can exist orbital properties of charges occupying orbits (or spectral-flows), or of condensed material components occupying orbits within a higher-dimensional (stable) geometry, as well as the existence of condensed material which determine "free" components (in the higher-dimensional hyperbolic metric-space).

How stable spectra are organized in the current model (2012) of physical description

The decreasing sequence of the spectral values would require that the material components be contained in some 6-dimensional (or higher-dimension) hyperbolic metric-space, with no orbital structures (unless unstable, non-linear models of material interactions in a many-body orbital system), and perhaps some condensation of components of the same dimension.

This is more-or-less the model of materialism in 3-space (or in space-time), where material reduces to points, where, essentially, only non-linear models of material interactions exist (in a context of indefinable randomness), and stable orbits are supposed to be caused by a 1/r potential-energy term which is associated to a spherically symmetric force-field for each point-particle (ie for a many-body system the math of such a context will be non-linear).

This seems to not work, yet the idea of higher-dimensional spectral structures representing ever smaller geometric properties is the conclusion (or the underlying assumption) of particle-physics and its derived string theory viewpoint.

The existence of both increasing and decreasing sequences of spectral values is not necessary in regard to allowing the atomic-sized spectra to exist. Though the existence of the decreasing sizes of spectral structures, as the dimension increases, makes it easier to identify smaller spectral-values for "discrete hyperbolic 3-shapes," and thus it is easier to imagine the spectra of atoms to be stable. However, of the 56 different possible sequences of spectral values, even for all the sequences characterized by increases of spectral values as dimension increases, many of these sequences could have the spectral values associated to the spectral values (sizes) of atoms.

However, if there were no hyperbolic 3-spaces whose spectral values were not the size of atomic components, then the (relatively stable) atoms would have to be resonant with the lower-dimension 2-spectral flows, associated to the "discrete hyperbolic 2-shapes" which are associated to the interaction structures of interacting stable components of charge, namely, the "discrete hyperbolic 3-shapes," of which the "discrete hyperbolic 2-shapes" are faces (or spectral 2-flows), where charges, nuclei, and electron clouds are all (stable) 2-dimensional components, though the electron clouds may not be bounded "discrete hyperbolic shapes."

The [11C2]=55 set of "discrete hyperbolic 2-shapes," which model the 2-dimensional hyperbolic metric-spaces and which define, part of, the 2-spectra of the 11-dimensional containment metric-space (set), ... where the spectral set of the 11-dimensional hyperbolic metric-space is a spectral set which is increased (added to) by the higher-dimensional spectra of the discrete hyperbolic shapes, which model metric-spaces, and this set of 2-spectra (which are sub-spectra to higher-dimensional spectral shapes) can be added to the 55, spectra of the 2-dimensional separate subspaces of 11-space, ... , would be small (in spectral value) in the "three 2-subspaces" of the particular 3-space (defined by the particular "discrete hyperbolic 3-shape"), due to the fact that constant factors defined between subspaces can (would) cause these particular subspace 2-spectra to possess relatively small spectral values.

as opposed to a spectral sequence whose ****

New material can be defined, which is different from both inertial material and charged material

Consider that, higher-dimensional material can exist in R(s,t), where s is the spatial subspace dimension and t is the temporal subspace dimension, and where s+t=n,

However, the interaction of any type of material would be spatially related to R(s,0), by the "new discrete shape's" distinguished point.

In such a pattern one can assume that t increases with each new material type which can be defined.

For example:

The new material types are the "odd-dimension" and "odd-genus" discrete hyperbolic shapes, whose stable spectral flows (or orbits) are filled with charge, but the geometry of these types of discrete hyperbolic shapes (will) have a geometric charge imbalance, and thus these shapes would naturally begin to oscillate and subsequently they would generate their own energy.

So the two types of (relatively) stable discrete hyperbolic shapes of material components exist in hyperbolic 4-space, ie "the stable components" and "the oscillating components."

Thus, there can be defined a R(4,2) space which: either contains both material types, or it contains the oscillating type, while R(4,1) contains the stable type, etc.

SO(4)

Mass can be defined as a circle in Euclidean space, where the mass of the circle is associated to k/r for the circle's radius, r, and some constant, k.

This defines a context for 1-dimensional inertial interaction structure within a 2-plane, but it can also be naturally related to an R(4,0) and SO(4), construct for changes of material positions. This is because of the geometry of SO(4).

Let R(4,0) have an (x,y,z,w)-variable coordinate structure, where SO(4) = SO(3) x SO(3), then consider labeling SO(3)1 x SO(3)2, and a context in which the description is organized within a particular 3-subspace structure, so that the SO(3)1 can be related to (x,y,z) subspace of R(4,0), and SO(3)2 is related to (x,y,w) subspace, and then, in regard to a description of material component interactions in the fixed (x,y,z)-space, which is the 3-space within which the interacting material is assumed to be contained, then [so that] SO(3)2 only affects (x,y) coordinates of the (x,y,z)-space.

In this context, in regard to the descriptions of electromagnetism, a "circle" created by conducting material is defined in regard to a (deformed) 2-plane, where this 1-hole constructed by material, can be used to define the magnetic field aspects of the electromagnetic 2-form defined in 4-space.

Electromagnetism has its "2-form, force-field" properties identified in space-time related to the geometry of charge and (usually) 1-currents . . . . , in relation to a solution to a wave-function (defined for both (+t) and (-t)) whose solution 1-form (supposedly related to potential energy of the electromagnetic system) is assumed to be single valued . . . . , but this space-time structure could just as well be 4-space, R(4,0), which can be adjusted to accommodate the existence of the two 3-dimensional force-fields, ie E and B, of electromagnetism. The Lorentz force which relates the electromagnetic force-field (which is defined in a 4-dimensional space) to inertial properties, defined in regard to point-vertices of material components, naturally belongs to R(4,0), in a similar way in which a "discrete hyperbolic 3-shape" belongs to hyperbolic 4-space.

While, in regard to mass, in this (same) context, there seems to be a "true (x,y)-plane" associated to gravitational-inertial interactions, which are defined by (in relation to) an interaction 2-dimensional discrete Euclidean shape (a 2-torus) contained in 3-space, so as to yield the math structures of Newton's gravitational laws, where the geometry of the local transformation structure of the SO(3) fiber group in relation to the geometry of the 2-form defined on the (Euclidean) interaction 2-torus, is consistent with the spherically symmetric geometry of gravitational interactions in 3-space.

This would also account for the plane geometry of the solar system, as well as Saturn's rings.

That is, it is the geometric structure of SO(4) which allows us to perceive both gravity and electromagnetism as having particular geometric properties in R(3,0) and/or R(3,1).

Note: There is also the possibility that the 2-plane related to SO(3)2 is what causes the inverse-square force field of free, and relatively motionless, charges, modeled as circles (or figure-eights) in a plane, to possess a spherically symmetric force field (associated to SO(3)2), whereas the geometry of magnetic force-fields would in this case be related to charged components possessing "discrete hyperbolic 2-shapes" being related to SO(3)1, and the (x,y,z) coordinates.

I do not claim to have identified any absolute, precise math structure within which all physical patterns can be described, but this structure "is the structure of stability," for both mathematical and physical description.

It is a simple context which leads to an even richer context for interpreting and modeling observed stable patterns.

These patterns give many possibilities and thus there are many choices needed to be made, in order to determine a more accurate, and sufficiently precise, description, but it is a geometric and stable descriptive structure, thus it is useful for practical creativity.

Thus, the project of finding the finite spectral set for an (or for our) 11-dimensional space is of value, though there may be many such, different, finite spectral sets, for many different 11-dimension hyperbolic metric-spaces. However, coupling between such spaces is (would seem to be) a solvable problem.

This is about the description, the knowledge, of existence, so that we, and our knowledge, are related to the creative process focusing on the creative, manipulate-able aspects of existence which could extend the properties of existence, so as to form new properties of an expanded existence.

For example,

"Can an 11-dimensional "discrete hyperbolic shape" be created?"

"Can the SO(3), SO(4) inter-related patterns be used in a practical context?" etc.

One wants high academic discussions to be about assumptions, contexts, interpretations associated to simple models of quantity and shape, and dependent on the properties of stability, and related to practical creative development.

The complex development of instruments, and/or fixed narrow descriptive contexts, should be the province of small groups, but the whole enterprise needs to be equal, and based on equal free-inquiry, motivated by visions of creative possibility.

Language can be manipulated in a great variety of ways, and because of the obvious failure of today's technical descriptive languages, new contexts of description should be taken seriously and encouraged, rather than measuring a person's "intellectual value" by their dogmatic purity within a preposterous, and failed, fixed descriptive context, which is primarily associated to only a few complicated instruments (most notably engineering of nuclear weapons).

Education based on developing incomprehensible complexity of descriptive languages, is mostly about the claim that people are not equal.

For example, the discussion of law needs to be about the nature of being human and the types of society one wants to develop, where inequality is about selfishness and extreme violence; while equality is about creativity, knowledge, and selflessness, etc.

Everyone must be considered to be an equal creator.

Part II
The speech by concoyle, at the San Diego Math conference (2013), about using geometrization to establish a new descriptive context for the physical world

Chapter 16

Abstracts

Abstract I

This relatively new (since 2002) and relatively simple context of math containment provides the setting for a solution to the problem of finding the math structure for the observed stable material systems which are so fundamental and so prevalent. It also provides a basis for a quantitative structure which is defined on a finite set. However, these stable physical systems go without any valid math structure (for these systems) in a currently accepted math context of indefinable randomness (eg improperly defined elementary event spaces), non-linearity, (global) non-commutativity, or only locally commutative, (eg quantitative inconsistency, eg chaos) all defined by a contrived descriptive structure of convergence and divergence onto a continuum. Where these math structure are together used to explain (or identify) the (stable) properties of physical systems. But such a math context really only applies to physical systems in a chaotic transitioning process (eg reactions in weapons) and for feedback systems (eg guided missiles) whose range of applicability is difficult to define and it is a context which applies to quantitative complexity (eg secret codes). But it is also used by the media to create an illusion of expert "mastery" and "expert complexity."

Due to these new ideas, many difficult problems now have solutions: nuclei, general atoms, molecules, a new way in which to analyze crystals, and the stable solar system, this means that these relatively new ideas should be dominating the attention of the professional mathematicians and physicists, but they are not.

Social forces which affect communication of ideas

Apparently there are stronger social forces involved in an inability of a public, or of an expert-class, "to discern truth."

The US propaganda system is the sole authoritative voice for all of society and it is the propaganda system which directs the attention of the researchers. These researchers are dependent on a funding process. However, these same researchers claim to be the personifications of the highest cultural attainments in the society, nonetheless they have social positions of being both wage-slaves and society's, so called, top intellects in regard to a religious personality-cult, expressed through the media, so public-worship consolidates their belief in their far too authoritative mathematics and physics dogmas, which has failed to solve the problem of the cause of physical stability for nearly 100 years (ie it is a failed dogma), ie the media turns "top intellectualism" and the dogmas upon which such a "measure of intellect" rests (the "intellectual winners" of the competition, whose rules, in the education system, are defined by an, essentially, absolute authority) have been turned into a religion, this is really a deep religious belief in what the media labels as science [Copernicus would have a more difficult time persuading others to consider an alternative way in which to organize and fashion language within such a current religion-of-an-absolute-science-dogma (2013) of expert authority, than the difficulties he had in regard to the authoritative religion of his time].

The professionals are following their "deep beliefs" as dictated to them by the propaganda system. Apparently these professionals can rigorously prove properties which are contained in a world of illusion, eg where a description based on randomness also possesses well defined geometric properties, eg particle-collisions.

It should be noted that the best interpretation of the Godel's incompleteness theorem is that precise languages can be very changeable when reduced to the elementary levels of assumption, context, containment, organization, interpretation, etc. Yet the failure to describe the stable underpinnings of physical existence has not been seen as a crisis of the knowledge which is being derived from the currently accepted authoritative dogmas of math and physical description.

There is other social organizational properties which manage society, and with which one must deal with, there is a vast social organization in regard to management which manage the math and physics (or science) communities, managing both the topics of interests and managing the nature of the inquiring language, where this is done in both a context of funding of wage-slaves and managing (autistic) personality-types, eg managing personality types similar to the management of personality types in politics and the justice system.

Note: The commercial world is related to a fixed stationary way of behaving or acting, a commercial structure is a very narrow context, based on a limited range of creativity and a fixed way in which to use material resources. The power of business monopolies depend on society not changing how it uses the material resources a business monopoly supplies to a society. The law is supporting this type of narrowness, essentially based on property rights and minority rule (creditor vs. debtor, smart vs. stupid, etc), and it supports such selfish actions with great violence. In fact, the economy is tied to a fixed narrow way in which to live and create, and this model of monopolistic economies is being used as a means to conquer ever larger populations, but it is being put into-place by means of extreme violence and coercion (often an economic coercion).

Does one want a society to be based on a fixed way to use material, and a fixed way in which one is to serve the material based, and fixed structure of society, and a fixed overly authoritative organization of descriptive knowledge, so that this type of power, and associated narrowly defined knowledge, depends on expansion in the form of an ever greater exploitation of particular types of material (usage)?

In the new context of containment one uses the most prevalent of the stable geometric patterns identified in the Thurston-Perelman geometrization, namely, the discrete hyperbolic shapes and the properties that these shapes possess as identified by Coxeter.

Furthermore the ability to "surround" a "hole" by a closed shape, so that a continuous deformation is limited, ie the "holes" introduce stable properties into the context of the continuity of shape.

The discrete hyperbolic shapes [with component interactions mediated by discrete Euclidean shapes (tori)] are also very rigid shapes with very stable spectral properties.

That the solar system is stable is evidence, which can be interpreted, to prove this new context for mathematical descriptions of the physical world is true, especially, since the professionals have no valid model of stability for these stable systems.

Abstract II

A new context in which to apply geometry to: math, quantum physics, and the solar system, etc

Quantum physics assumes the global and descends to the local (ie random particle-spectral measures).

Is geometry a better vehicle to define the stability of quantum systems rather than function spaces?

Is the stable construct to be the very stable discrete hyperbolic shapes, in a many-dimensional context?

A geometrically stable and spectrally finite math construct, where, in adjacent dimensional levels, the bounding discrete hyperbolic and Euclidean shapes are defined, and then mixed as "metric-space states" in a Hermitian (or unitary) context, can provide a structure for stable properties.

Assume that math be consistent with (local) geometric-measures of stable shapes, which define finite spectral sets, contained in higher-dimensions.
The stable shapes in the different dimensional levels are con-formally similar, and resonate with a finite geometric-spectral set contained in a high-dimension space.

A new interaction type consists of a combination of hyperbolic and Euclidean components, but when in an "energy-size range" the system can resonate with the spectra of the containing space, and thus it can change to a new stable, discrete shape.

Abstract III

There are (moded-out) "cubical" simplexes in a many-dimensional context, whose structure is determined by hyperbolic metric-spaces, which can, themselves, be modeled as moded-out "cubical" simplexes. Transition between the different dimensions determine physical constants, and the value of these physical constants can imply that:

1. the different dimensional levels can be hidden from one another, ie the size of the interacting materials change from dimensional level to dimensional level and the geometry of the interaction can also change, ie material interactions are not usually spherically symmetric.
2. The over-all high-dimension containing space can be defined as having a finite spectral set.
3. The descriptions of both mathematical and the physical systems are (or can be) stable, because the "cubical" simplexes can define discrete hyperbolic shapes.

Metric-spaces have properties and subsequently an associated metric-space state, and this determines the dimensional distinction between Fermions and Bosons, as well as determining the unitary (invariant) mixing of (metric-space) states in subsets of complex-coordinates.

Unitary invariance implies continuity, or the conservation laws, eg the conservation of energy and material, etc.

Each dimensional level is a discrete hyperbolic shape, and this implies such a set can define a finite spectral set for the entire space.

A new interaction type consists of a combination of hyperbolic and Euclidean components which are one dimension less that the dimension of their containing metric-space. A 2-form construct emerges from this geometric context which is the same dimension as the adjacent (higher) dimension Euclidean base space of its fiber group which determines discrete spatial displacements. This interaction construct is either chaotic or it could begin to resonate during the interaction and, subsequently, to become a new stable spectral-orbital (discrete hyperbolic shape) structure, by means of its resonance with the spectra of the many-dimension containing space.

Chapter 17

Blurbs

Blurbs: I

If a fiber group's (primarily isometry, or unitary) local matrices are always diagonal (or commutative) on the global coordinates of a shape's locally-identified vector-field then the geometry is stable, quantitatively consistent, and simple enough to be considered to be a good candidate to be an element in a set of shapes which is to be used to model existence in a many-dimensional, macroscopically-geometric context so that stable spectral-orbital properties of physical systems [which exist at all size scales, eg nuclei to solar systems] can: generally, accurately (with sufficient precision), and practically usefully; be described (where a stable geometric description is a practically useful description).

Basically these simple commutative shapes are the stable circle-spaces, defined within metric-spaces of non-positive constant curvature, where (but) they are also primarily the "discrete hyperbolic shapes," though metric-spaces which are different from R(n-1,1) [which is either a general space-time or an (n-1)-hyperbolic metric-space] can be considered, such as R(s,t), wherein new higher-dimensional material can be (newly) defined, and a higher-dimensional model of a life-form can be defined.

Furthermore, the metric-spaces possess intrinsic properties, which exist as opposite pairs of metric-space states. This leads to complex coordinates and unitary fiber groups, as well as fiber spin-groups and the spin-rotations of opposite metric-space states.

The different dimensional levels are to be modeled as discrete hyperbolic shapes, where up to and including hyperbolic dimension-5, these shapes may be assumed to be bounded, and assume there is a shape which is maximal for each dimensional level and for each subspace of that same dimensional value. This is the basis for defining a finite spectral set, associated to the over-all high-dimension containing space, ie an 11-dimensional hyperbolic metric-space, since the last existing infinite-extent discrete hyperbolic shape has hyperbolic dimension-10.

Thus, the different dimensional levels of the higher-dimensional (over-all) containment space are partitioned by open-closed shapes, which are associated to "rectangular-like" fundamental domains in hyperbolic space.

The "rectangular" simplex fundamental domains for the metric-space within which we and our solar system are (both) contained are 4-dimensions, upon which our metric-space is a 3-flow, and this fundamental domain would be the size of the solar-system, these block-like fundamental domains (or equivalently circle-space metric-spaces) would appear open, and thus the incoming light (from outside the solar system), where light is modeled as an infinite-extent discrete hyperbolic shape, would pass through the blocks from far away, while on-the-other-hand the apparently infinite extent neutrino discrete hyperbolic shapes which define the 2-faces of the "discrete hyperbolic 3-shapes," which model electron clouds, would (could) have its infinite extent defined by being bounded by the bounded discrete hyperbolic 4-shape metric-space which is also a model of the discrete hyperbolic 4-shape of the solar system. Thus, one has that when one looks away from the rectangular 4-simplex of the solar system (or looking out to the universe) one would see light coming in from the distant places, while looking within the rectangular 4-simplex one would see the closed boundaries of bounded discrete hyperbolic 2-shapes (and/or possibly bounded discrete hyperbolic 3-shapes) modeling the material components contained in the 3-flow model of our containing metric-space

In this context, the way to use higher-dimensions is to model the different dimensional-levels as the stable circle-spaces, and the way in which to get a (math) pattern which is associated to the existence of stable spectral-orbits is to multiply the different dimensional levels by constants, ie the nature of physical constants, so that the spectra . . . , of the adjacent next higher dimensional level . . . . , increases in value, where this allows the lower-dimensional shapes to be contained within a stable shape, where, in turn, this shape can define a relatively stable orbital geometry, for the material which is contained within the shape (ie the metric-space is a shape).

The idea of materialism as well as of quantum physics and particle-physics (both consistent with the idea of materialism) assumes that the higher dimensions are either continuous, in an open context, and that material reduces to point-particles and the spectra (associated to the physical systems which these point-particles occupy) decrease in value (spectral size) as the dimensions of containment increase, where in both cases (continuous and discrete particle context) the force-fields are assumed to be spherically symmetric (though perhaps not always inverse square), but this spectral structure implies that no stable spectral-orbits exist (that is, where does one now (2013) find valid descriptions of the general and spectrally relatively stable: nuclei, atoms, molecules, crystals, and solar-systems?).

The derivative operator can be re-defined as a discrete operator, defined between: (1) dimensional levels (2) time intervals, and (3) toral components, wherein Weyl-transformations can be used to identify stable orbital shapes associated to a set of angularly-deformed "discrete hyperbolic shapes," which in turn, define envelopes of orbital stability for the condensed material components, which these metric-space shapes might contain.

Blurb 2

Topic list

The stable properties of general sets are related to precisely identifiable properties of physical systems where these precisely identifiable properties of physical systems exist at all size scales; from nuclei to solar systems, and these fundamental systems have no valid descriptions, eg Hartree-Fock etc.

List of fundamental topics concerning these new math-science ideas:

From

1. precise language (build new languages at an elementary level of assumption, and context, and interpretation, in order to broaden the capacity to create; Godel's incompleteness theorem can be interpreted to mean add more assumptions or it can be interpreted to mean review and alter one's precise language at the level of assumptions),

   Does language fit into a "fixed scene" (which is implicitly assumed to be moving toward some absolute truth) which is similar to the idea that an authoritative math language is always relatable to ever more complicated instruments, but in an industrial society all the instruments are built so as to be based on the same principles (electromagnetism or other classical theories which allow stable patterns which are controllable), so that when the instruments are adjusted in their "complicated context," they either improve or they reach their limits of performance, likewise physical theory and math patterns are carefully adjusted within the realm of fixed principles, but the math and the physics do not work, and the careful adjustments either do not work or they have already reached their limits of performance (and thus, they have lost their relevance), to
2. quantitative structure (continuum, quantitative consistency, comparisons [length, time, material {particle-number or density}] or spectral values [momentum, energy {single-valued in regard to holes in the shape of the domain}, as well as an assumption of fundamental randomness [which apparently, for physics, cannot be defined as a valid elementary set of random events]]), to
3. failing descriptions (for valid descriptions of the stable spectral-orbital physical systems which exist at all size scales, rather the descriptions are of fleeting and unstable patterns), to
4. the stability of mathematical patterns (how can their stability be established), to

5. the proper role of geometry (stability of patterns, a measuring context, a useable context), to
6. interpreting observed patterns (Are the incessant examination of the properties of elementary-particles best interpreted to mean that existence is higher-dimensional and unitary? Furthermore, the existence of high-energy cosmic-rays, as well as an apparent property of dark-matter, are best interpreted to mean the existence of a large-scale spectral structure which exists in higher-dimensions), to
7. fundamental structure of math, functions vs. numbers, (algebraic equations and partial differential equations, is math really about finding the stable geometric confines related to the existence of stable describable math patterns or stable measurable properties).

   For example, How to determine the structure of the derivative?: {From a derivative interpreted as

   1. an operator on a function space, to
   2. a model of locally measured properties within a sufficiently determined containing space, to
   3. a discrete operator in a finite math structure determined by (discrete) stable geometry in a newly organized, many-dimensional, containing space with new interpretations, which can lead to many more possibilities (the functions in the function space are [now] the discrete shapes).}

Math is about quantity and shape, . . . , but when shape is considered primarily in terms of measurable quantities and functions (or functions and their coordinate domain spaces), it (the shape which is being described) is most often non-linear and unsolvable, except locally (the fiber diffeomorphism group is locally invertible), but non-linear quantitative properties are chaotic, so the local pieces of shapes cannot be put together in a quantitatively consistent manner, and . . . , thus it (shape) fits into an indefinable random structure, in turn, to be fit into function spaces which, in turn, depend on indefinable sets of spectral functions (whose associated operator structures, or measurable properties, usually do not commute).

However, Thurston's geometrization finds that the variety of stable geometric systems depends most strongly on the discrete hyperbolic shapes, while Coxeter found that the last bounded discrete hyperbolic shapes are 5-dimensional, and the last set of infinite-extent discrete hyperbolic shapes are 10-dimensional.

Consider:

1. The stable properties of general sets of precisely identifiable physical systems which exist at all size scales; from nuclei to solar-systems, and these fundamental systems have no valid descriptions, eg Hartree-Fock, general relativity, etc.

2. Basing measurable descriptions of systems which possess stable properties on a quantitative set which is determined by a finite spectral set, put into a context of stable (linear, solvable) geometries, contained in a higher-dimensional set which is also organized around stable shapes, ie organized around stable circle-spaces. Namely, a finite number of discrete hyperbolic shapes contained in an 11-dimensional hyperbolic metric-space.

Where as consider the following patterns of spectral values associated to a containment space, as the dimension increases there is either:

(a) an increasing finite sequence of spectral values with an upper bound, or
(b) a decreasing finite sequence of spectral values with a lower bound.

That is, (a) is the new proposal, while (b) is essentially what is assumed today (2012) when guided by the principle of materialism and the principle that material reduces to a set of fundamentally random elementary-particles, where (b) is helpful in regard to building nuclear weapons.

The consideration of (a) is about circle-spaces, which in turn, are associated to complex-numbers and subsequently complex-coordinates, and, in turn, the relation of circle-spaces to: quantitative consistency, stability of measured patterns, and the relation of the shape of a circle-space to both spectral values and the geometric properties of spectral constructs (multiplicative constants, and Weyl-transformations [or allowable folds] on the lattice of a discrete shape, as well as action-at-a-distance {or non-local} structures of material interactions) and the subsequent rigidity of measurable structure (eg analytic [complex] function structure), and its relation to the stability of pattern, and the existence of either analytic continuation, or the controllability of linear solvable systems.

Though the apparent randomness of point-particles, may appear to dominate observed material phenomenon, it is really the confinement to a set structure which allows for stable patterns to: exist, be measured, and used, ie controlled (and/or formed). This is about how material components relate to either a stable math structure of their own, or a stable math context needed for reliable measurements (existing in stable orbits or existing as free components in a metric-space, which in turn is about the math structure's, eg spectral values of the different dimensional levels, as well as the energies of an interacting context, which determine condensation vs. resonances with the existing constructs of the stable circle-spaces, where resonances allow for a system's stable math structure to form within itself).

3. Life: The odd-dimensional discrete hyperbolic shapes which also possess an odd-genus, when their faces (or spectral-flows) are occupied then they are charge unbalanced, and would naturally oscillate and generate energy, ie a simple model of life.

4. Mind (related to the spectra which can be contained within a maximal torus of the fiber group)
5. Intent (directing the flow of energy within a cognizant system)
6. Creativity (creating and expanding the possibilities of existence, itself, in a direct manner, or how the instrument of life can be used) Etc.

The truth of a precisely identified pattern should be determined by the relation that the pattern has to practical usefulness, generality of application, so as to provide a wide range of accurate descriptions, made to a sufficient level of precision, based on simple easily applicable laws, or based on the "correct" context which can be seen to limit possibilities, so the information is accurate and the context is practically useful. That is, math truth, not necessarily a context about so many independent variables defining containment so that the description needs the same number of independent equations. That is understanding the context of existence in the context of its stable shapes which organize the different dimensional levels of existence.

There are all types of social issues concerning the expression, and consideration of new ideas, where these issues get mixed up in the way in which tradition and authority dominate the published expressions of a society, but more strikingly the structures of investment and the condition of wage-slavery which afflicts the development to of knowledge, and the subsequent set of lies about personal value and worth as well as competition, where both worth and the competitive game are narrowly defined by the investors, and these forces constitute the social context, and it is a context which opposes new ideas and new expressions, the investors require that knowledge serve the creative interests of the investors, thus such a system grinds itself to a halt, because of the highly enforced narrow viewpoints. It should be noted that, these social structures, which are based on inequality, are created and maintained by means of extreme violence, this emanates from the justice system, and from a militarized management system.

One needs the math properties of stability:

1. Linear (partial differential equations)
2. Metric-invariant, with non-positive constant curvature, where the metric-function has constant coefficients,
3. The shape must be parallelizable and orthogonal at each point of the global shape of the containing coordinate system, ie the partial differential equation is separable so that the locally linear matrix structure associated to derivatives and differential equations is always commutative, or diagonal.

These properties of stability are about the properties which the discrete Euclidean shapes and the discrete hyperbolic shapes, possess, as well as being possessed by the (discrete) shapes in R(s,t), where space-time is R(3,1).

Other properties:

I. principle fiber bundle with metric-space base spaces and isometry and unitary fiber groups
II. the metric-spaces possess properties, eg properties of position and the property of a stable pattern. This leads beyond the isometry Lie groups to both spin-groups and unitary groups
III. Both the metric-spaces and the material components are discrete hyperbolic shapes, essentially modeling adjacent dimensional metric-spaces, ie a higher-dimensional context is not based on continuity of the lower dimensions until one reaches the discrete shape which defines a particular dimensional level, a 3-shape does not see a 7-shape as part of its containment context.
IV. The containing space is an 11-dimension hyperbolic metric-space, a hyperbolic metric-space is chosen since the discrete hyperbolic shapes are so stable in both their shapes and their spectral properties. There are constant factors defined between dimensional levels and between toral components of discrete hyperbolic shapes
V. The derivative can be defined as a discrete operator between

    (a) dimensional levels, between
    (b) time intervals, and between
    (c) toral components of discrete hyperbolic shapes by way of the Weyl-transformations.

VI. Life, mind, etc

As well as:
Holes in metric-spaces; material either resonates with a shape so as to occupy a hole, or spectral-flow, defined within a discrete hyperbolic shape, or it condenses and orbits around the holes (spectral flows) defined by a discrete hyperbolic shape which is much larger than the size of the condensed material.

Increasing or decreasing spectral-size sequences
[The properties of a physical/mathematical construct of an increasing spectral sequence, as the dimension increases, can be organized, based on: shapes, sizes, and formation processes (or formation structures) so as to cause an observer within a particular dimensional level to not perceive the higher dimensions. Furthermore, in an n-metric-space the observer mostly sees only the (n-1)-material components. In a decreasing spectral sequence, as the dimension increases, which is the assumption of both particle-physics and string-theory, it is assumed that there exists the property of continuity between all dimensional levels, and all subspaces, so the spectral sequence must decrease, as the dimension increases, so the observer cannot see the affects of the

higher dimensions, thus the higher-dimensions are curled into small shapes which possess small spectral values, but such an assumption of continuity between dimensional levels is not needed, and it is certainly not necessary.]

Material condensation (often due to the material sizes which are defined in a particular subspace of a particular dimensional level, where the condensed material is smaller than the material-component sizes, ie discrete hyperbolic shapes, of the particular dimension and subspace [of that same dimension])

Only valid model expressing the principle of inertia, as identified in general relativity, is "orbits of condensed material on the discrete hyperbolic shapes of the condensed material's containing metric-space."

A main issue is: "finding one of the finite spectral sets, which identify the stable context of an 11-dimension hyperbolic metric-space."

Blurb 3

Measuring and circle-spaces

Measuring is about relating variations of a system, eg curved-shapes or changes in relative form, to stable uniform attributes, eg measuring rods ie the ability to compare different material structures, linear rods (or clocks, or counting material components) compared to the tangents of either shapes or motion-graphs (of material components), so as to define a measuring process.

But it turns out that both circles and lines are quantitatively consistent geometries, eg the real-number line and the complex-number plane, have the same set of algebraic properties, where algebraic properties of quantities are about the properties of math operations (multiplication and addition):

(1) the existence of inverses, and
(2) about the order of operations (associativity and commutativity), and
(3) the distributive property (which indicates that polynomials are consistent with our decimal (or any base) number system), where the property of commutativity is central to the solvability (or invertibility) of matrix equations, in relation to functions defined on domain space with a many-variable coordinate system, ie many-dimensional, where matrices model (either quantitative or local) linear function relations.

The local vector structure of curved coordinates defined for the local tangent structure of a function's graph, are determined by derivatives, ie this (property of commutativity) is a statement about solving differential equations, and local commutativity is related to a (local) "one-to-one and onto" relationship between the function values and the coordinate values.

Partial differential equations are solvable when local commutativity is a global property, defined everywhere for the local tangents to the graphs of the function values, defined by formulas on the coordinate variables. The conditions of such solvable function-coordinate constructs are satisfied by the circle-spaces, where the function's formula defines a circle-space, eg a torus.

In considering the properties of existence, one might ask, "Which of the following is more fundamental?"

1. The measurable properties of material in space and time, which are relatively stable, or
2. The stable spectral (and orbital(?)) properties of material systems?

The answer might well be found in a context about circle-spaces, the holes in space associated to the circle-space structure of existence, which, in turn, determine the properties of stability in both cases ((1) spectra and (2) measuring in metric-spaces).

The circle-space shapes, which "surround" the holes "which exist in these same circle-space shapes," identify stable spectral values, as well as stable orbital structures, and it is within the context of stable spectra and stable planetary orbits that the measurable properties of material systems have stability in regard to detecting definite spectral values and, performing measurements consistent with a metric-space's measuring properties, and subsequently, resulting in the reliability of measurements.

This leads directly to the hypothesis that . . . , "existence is many-dimensional, (where the holes in space at the different dimensional levels also depend on the size of the circle-shapes which define a dimensional level) where different dimension circle-spaces define (or partition the high-dimensional containment set) the different dimensional levels of the high-dimensional containment set," . . . , where the circle-spaces can also determine the inter-relationships which exist between the material components, which a circle-space contains . . . , (where circle-spaces model both (1) any particular dimensional-level, and (2) a material component [where the material component is also a metric-space but possessing a dimension at least one less than the dimension of the {lowest dimension} metric-space in which the material component is contained).

The circle-space models of both material and their containing metric-spaces determine the properties of material interactions, where the size and stability of the material components which are either a part of the material interactions or which emerge from material interactions depend on the properties of size and stability of the circle-space models of both material and space. It needs to be noted that considerations about dimensional level, and the actual subspace of containment, must be taken into account, since different subspaces might contain within themselves circle-space-shapes which have different sizes, or which have material circle-spaces characterized by different (defining) sizes.

That is, if one wants to partition a high-dimension space by means of "circle-spaces of different dimensions" then it is natural to consider an increasing spectral sequence, as the dimension of the circle-spaces also increases, so as to allow for the containment of the lower spectral set within the higher-dimension (containing) circle-spaces.

Though a decreasing spectral sequence, as dimension increases, is also allowable, but there will not be an ordered structure in regard to the existence of stable orbits for the containment of the lower-dimension material (for such a deceasing spectral sequence) [this is mentioned again below].

That is, (it is only when) both material and space both have stable properties, where these stable properties (associated to these circle-spaces) are relatable to uniform models of quantitative sets (or uniform models of measuring, ie measures and measuring processes) because both material and space have the structure of circle-spaces. That is, both stable systems as well as the property of stable, reliable measurements of "material system properties" can both exist. Namely, the very stable properties, in regard to both geometric-measures in a metric-space and spectral properties, of the discrete hyperbolic shapes.

Why is there stability (eg the relative stability of very many fundamental material systems, eg general nuclei, atoms with more than five charged components, etc)?

Why is there quantitative consistency (why is [macroscopic and microscopic] measuring so reliable and stable?)?

Both questions have the same answer, "It is because of their relation to the structures of circle-spaces."

When motion and position change, but the form of the system remains the same, this means that the energy of the system remains constant. This is a property of the very stable discrete hyperbolic shapes.

Inertia is related to force-fields, and force-fields are related to both material positions and material motions in space, but the motions, which are related to force-fields, tend to be uniform motions (or uniform patterns of motion) which exist within a stable (material) structure, eg the relation of electric currents to the magnetic field. In turn, inertia is related, by integration, to both energy of motion and energy of position, where these functions of energy are (assumed to be) single-valued properties, which are identified by integration, a linear math operator with which one can invert the locally linear differential structure of local inertial descriptions (measures), if there is a way to determine the constant of integration, an initial (or boundary) condition, (and if there is a region [with a boundary, or a set of bounding properties] upon which the equations of inertia are defined). This single-valued-ness of energy function can only be believed to be true if the integration is defined in regard to a very small region of the metric-space so that the holes of the circle-space model of the metric-space are not considered to be incorporated into (or a part of) the integration process. The distant regions of the metric-space are not seen to possess any holes, since the metric-space is, in reality, seen by the observer within the metric-space to be the lattice structure of the metric-space and not the circle-space-shape (of the metric-space).

That is, if things (properties) stay fixed, ie the formulas which define the measurable properties stay applicable (and are single-valued), and if these properties are confined to a spatial region (or temporal, region), which is "small enough" then the quantitative inter-relationships between the (single-valued) function values and the domain values, eg functions defined by either position or local motion, are found to be very rigidly related to one another.

What allows such rigidity of both physical attributes and spatial uniformity (ie measurability of properties, within the domain, which remain relevant to a system's properties, the function's values)?

This is essentially the same context of the question concerning the existence of (say) the maximum and minimum values which exist for a continuous function defined on a closed and bounded interval (domain region). Measuring stays consistent if a formula places consistent bounds on the local structure of the continuum (the domain space).

But, "Is this truly (or in reality) an issue about the over-all stability of the domain space?" and "Not necessarily a property of the continuum (or of a formula defined on a continuum)?" Consider the further context, "Contained in a closed bounded region" or is this really about "contained in a stable circle-space."?

That is, the discrete hyperbolic shapes are so very rigid in their geometric properties that this property allows action-at-a-distance to be a relevant part of the material interaction process.

Ultimately, one may ask, "How are shape and quantity related?"

Can quantitative consistency be related to a finite spectral set, defined in regard to a many-dimensional set partitioned by circle-spaces defined for each different dimensional level (and each subspace for all the different dimensional levels)?

The vast variety of stable shapes are the discrete hyperbolic shapes, which are essentially built up of toral components, ie the discrete Euclidean shapes, and together, this is about the circle-spaces and their relation to stability, and how to relate these spaces to Newton's inertial law, where Euclidean space allows action-at-a-distance (or non-locality).

Coxeter has shown, in regard to the discrete hyperbolic reflection groups, that there are bounded discrete hyperbolic shapes which can be defined up-to, and including, five-hyperbolic dimensions, yet the unbounded discrete hyperbolic shapes exist up-to, and including, hyperbolic dimension-10.

Thus one can consider that all of these discrete hyperbolic shapes can be fit into a hyperbolic metric-space of dimension-11.

The mathematical context, is that of metric-spaces, with fiber groups which are the classical Lie groups, and the partitioning of the dimensional levels by discrete hyperbolic shapes identify within any particular dimensional level an open-closed topology, while when a shape is observed from the adjacent higher-dimensional containing metric-space (if the metric-space's shape is big enough to contain the lower dimensional shape) the lower dimension shape forms a lower-dimension boundary, which is interpreted to be a material component.

It needs to be noted that discrete hyperbolic shapes possess distinguished points on their shapes, which relate to the vertices of the shape's right-rectangular (or "cubical") fundamental domain.

Inertia and force-fields relate to a constant energy-value for a system, mostly, when the material components of the system are in (stable) orbits, on circle-spaces, where inertial interactions are defined as a set of inter-dimensional relationships.

Quantitative models of existence are often given in regard to probabilities of random, but distinguishable, events, which are distinguishable well-defined properties associated to a process which is always disturbing or changing the properties of the finite set of elementary events, or distinguishable and countable set of properties, where general random events are subsets of the elementary event space, eg the six-faces of a (fair) dice which is tossed, where these countable events are supposed to define a "finite" set of: stable, distinguishable, well-defined, and calculable set of properties, associated to (?) properties of things or simply distinguishable measurable (or countable) properties? These sets are only well-defined if these events are distinguishable properties of stable material things. That is, if one distinguishes an unstable pattern then it is not clear that the counting process needed to identify such a random event's probability is well defined, and thus the probabilities are not reliable measures of such a distinguishable-property's randomness.

Consider:

The existence of high-energy cosmic-rays, as well as the possibility of "dark matter" (etc), are better interpreted to mean that there exist large-scale spectral structures in higher-dimensions.

That is, we should not assume that the spectral sequence decreases as the dimension of containment increases, which is the idea of both particle-physics and string-theory etc, since this type of descriptive context implies that there do not exist any stable spectral-orbital structures anywhere.

Also consider, "Where are the descriptions based on the laws of particle-physics which can describe the stable spectra of general nuclei?"

That is, random particle-collision probabilities are only applicable to the rates of reactions in nuclear weapons, where nuclear reactions are only defined in a system which is in an unstable context of "system transition," the probabilities of elementary particle-collisions are irrelevant to describing the stable structures of existence.

The existence of stable planetary orbits is evidence that there do exist high-dimension, large-scale (macroscopic), spectral-orbital shapes, which affect the properties of material (in a manner consistent with the idea of inertia as defined general relativity) and which are important in regard to how material structures are organized.

That stable planetary orbits exist can also be interpreted to mean that large-scale spectral structures exist in higher-dimensions.

Blurb 4

These books contain both old and new essays, with many discussions about the social context of "knowledge," and its relation to monopolistic economic social domination, within a context of wage-slavery. In our society, authority (associated to "high social position") is interpreted to be truth, and that authority is expressed by the propaganda system.

But these essays are mostly about, the subject matter discussed by M Concoyle at the San Diego, 2013, math conference. Namely, the existence of a finite spectral set (associated to the stable properties of material systems) that stable geometries can identify as a dimensional partition of a many-dimensional containment set.

These stable geometries are identified by Thurston's geometrization ideas, concerning the circle-spaces, where these circle-spaces have a strong relation to analysis in complex-coordinates, ie circles in the complex-number plane are quantitatively consistent shapes with the real line.

These circle-spaces are macroscopically related to a new geometric organization of "material" within a high-dimension containing space [for (all) existence of], an 11-dimension hyperbolic metric-space, so as to define a finite spectral-orbital set for "material" within this containing space, where each dimensional level is modeled as a discrete hyperbolic shape, ie a very stable circle-space, and each material component is also modeled as a discrete hyperbolic shape.

These shapes (of both "material" and metric-spaces, which exist at [or are defined in relation to] adjacent dimensional levels) determine the relatively stable spectral-orbital properties of stable material components, as well as condensed material's relation to stable orbits in the various dimensional levels of existence.

The question is, "Why define a decreasing spectral-orbital sequence, as the dimension increases, as is now done in both particle-physics and string-theory?" Where it needs to be noted that particle-physics is not capable of describing the relatively stable spectral structures (or properties) of general nuclei.

Instead, consider an increasing spectral orbital sequence, as dimension increases, so that the stable spectra can define the stable orbital envelopes within which condensed material can define relatively stable planetary orbits (ie the lower dimensional stable-shapes can fit into the higher-dimensional stable shapes).

In the new context the high-dimension containing space is partitioned into shapes so that the space is not continuous, this allows for the existence of an increasing spectral sequence as the dimension increases.

The stability of the solar system cannot be described using general relativity, but the stability of the solar system is describable in the new context.

Blurb 5

The books are about using discrete hyperbolic shapes (circle-spaces, eg a torus, or shapes with toral components) to partition higher-dimensional containing space to both describe physical properties and to define a finite spectral set upon which a stable context for quantitatively consistent descriptions can be based, where a description is about either a shape or a function defined on a containing coordinate metric-space, so that the function defines either a shape or a spectral set (or both).

By partitioning a many-dimensional containing space by open-closed stable discrete metric-space shapes (open-closed when within the metric-space) of different dimensions (the dimensional levels are shapes) the many-dimensional containment construct is not continuous in regard to dimension (or throughout its dimensional structure).

It is only the 5-dimensional discrete hyperbolic shapes which are continuous in a 6-dimensional metric-space, which the 6-dimension (open-closed) shape defines.

It is "only" (primarily) the motions of the stable 4-shapes, which are continuous within the 5-space. It should be noted that 4-shapes which are part of a particular 5-dimensional subspace identify 4-dimensional boundaries when within a 5-dimensional shape. Furthermore, a closed 5-dimensional shape, contained in 6-space, defines a 4-dimensional boundary.

That is, it is, essentially, the motions of these 4-dimensional boundaries which identify continuity (or the continuous motions of objects) in the 5-dimensional level.

As dimensions increase one can define an increasing spectral-orbital sequence . . . , [as opposed to the current construct of defining a decreasing spectral sequence as the dimension of containment increases, as assumed by particle-physics and string-theory, which cannot be used to describe the stable properties of material systems] . . . , so that both lower-dimensional and smaller discrete hyperbolic shapes which can fit into, and follow orbits associated with, the higher-dimensional discrete hyperbolic shapes. Thus, both physical phenomenon of material in regard to either spectra or relatively stable material dynamics can be consistent with the structure of space and material in a sequence of dimensionally dependent spectral-orbital values, ie measured properties can possess stable, consistent, reliably measurable, and practically useable properties.

Discrete hyperbolic shapes can be determined by linear, metric-invariant, and separable partial differential equations. They have very stable geometric and spectral properties. They can be thought of as circle-spaces with right-rectangular-like ("cubical-like") fundamental domains, where the 'cubes" can be thought of as being attached at their vertices, and the circle-spaces can be composed of toral components, as many toral components are the genus of the shape (or the number of holes which can be identified on the shape).

Circle-spaces are related to non-positive constant curvature metric-spaces, R(s,t), where R(3,1) is space-time and R(3,0) is Euclidean 3-space.

Each type of metric-space, characterized by the dimension of the temporal subspace, is associated with mathematical properties and physical properties.

The first two such properties defined: on R(n,0), upon which position is identified, in regard to the distant fixed stars, while on R(n,1) [either generalized (n+1)-space-time, or an n-dimensional hyperbolic metric-space], upon which stable patterns of shape can be identified, ie patterns which are continuous in time, or conserved properties.

Blurb 6

These new books

These are science and math books about new ways in which to describe the observed patterns of existence, they are new ideas which are based on partitioning a many-dimensional space

(12-space-time-dimensions) by modeling each different subspace of each different dimensional level by a stable circle-space.

This is a simple model, but the new context it identifies allows for new ways in which to consider very complicated systems, which are relatively stable, eg life-forms. These complicated systems can be placed in what appears to be quite a simple measurable construct, but which identifies pathways of organization and control.

A circle-space is a shape composed of toral components, where a torus is a doughnut-shape. The circle-spaces, which form the dimensional partition of 12-space, have sizes, so as to form a finite sequence (of circle-space sizes) which is increasing as the dimension increases.

Thus, continuity to an observer would depend on the: dimension, subspace, and size of the circle-space within which they are contained. When within a circle-space an observer would not see the shape of the space which they are in (they would see the full space of the lattice, which defines the circle-space). That is, we cannot see the higher-dimensions.

This set-up allows for a finite spectral set to be identified (where the spectral values are associated to the radii of the circles) upon which the spectra of material-component systems are based, ie where material-components are lower dimension circle-spaces which are contained in any (particular) dimensional level, and they resonate with some subset of values of the finite spectral set, where a relatively stable nucleus would be one of these stable (low-dimension) circle-spaces.

It is a new model which also allows one to understand the basis for the stable large-size orbital-systems which also exist, ie the relatively stable solar-system. In fact, the stability of the solar-system can be interpreted to mean that such dimension-size structures associated to very stable circle-spaces do exist.

This (new) construct, in fact, does quite a lot for science and math, but it also allows for new simple, higher-dimension, models of life, and mind, which can be organized around a simple hierarchy, and it identifies a new context (beyond the idea of the material world, though the material world is a subset of this construct) within which life can be creative.

It also accounts for quantum randomness.

It is a description which is consistent with the fact that non-linearity and instability are so very prevalent, but it identifies the reason for an underlying stability of the very many well defined forms of material systems.

This is something which the current descriptive construct of the physical world cannot do.

Chapter 18

Introductions to math meeting talk

Introduction 1

Math flyer (1), San Diego joint math meeting (2013)
(1-12-13, rm 6E (main bldg) 3:30 pm, #1086-VR-413, Assorted Topics II)
A new context in which to apply Geometry to: Math, Quantum Physics, and the Solar System, etc. By M Concoyle Ph. D.

I. This new "math construct" addresses unsolved problems, as well as unrealized models of quantitative containment (within a finite quantitative construct). This is done by using the simplest of math structures, and elementary ideas about quantity and shape.

II. The unsolved physical problems are about finding the context within which one can describe very stable fundamental physical systems: relatively stable nuclei, general atoms (atomic number greater than five), molecules and their shapes, crystals (BCS predicted a critical temperature which has been exceeded), and envelopes of orbital stability for orbital planetary systems, [and dark matter, dark energy] etc.

    These problems cannot be solved in the math structures (dogmas) based on a non-linear, and an indefinably random context. Nonetheless, this is the context which is used by today's math and physics professions. Such a context leads to the statement, that "the stable properties of fundamental (physical) systems are 'too complicated' to describe," and the focus of the description is on describing fleeting, unstable patterns, in a quantitatively inconsistent manner, and this is done within sets which are "too big," eg the continuum, so as to possess the capacity to be logically inconsistent (eg geometry defined in a random context, allowed by defining various convergences), ie they focus on non-descriptions and irrelevant issues, which they try to describe in too great detail.

III. The new math context is ultimately about, "How to guarantee: stability, quantitative consistency, and to define finitely generated quantitative sets," and about "How to guarantee the stability of a uniform unit of measurement, within descriptions of fundamental systems, which possess stable measurable properties."

IV. The new math context is about basing physical description, as well as stable quantitative structure, on stable geometry. Namely, the geometry based on the discrete hyperbolic shapes, in conjunction with discrete Euclidean shapes, as well as with other metric-space "discrete shapes" associated to non-positive constant curvature spaces, with metric-functions with constant coefficients, eg the R(s,t) metric-spaces and C(s,t) Hermitian-spaces (of finite dimension), (C(s,t) is a result of (4) below), where, s, is the dimension of the spatial subspace, and, t, is the dimension of the temporal subspace, and s+t=n is the dimension of the metric-space. [and where R = real, C = complex numbers]

These are the circle-spaces ie spaces related to "cubical" simplexes or rectangular simplexes (where cubical simplexes are related to circle-spaces by equivalence topologies, ie a moding-out processes).

V. Just as Copernicus and Kepler provided the correct quantitative-geometric context for the properties of the solar system, which could be fitted by Galileo's law and Newton's global solutions to (Newton's) differential equation models of the 2-body, "center-of-mass coordinate" model of the solar system . . . . , which is a more useful context, in regard to feedback systems in gravitating systems, than is the "practically useless," 1-body, spherically-symmetric model of gravity, given by an unrealistic (1-body) non-linear, general relativity theory, ... , this new math context provides the answers (the types of shapes for a solution function) for stable material systems (as well as metric-spaces) modeled in a stable, geometric context, where the assumptions provide the context for the existence of solutions to stable systems, and by modeling metric-spaces as stable shapes, this allows for a finite spectral set to be defined.

But the new context also organizes math patterns, and provides new types of quantitative processes through which fundamental stability can be placed into a general descriptive context, in relation to wide ranging applications, which can lead to sufficiently precise "stable spectral-orbital" descriptions (constructs), which are geometric and thus the information provided can be used in a practically creative manner.

VI. The stable spectral properties of general atoms and nuclei (as well as the other stable material systems) are related to finite (integer) values, ie number of charged components (atomic number), and a number of (uniform) nuclear components (atomic weight), respectively, but they do not have valid descriptions based on physical law, and it has (also) not been understood "how envelopes of orbital-stability form in a macroscopic solar system," but all of these systems can now be modeled, based on stable geometry, ie similar stable geometric constructs hold for both stable microscopic spectral systems, and macroscopic orbital-envelopes of stability.

VII. There are 6 fundamental aspects to the description:

1. The metric-spaces and their associated isometry groups fit into a principle fiber bundle (one ends-up using this construct more later, and it is just as easy to introduce it first).
2. The fundamental shapes of existence: both metric-spaces, and material systems, are to be modeled in the context of discrete hyperbolic shapes and discrete Euclidean shapes (tori), or in non-positive constant curvature spaces for metric-functions with constant coefficients. This is related to the classical Lie groups of both SO(s,t), and spin groups, and SU(s,t) [see 4. below]
3. The containing space is many-dimensional (11-dimensional hyperbolic metric-space) so each dimensional level, as well as each subspace of the same dimension, is to be identified with, either a macroscopic or a microscopic, stable discrete hyperbolic shape.
4. There are (physical) properties associated to metric-spaces of various dimensions, and they are associated to various metric-spaces of the type R(s,t). This leads to metric-space states which come in opposite-pairs. These pairs of opposite metric-space states can be fit into both the real and i(real) [or pure imaginary] subsets of complex-coordinates. Thus the description is (can be) related to SU (unitary) fiber groups. Such descriptions, of opposite metric-space states, also relate to spin-groups, and Dirac operators.

   These properties of metric-spaces can be related to both math and physical patterns since the properties deal fundamentally with "position in space" (Euclidean space), and "a stable existing pattern" (continuity of stable patterns in time, an implied assumption in mathematics), eg energy, mass, and charge conservation assumptions.
5. The derivative as a discrete operator which can be related to: (1) material interactions as discrete operations, (2) dimensional levels involved in dynamics (also physical constants can be modeled as discrete operators defined between dimensional levels) (3) Weyl-transformations are discrete angular transformations, and deal with the shapes of discrete hyperbolic geometries of a physical system's related orbital-spectral properties.
6. A new way in which to model both life and mind.

Introduction 2

Describe the properties of the stable spectral-orbital systems, How does one describe the very stable, many-body, spectral-orbital physical systems which exist at all size scales from nuclei to solar systems, as well as life and mind?

The subject of stability is mostly about using very confining, but very simple set of geometric shapes, and some simple calculus, as well as a few other (local) geometric properties defined in a context of a principle fiber bundle with a many-dimensional base-space, where the base-space is (dimensionally) partitioned into sub-metric-spaces which are identified by discrete shapes, which are open-closed (when observed within the metric-space which possesses the discrete shape), but these metric-spaces (with shapes) form a boundary when viewed from an adjacent higher-dimensional metric-space which contains the discrete shape.

The math properties for stable solvable physical systems, The simple math properties which allow partial differential equations (or differential equations) . . . , which are used to model physical systems (either geometric-inertial or spectral) . . . , to be solved can be listed.

That is, solvability is related to the set of properties: linear, metric-invariant, separable (locally linear and commutative (diagonal matrices at each point in the global coordinate system)), where the metric-functions can only have constant coefficients, (for the various R(s,t)-metric-spaces {where s-space, and t-time; dimensions}, and associated metric-function signatures, where these different metric-spaces are, in turn, related to new types of materials).

[These new material-types are characterized by their properties of being odd-dimensional spatial subspaces, with an odd genus-number, (analogous to discrete hyperbolic shapes) so as to be charge-unbalanced, and thus, naturally oscillating and energy-generating material systems, which model life. Life is defined as new material-types, which are contained in either: R(4,0), R(6,0), R(8,0), or R(10,0) Euclidean spaces.]

The properties of stability are properties which are possessed by the circle-spaces. One can note that, circle-spaces can be modeled as metric-spaces of non-positive constant curvature, whose discrete shapes (or discrete isometry subgroups) are determined from lattices (with an associated fundamental domain, which is related to right-rectangular (or "cubical") simplexes, by a moding-out process (or a process of defining an equivalence topology on the cubical shapes of the fundamental domains).

[Note: The quasi-spectral-geometric properties used to describe particle-physics, ie the non-linear-random-geometric model of particle-physics . . . . , which deals primarily with probabilities of particle-collisions {which, in turn, only model (nuclear) reactions}, and which is a completely irrelevant and practically useless descriptive construct . . . . , are excluded from consideration, since these descriptive constructs only provide . . . , in a chaotic or random fashion . . . , brief and fleeting descriptions of unstable patterns. It is a model which is only applied by means of its relation to reaction-rates.]

Holes-in-space, spectra, force, and the circle-spaces, are all defined within a many-dimensional space. The simple math structures of circle-spaces, as the basis for the stable (geometric) properties of existence, causes both spectral properties and force-fields, to have an analogous (or parallel) math structure . . . , [in relation to holes in space caused by either the shape of a metric-space or by rigid material shapes, eg (usually) 1-dimensional currents defined by rigid material defining a closed curve] . . . , than is usually believed to be true (ie the material-shapes are equivalent to spatial [or metric-space] shapes).

The very stable "discrete hyperbolic shapes" are used as models for both material components and metric-spaces, so as to construct a very rigid geometric structure, in a many-dimensional context, so that Newton's laws still define inertial dynamics. Discrete hyperbolic shapes are stable, with stable spectral properties, and they provide a constrained and stable and very rigid set of boundary conditions for both material containment (confinement) and material interactions, so that differential equations, at first, appear to have a very restrictive containment structure, but this is needed to model stable systems, and it can also be used to model living systems in new ways.

Most material components, as well as metric-spaces, which are contained in this new many-dimensional context, have the shapes of circles-spaces, in particular the "discrete hyperbolic shapes" define a very rigid set of both constraining contexts and boundaries for measurable descriptions. In this new context, descriptions which are related to (partial) differential equations, require that material interactions be mediated by discrete Euclidean shapes, which in turn, are related to the differential equations associated to metric-invariant differential-forms of the force-fields, ie the descriptive structure is based on the geometry of circle-spaces (and holes in the shape of space). Force-fields are applied to material shapes by Newton's law of inertia, which is defined in absolute Euclidean space (which also allows for action-at-a-distance), in this context of very rigid, and confining, geometry (but Newton's universal law of gravitation is modified).

Stable spectral shapes, and material's orbital properties, and the properties of (condensed) "free" material components in space, There are "free" material systems, and orbital material systems, where "free" material components can condense into limited orbital structures, or condensed material can be guided by highly confining orbital constructs, because these (condensed material) components do not have the "correct" size to be stable material components, in their (particular containing) dimensional level, as well as particular subspace, so that the resonances . . . , which allows the existence of material components . . . , are either from another subspace (of the same dimension), or are determined to be from resonances which are defined on the facial structure of the stable simplex structure of the condensed material, in order to have a stable spectral structure for the condensed system. If a higher-dimensional material component is not "big enough" to be a material component, in its particular subspace (of some particular dimension), but nonetheless this higher-dimensional material component does (can) interact. However, the property of "the faces of the material component," which is interacting on a higher-dimensional context, are contained in a closed metric-space. This causes the dynamics of the interaction (in a higher-dimensional level) to be defined on the entire material-containing metric-space. The entire, rigid, metric-space being pushed by the interaction is not noticeable within the rigid geometry defined within the metric-space, ie a perfectly rigid-rod can transmit a "push" instantaneously across the space it occupies.

An example of a true manifestation of inertial affects defined in general relativity, "Free" material components can also be related to orbital structures, where most often condensed material constrained by the orbital structures of discrete hyperbolic shapes, wherein geodesics can (now) affect inertial properties of these "free" material components, and can be described in an explicit manner in a stable (linear) context, ie general relativity is being defined beyond the 1-body problem with spherical symmetry, so that the description is stable, since the geometry in the new descriptive structure is linear, metric-invariant, and separable, so as to define a stable orbit.

[Note: An orbit defined on a discrete hyperbolic shape is pushed . . . , to the limited and rigid structure associated with the shape's geodesics . . . , by the coordinate structure of hyperbolas, which exists away from the geodesic paths (where geodesics will be contained on the faces of the simplex of the discrete hyperbolic shape's fundamental domain)].

The interaction "discrete Euclidean shape," or the interaction torus, Euclidean space-forms (a synonym for a "discrete Euclidean shape") mediate material interactions within the rigid constraints of the "discrete hyperbolic shapes," which model both metric-space existence and material existence. During a material interaction a differential-form is defined upon the geometry of an interaction torus, ie a differential 2-form model of a force-field is defined on the interaction torus. Yet, a "discrete Euclidean shape" of an interaction can also transform, so as to become a toral component of a newly formed discrete hyperbolic shape, where this can happen if both (1) resonances exist for the interaction, and (2) the energy and size of the interaction is within the "correct" energy and size ranges. The geometry of the 2-form is related to the geometry of the fiber SO(n) group in order to determine the direction of the force-field's push [in the base-space containing the interaction torus]. Then the spatial positions of the interacting material's vertices are locally transformed, in a (local) context of opposite metric-space states.

A finite spectral-orbital set, defined upon a many-dimensional containing space, A finite spectral set can be defined on an 11-dimensional hyperbolic metric-space, which is an over-all high-dimension containing space for (of) a model of existence (where the existence is defined by the finite spectral set). This is possible because in each dimensional level (and for each subspace of the same dimension) there is a maximally-sized discrete hyperbolic shape in regard to the finite set of discrete hyperbolic shapes which are used to model the subspace metric-spaces (subspaces of the 11-dimensional space), so that the spectra defined for all the dimensional-levels and subspaces of each of those different dimensional-levels is finite, and thus, it can determine a finite spectral set, upon which all material properties . . . , which are allowed to exist in this 11-dimensional hyperbolic metric-space . . . , depend.

Defining angles between toral components, Weyl-transformations of angles between a discrete hyperbolic shape's toral components, (or a set of folds allowed on the base-space lattice structure), so that there are a finite set of angular relationships which can be defined between the toral components of a discrete hyperbolic shape. These folds between toral components allow envelopes of orbital stability to be defined, based on the orbital (or metric-space shape) structure which a discrete hyperbolic shape can have after its toral components are transformed by certain angular values.

The operation of multiplying by a constant factor, Constant multiplicative factors can be defined . . . , so as to affect the properties of: shapes, sizes, orbits, and the stability of "discrete hyperbolic shapes" :

. . . , between dimensional levels,
. . . , between subspaces of the same dimension, and
. . . , between toral components of a discrete hyperbolic shape.

Physical properties (and math properties) attributed to metric-spaces, There are physical properties associated to metric-spaces. Two examples of physical properties are: (1) position in space of a system's vertex, in regard to the distant stars, (2) the stability of a system's (or a mathematical) pattern.

This results in the definition of "metric-space states" of opposite physical properties, eg (+t) and (-t). These opposite metric-space states are a part of the dynamic processes of material interactions, eg fixed stars, rotating stars (Euclidean); forward time, backward time (hyperbolic space) etc.

In turn, this implies unitary containment, in regard to the containment (of opposite metric-space states), within both real and pure imaginary subsets of finite-dimension Hermitian containing set of coordinates, as well as allowing the definition of the spin-rotation of metric-space states, so that this spin-rotation of metric-space states is defined on opposite metric-spaces states so that these opposite states are a part of the dynamic process. The (time interval of the) period of the spin-rotation of opposite metric-space states is a property used in the dynamic (or inertial) material interaction process.

In this description there is no need of a continuum, instead rigid geometric stability is used rather than using (indefinable) randomness and non-linearity as a basis for (physical, or mathematical) description.

Indefinable randomness and non-linearity seem to possess the properties which are needed to briefly describe patterns which are unstable and fleeting in duration, which, at best, are relatable to feedback constructs, ie it is a description which depends on the validity of the fleeting pattern of a system modeled as a partial differential equation, which has limited descriptive value, and which is used in a relatively unimportant contexts, ie it is a flawed viewpoint which cannot be used describe the observed stable spectral-orbital properties of physical systems at all size scales.

Introduction 3

Outline of ideas

Describe the properties of the stable spectral-orbital systems, How does one describe the very stable, many-body, spectral-orbital physical systems which exist at all size scales from nuclei to solar systems, as well as life and mind?

The subject of stability is mostly about using very confining, but very simple set of geometric shapes, and some simple calculus, as well as a few other (local) geometric properties defined in a context of a principle fiber bundle with a many-dimensional base-space, where the base-space is (dimensionally) partitioned into sub-metric-spaces which are identified by discrete shapes, which are open-closed (when observed within the metric-space which possesses the discrete shape), but these metric-spaces (with shapes) form a boundary when viewed from an adjacent higher-dimensional metric-space which contains the discrete shape.

The math properties for stable solvable physical systems, The simple math properties which allow partial differential equations (or differential equations) . . . , which are used to model physical systems (either geometric-inertial or spectral) . . . , to be solved can be listed.

That is, solvability is related to the set of properties: linear, metric-invariant, separable (locally linear and commutative (diagonal matrices at each point in the global coordinate system)), where the metric-functions can only have constant coefficients, (for the various R(s,t)-metric-spaces {where s-space, and t-time; dimensions}, and associated metric-function signatures, where these different metric-spaces are, in turn, related to new types of materials).

[These new material-types are characterized by their properties of being odd-dimensional spatial subspaces, with an odd genus-number, (analogous to discrete hyperbolic shapes) so as to be charge-unbalanced, and thus, naturally oscillating and energy-generating material systems, which model life. Life is defined as new material-types, which are contained in either: R(4,0), R(6,0), R(8,0), or R(10,0) Euclidean spaces.]

The properties of stability are properties which are possessed by the circle-spaces. One can note that, circle-spaces can be modeled as metric-spaces of non-positive constant curvature, whose discrete shapes (or discrete isometry subgroups) are determined from lattices (with an associated fundamental domain, which is related to right-rectangular (or "cubical") simplexes, by a moding-out process (or a process of defining an equivalence topology on the cubical shapes of the fundamental domains).

[Note: The quasi-spectral-geometric properties used to describe particle-physics, ie the non-linear-random-geometric model of particle-physics . . . . , which deals primarily with probabilities of particle-collisions {which, in turn, only model (nuclear) reactions}, and which is a completely irrelevant and practically useless descriptive construct . . . . , are excluded from consideration, since these descriptive constructs only provide . . . , in a chaotic or random fashion . . . , brief and fleeting descriptions of unstable patterns. It is a model which is only applied by means of its relation to reaction-rates.]

Holes-in-space, spectra, force, and the circle-spaces, are all defined within a many-dimensional space The simple math structures of circle-spaces, as the basis for the stable (geometric) properties of existence, causes both spectral properties and force-fields, to have an analogous (or parallel) math structure . . . , [in relation to holes in space caused by either the shape of a metric-space or by rigid material shapes, eg (usually) 1-dimensional currents defined by rigid material defining a closed curve] . . . , than is usually believed to be true (ie the material-shapes are equivalent to spatial [or metric-space] shapes).

The very stable "discrete hyperbolic shapes" are used as models for both material components and metric-spaces, so as to construct a very rigid geometric structure, in a many-dimensional context, so that Newton's laws still define inertial dynamics. Discrete hyperbolic shapes are stable, with stable spectral properties, and they provide a constrained and stable and very rigid set of boundary conditions for both material containment (confinement) and material interactions, so that differential equations, at first, appear to have a very restrictive containment structure, but this is needed to model stable systems, and it can also be used to model living systems in new ways.

Most material components, as well as metric-spaces, which are contained in this new many-dimensional context, have the shapes of circles-spaces, in particular the "discrete hyperbolic shapes" define a very rigid set of both constraining contexts and boundaries for measurable descriptions. In this new context, descriptions which are related to (partial) differential equations, require that material interactions be mediated by discrete Euclidean shapes, which in turn, are related to the differential equations associated to metric-invariant differential-forms of the force-fields, ie the descriptive structure is based on the geometry of circle-spaces (and holes in the shape of space). Force-fields are applied to material shapes by Newton's law of inertia, which is defined in absolute Euclidean space (which also allows for action-at-a-distance), in this context of very rigid, and confining, geometry (but Newton's universal law of gravitation is modified).

Stable spectral shapes, and material's orbital properties, and the properties of (condensed) "free" material components in space, There are "free" material systems, and orbital material systems, where "free" material components can condense into limited orbital structures, or condensed material can be guided by highly confining orbital constructs, because these (condensed material) components do not have the "correct" size to be stable material components, in their (particular containing) dimensional level, as well as particular subspace, so that the resonances . . . , which allows the existence of material components . . . , are either from another subspace (of the same dimension), or are determined to be from resonances which are defined on the facial structure of the stable simplex structure of the condensed material, in order to have a stable spectral structure for the condensed system. If a higher-dimensional material component is not "big enough" to be a material component, in its particular subspace (of some particular dimension), but nonetheless this higher-dimensional material component does (can) interact. However, the property of "the faces of the material component," which is interacting on a higher-dimensional context, are contained in a closed metric-space. This causes the dynamics of the interaction (in a higher-dimensional level) to be defined on the entire material-containing metric-space. The entire, rigid, metric-space being pushed by the interaction is not noticeable within the rigid geometry defined within the metric-space, ie a perfectly rigid-rod can transmit a "push" instantaneously across the space it occupies.

An example of a true manifestation of inertial affects defined in general relativity, "Free" material components can also be related to orbital structures, where most often condensed material constrained by the orbital structures of discrete hyperbolic shapes, wherein geodesics can (now) affect inertial properties of these "free" material components, and can be described in an explicit manner in a stable (linear) context, ie general relativity is being defined beyond the 1-body problem with spherical symmetry, so that the description is stable, since the geometry in the new descriptive structure is linear, metric-invariant, and separable, so as to define a stable orbit.

[Note: An orbit defined on a discrete hyperbolic shape is pushed . . . , to the limited and rigid structure associated with the shape's geodesics . . . , by the coordinate structure of hyperbolas, which exists away from the geodesic paths (where geodesics will be contained on the faces of the simplex of the discrete hyperbolic shape's fundamental domain)].

The interaction "discrete Euclidean shape," or the interaction torus, Euclidean space-forms (a synonym for a "discrete Euclidean shape") mediate material interactions within the rigid constraints of the "discrete hyperbolic shapes," which model both metric-space existence and material existence. During a material interaction a differential-form is defined upon the geometry of an interaction torus, ie a differential 2-form model of a force-field is defined on the interaction torus. Yet, a "discrete Euclidean shape" of an interaction can also transform, so as to become a toral component of a newly formed discrete hyperbolic shape, where this can happen if both (1) resonances exist for the interaction, and (2) the energy and size of the interaction is within the "correct" energy and size ranges. The geometry of the 2-form is related to the geometry of the fiber SO(n) group in order to determine the direction of the force-field's push [in the base-space containing the interaction torus]. Then the spatial positions of the interacting material's vertices are locally transformed, in a (local) context of opposite metric-space states.

A finite spectral-orbital set, defined upon a many-dimensional containing space, A finite spectral set can be defined on an 11-dimensional hyperbolic metric-space, which is an over-all high-dimension containing space for (of) a model of existence (where the existence is defined by the finite spectral set). This is possible because in each dimensional level (and for each subspace of the same dimension) there is a maximally-sized discrete hyperbolic shape in regard to the finite set of discrete hyperbolic shapes which are used to model the subspace metric-spaces (subspaces of the 11-dimensional space), so that the spectra defined for all the dimensional-levels and subspaces of each of those different dimensional-levels is finite, and thus, it can determine a finite spectral set, upon which all material properties . . . , which are allowed to exist in this 11-dimensional hyperbolic metric-space . . . , depend.

Defining angles between toral components, Weyl-transformations of angles between a discrete hyperbolic shape's toral components, (or a set of folds allowed on the base-space lattice structure), so that there are a finite set of angular relationships which can be defined between the toral components of a discrete hyperbolic shape. These folds between toral components allow envelopes of orbital stability to be defined, based on the orbital (or metric-space shape) structure which a discrete hyperbolic shape can have after its toral components are transformed by certain angular values.

The operation of multiplying by a constant factor, Constant multiplicative factors can be defined . . . , so as to affect the properties of: shapes, sizes, orbits, and the stability of "discrete hyperbolic shapes" :

. . . , between dimensional levels,
. . . , between subspaces of the same dimension, and
. . . , between toral components of a discrete hyperbolic shape.

Physical properties (and math properties) attributed to metric-spaces, There are physical properties associated to metric-spaces. Two examples of physical properties are: (1) position in space of a system's vertex, in regard to the distant stars, (2) the stability of a system's (or a mathematical) pattern.

This results in the definition of "metric-space states" of opposite physical properties, eg (+t) and (-t). These opposite metric-space states are a part of the dynamic processes of material interactions, eg fixed stars, rotating stars (Euclidean); forward time, backward time (hyperbolic space) etc.

In turn, this implies unitary containment, in regard to the containment (of opposite metric-space states), within both real and pure imaginary subsets of finite-dimension Hermitian containing set of coordinates, as well as allowing the definition of the spin-rotation of metric-space states, so that this spin-rotation of metric-space states is defined on opposite metric-spaces states so that these opposite states are a part of the dynamic process. The (time interval of the) period of the spin-rotation of opposite metric-space states is a property used in the dynamic (or inertial) material interaction process.

In this description there is no need of a continuum, instead rigid geometric stability is used rather than using (indefinable) randomness and non-linearity as a basis for (physical, or mathematical) description.

Indefinable randomness and non-linearity seem to possess the properties which are needed to briefly describe patterns which are unstable and fleeting in duration, which, at best, are relatable to feedback constructs, ie it is a description which depends on the validity of the fleeting pattern of a system modeled as a partial differential equation, which has limited descriptive value, and which is used in a relatively unimportant contexts, ie it is a flawed viewpoint which cannot be used describe the observed stable spectral-orbital properties of physical systems at all size scales.

Social comment

The social structure which is so destructive to science, math, and equal free-speech, is that the US society has adopted the western tradition . . . , which has existed since Rome . . . , of a society being based upon the public supporting an emperor, exactly as the Emperors were upheld within the Holy-Roman-Empire, where religion was added as a propaganda-wing, espousing high-value to support the tyrant Emperor. Today those who support the high-values, which our culture is supposed to represent are, in fact, supporting the structures which uphold the Emperor and oppose equality and creativity.

Social comment

The US society today is the same type of society as was the Holy-Roman-Empire, and this is based on property rights and minority rule, which is the law, if the Bill of Rights and the Declaration of Independence are not upheld.

**** The small group of people, who own (the controlling interest in) the US society, is claimed to be so adamantly "pro-life," but "the reality" (if examined) shows that this is a cruel joke.

For example, if a person is to be "formed from a lump of clay," then this will cost, due to property rights, enforced by a "justice" system with such extreme violence, so as to terrorize the public (where the public is expected to take a vow of poverty, and to only serve the creative interests of the owners of society), and the so called "value-added to the clay" can be priced (by the monopolistic owners) at any arbitrary amount, to cover the cost of the extreme violence (which is really covered by the tax-payer), and which is also associated to "maintaining intellectual property rights," "the right of the owners of society to steal the ideas of others" for the owners of society, which is again protected for them by the justice system (maintained by the tax-payer), and then the need to keep the society fixed, or conditioned to live in a traditional structure (or organization) "in the way which assures that certain materials are used by the public 'to live,'" so that the society is dependent on the particular way in which those who own society have organized material and its "value-added" products, so as to be able to monopolize all the human-life within society, so the people will serve their pay-masters. That is, the society is opposed to human-life, where human-life is all about developing knowledge and using it for practical creative purposes, where creativity is an expression of life and not an expression of selfish gain, and a proof of one's own superiority.

That is, the inequality of the people in the US society "places no value on-life (except for the lives of the owners of society)" because this inequality depends on extreme violence to maintain.

Thus, the public is led to believe that they are not allowed to have any access, or any control over material, as well as no access to useful knowledge. They are filtered out of the knowledge "used to maintain the narrow domineering monopolistic social structure," by a process which supports obsessive memorization of a language which is incomprehensible, and irrelevant (except the little knowledge needed to maintain the traditions upon which the monopolistic economy-society depends) [and the tax-payers also pay for this affront to their knowledgeable and creative capacities]. The extreme violence (used to maintain such a narrowly defined range of value for society) expresses the idea that only "the life of the owners of society" has any value. The public is deceived into believing in narrowness associated to a great obsession over irrelevant matters (and this is called education) and to serve their pay-masters, and to bow-down to the palace guard (the justice system and the militarized state).

As the military becomes bigger the few owners of society can both become fewer, and become more separate from society.

The monopolies (of the few owners of society) claim to be the great successes, but their success is all about, how to apply violence, and has nothing to do with their use and development of knowledge, which if examined at all, one sees that "they are complete and total failures." "They are complete incompetents," and they want to take-over the education system, where their propaganda is all about their lies . . . . (which dominate the propaganda system, namely, that people are not equal, and that the superior people do [and should] run the country) . . . . , about "identifying superior teachers" so as to be able to teach the (very narrow concept of an) "absolute truth," namely, and absolute truth which is used to maintain the power of the owners of society, (the knowledge which is leading to ever greater failures). However, it should be noted that it is the same type of power (and power structure) as was the power structure of the Roman-Empire, and this means that it is power which is based on extreme violence and extreme inequality (and a disregard for human life).

There is the mistaken idea that equality is all about material (dividing up material, where control of material is the basis for western oligarchy), but this is far from being true, though people should have equal access to material, and equal access to knowledge, where "people most want" a relation within themselves "between knowledge and creativity" ie they want to develop knowledge which is practically useful in regard to creativity.

This is the defining idea of human-life, and it is the idea most opposed by a society based on extreme violence which upholds property rights and minority rule.

The result of basing society around the idea of property and inequality (ie minority rule) is that there is only a focus on materialism, and a capacity to concentrate and control power by the society's relation to material and its uses.

That is greater inventive-ness allows for greater diversity in regard to the ways in which to organize and use material within society.

The point of the US revolutionary war was to break from the "model of society" which is dominated by a very oppressive and anti-life "form of a society" governed by a Caesar-like emperor, an emperor of a Puritan-US-Empire, upheld (and expanded) by extreme violence, and a subsequent disregard for human-life (or any other life). At first it expanded across America, until (around) the 1920's, and then the US ended-up controlling the world after WWII.

The equality upon which the American Revolution was based is the "equality" represented by the members within the Quaker society, built by W Penn.

However, it is also the spirit of the entire Protestant movement, where people were to read an interpret the Bible for themselves. Unfortunately, the bible is the literature of "propaganda for Empire," which is what the Roman-Emperor Constantine provided to the western world (and expressed throughout the west by the Catholic church, until the 15th century, which is also when the scientific-revolution started).

The equality of the American Revolution was also about the spirit of "the scientific revolution," which was (perhaps) most freely expressed in the 18th and 19th centuries, (where the electronic industry is still expressing and developing the ideas of 19th century electromagnetism).

However, western knowledge has always been under the control of a de-facto emperor, but the American revolution was all about equality, "equality to have equal free-inquiry" in relation to finding (developing) knowledge, which can be used for practical creativity, where a creative notion is most often what drives one to seek knowledge or to develop knowledge.

Godel showed that knowledge is all about equality since it is mostly about elementary language development at the level of: assumptions, interpretations, contexts, containment sets, organization; all in regard to: material, organization, space, and change. The history of technical development has shown that when a precise (measurable and practically useable) language is at the simplest level, there is the greatest practical developments.

That is, though there can be great complications that can occur for a particular technical context, "in such a case the development is all about more of the same technical context," but wider diversity in regard to practical creativity emerges from a simple language, which relates stable measurable properties so as to form a context for coupling (between subsystems) and (subsequent) control of their stable measurable properties.

That quantum systems have such stable definitive and precise system properties does not indicate that the descriptive language of these systems should be based on randomness, rather it implies a controlled geometric structure (and not string theory).

A main focus on equality is not about focusing on materialism, rather "it is a main focus" on both useful knowledge and creativity, (ie a focus on the spirit), where practical creativity does not necessarily mean creativity in regard to materialism.

That is there already exist models of higher-dimensions, and these high dimensions are of macroscopic sizes, where material is a subset (where the higher-dimensions are hidden by both the way the process of material interaction is organized and the size of the interacting materials (which is controlled by physical constants) [eg there is a jump from the size of nuclei to the size of solar systems going from 2-dimensional material to 3-dimensional material]).

It is the violence of empire, where the emperors are those few people who (effectively) own society, and these emperors depend "for the maintenance of their social positions" on the extreme violence of the justice-system and the subsequent social domination of an artificially fixed (and traditional) society by monopolistic capitalism and money (which, essentially, is society's only measure of value), with societal law based on property-rights and minority-rule, and a set of rulers who focus (so much) on materialism, where materialism is (has come to be) the opposite of creativity and the spirit.

Marx was a "western (or oligarchical) materialist," and he analyzed material-based capitalism, and he identified an oligarchic social structure associated to his western idea about socialism (the type of oligarchy which most threatens the capitalists, since Marx's socialism also focuses on the idea of materialism), where the socialism of Marx is (also) focused on materialism (as capitalists are focused on materialism, since it is through material-control that society is controlled, and it is material from which they derive their social power) as opposed to focusing on the social attributes of knowledge, creativity (and spirit).

Marx's ideas were consistent with the west, since the west was influenced by the Holy-Roman-Empire and its violence was centered around the idea of controlling material, and that society should be organized around an oligarchy.

However, the original intent of the US Revolution was to oppose "rule by emperors," and it wanted to base the law of the new society on equality.

Unfortunately, the ruling class (of a society of European settlers who were stealing from the native peoples, resulting in the selfish gain for these settlers [within a European organization of society]) wanted law based on property rights and minority rule. The constitution was based on property rights, and slavery, and minority rule, but it was to depend on the Bill of Rights.

The constitution has had no validity other than through its relation to extreme violence because the Bill of Rights has never been upheld.

Instead extreme violence was channeled through the justice system, and it was used to support property rights, minority rule, and wage-slavery, which is essentially the same basis in law as the Holy-Roman-Empire, and again the power of the oligarchs depending, for maintenance, on extreme violence "aimed at the public," and as a side-show "it is about world domination."

In regard to material, equality means "life has a material right to its life," and human life is mostly about knowledge and practical creativity, but practical creativity may not be based on materialism.

A fixed aspect to society which does include material needs of life, must be organized, not for selfish domination and extreme violence, but for a balanced arrangement which allows (requires) the equality of people.

Note: The golden rule might best be re-stated in the simple terms that, "people are equal."
Though Constantine's Bible (of empire) expresses the golden rule in an almost incomprehensible manner, "Do to others as one wants others to do to you." (?)

Introduction 4

Math propaganda

Base the descriptions of practically useful math patterns on a finite set of quantities, namely, a finite set associated to spectral-orbital properties of stable physical systems, defined over a range of different dimensional levels, up to hyperbolic dimension-11.

Do professional mathematicians actually understand what a derivative operator, or a differential equation, is?

Interpret the derivative, in a new way, as a "discrete operator" which acts on stable geometries, so as to not be dependent on (or identify) either a continuum, or having a (necessary) relation to a function space, due to an assumed property of existence of fundamental randomness.

Stable spectral-orbital properties imply a: linear, separable, solvable, controllable, stable, geometric context.

Stability of shape is the only way in which to both ensure quantitative consistency, and provide valid constructs of stable (observed) measurable properties of (for) physical systems.

Stable, linear, and invertible, ie solvable, always means continuously commutative on the global coordinate space, in turn, this implies the geometry being related to discrete shapes which have "cubical" (or rectangular) fundamental domains (or simplexes), or equivalently, a coordinate space whose coordinate shapes are based on circles and lines (or line segments).

Does a uniform unit of measuring (used to measure an existing [or an observed] pattern), ultimately, depend on a stable geometric construct? {Yes}

Can a mathematician almost always assume that their math constructs stay quantitatively consistent when their quantitative basis depends on (or is) "a stable uniform unit of measurement?" {No}

Should descriptions of math patterns be practically useful, or are they intellectual games which are all about word agreement (Platonic truths), and which are unrelated to any practicality (other than forming a literary body of useless Platonic truths)? {Truth can only be discerned if it is related to practical creativity.}

Consider a new context for mathematics and physics, which is built from stable, simple, solvable geometries, a model which goes beyond materialism (because it is a model of existence which has many-macroscopic-dimensions), and which is quantitatively based on a finite spectral-orbital set, which in turn, can be used to model life-forms as high-dimensional, coherent shapes, where these shapes are associated to classical Lie groups, Both SO (metric-invariance), and SU (Hermitian invariance), and likely Sp (symplectic, or p^x-invariant [where the symplectic space is about combining Euclidean and hyperbolic spaces]).

Stability and quantitative consistency depend on discrete isometry subgroups of classical Lie groups, which, in turn, are based on the general expression of "cubical" (or limited sets of rectangular) simplexes (or fundamental domains). These stable shapes are the very stable discrete hyperbolic shapes, and the continuous, flat, and adaptable discrete Euclidean shapes.

Is a derivative best placed in a context in which it is a discrete operator, defining both discrete multiplication by constants, defined between dimensional levels and between toral components of a hyperbolic space-form, along with discrete angular changes, and discrete changes in material positions?

Math which is stable, and quantitatively consistent, and solvable, and controllable, then it is to be constrained to a "checker-board," or "cubical" lattice.

Generalizations concerning geometry, topology analysis, etc, which are based on the belief in a continuum, have not led to valid, or accurate, or practically useful math constructs.

Current descriptive languages are not working

The math constructs based on non-linearity and randomness, particularly indefinable randomness, do not lead (have not led) to a context of practical creativity, nor to accurate descriptions of stable, physically observed patterns, eg the stable spectral orbital properties of general nuclei, general atoms, molecules, molecular shape, crystals (the critical temperature predicted by BCS has been exceeded by high-temperature superconductivity), as well as the stable solar system, are all systems, which are not now being accurately described, for the wide range of these general (but fundamental) systems.

When a precise descriptive language does not work, ie either has very little accuracy or very little practical value, then the correct interpretation of the Godel's incompleteness theorem is that "the language should be revised at the level of assumptions, contexts, interpretations, containment, and definitions etc."

Furthermore, it is at the elementary level of assumptions, etc, that the language is best related to serve the interests of practical creativity.

Consider the elementary questions:

1. What are the axioms of quantity [(+, x), (+) counting things of the same type using a stable unit of measuring, and (x) rescaling, ie the linear functions, and (x) is also related to the changing of number-types],
2. what is a pattern (or system) which one wants to measure or carefully describe, (something continuous in time, or a coordinate grid of lines and circles)
3. what is the structure of the number system, (polynomials)
4. what is the set of quantitative properties associated to a containment set, and/or associated to the measurable properties of the system which is being contained,
5. what is the relation between shape and quantity (functions and coordinate domain-spaces; in the new descriptive structure the shapes of coordinate spaces and the properties of

functions are essentially the same as the shapes of coordinate spaces, though often defined on different dimensional levels),

6. what is the relation between dimension and shape, (one perceives 2-surfaces in 3-space, and 3-faces in 4-space etc.)
7. what is a derivative, [must derivatives be defined within a continuum] (local linear method of measuring formulas, or linear operators on function-spaces, or discrete operator defined between dimensional levels and between uniform, discrete time-intervals)
8. what is an equation,
   (different representations of the same thing, or an expression of how measuring sets (measured values) are changing, or a rule for an operator (function) on a set)
   [Equivalent measurable properties which are contained in a stable (math) structure, which allows and maintains a reliable measuring context.]
9. what is a physical law, (two representations of a system's properties, or "how a system's properties change")
10. Are metric-spaces associated to physical properties? (yes, eg position in space, stable pattern)
11. relating a containment set to a system's measurable properties, eg defining functions,
12. functions as geometric properties vs. a function as a spectral property, where such spectral-functions define function-spaces,
13. what is the set structure for inversion (of a system of operators acting on sets of functions) and its relation to solution of an equation, (continuously linear, and commutative, and metric-invariant, and invertible, as well as both one-to-one and onto)
14. what set of properties of a math construct allows quantitative consistency,
15. what is a function (a model of measurement within the set which contains the system, which the function is modeling as a measured value of the system),
16. how can it be consistently measured (in a linear manner), and
17. how is nearness to be determined {by a continuum, by (locally) linear geometric measures, by a set structure of a function (function-values and its domain values and the containment within the domain set in regard to an inverse-image set, where the image set is defined by the function-values)}? For some local operators the property of closeness is needed, but "How close does one need to get?" "Does one need a continuum?"
18. What is a spectral set? Are they averages and arbitrary vibrations? Do spectra correspond to sets of operators, eg sets of eigenvalue equations? or Do stable spectra possess a geometric structure? (They are associated to the stable geometric properties of discrete hyperbolic shapes.)
19. Is measuring about metric-invariant measures defined on coordinates whose shapes are quantitatively consistent, ie with shapes related to lines and circles, or is measuring about

the spectral properties of material components or systems, and associated to random harmonic properties?

Math papers should not be peer reviewed, rather their assumptions, contexts, and interpretations, etc, should be identified by the author and a computer should sort them out based on the fundamental attributes of the ideas (or patterns) being expressed, and they should be published openly, but the author needs to also identify the practical creative, or practical measurable pattern, or context to which the description (or the math constructs) apply, and they should be published openly

Being labeled a valuable intellect by an outside source is based on allowing the owners of society to define "what is valuable within society,"

Experts are people who seek to be high-paid wage-slaves, so as to help others unfairly dominate society by limiting markets, and limiting the creativity of the society.

Only in an equal society can people be creative.

People should live in a society as equal creators, ie people cannot (should not) be excluded (separated) from their knowledgeable and creative endeavors, both their creative aspirations and their equal free-inquiries.

If a precise language is not suitable for one's creative desires then create a new precise language.

An unequal society is a society held together by extreme violence, and it depends (for its stability) on great injustices.

Are metric-spaces associated to physical properties?

Inertia is defined within a Euclidean space, where spatial position can be determined within a Euclidean space, while stable energetic and electromagnetic systems are defined in a hyperbolic metric-space (note: a hyperbolic metric-space of dimension-3 is equivalent to space-time).

Note: Thus, $m=k/r$, where r is the smallest radii of a discrete hyperbolic shape [so that this is a property which relates the spectra of a discrete hyperbolic shape, by resonance, to a discrete Euclidean shape whose spectra is associated to the same value, r, so that, k/r, is the inertia (or mass) associated to the discrete Euclidean shape]. Where it might be noted that, m=k/r, is consistent with (or similar to) the relation, p=[h/(wave-length), where p is momentum].

Can sets be "too big?"

Is the continuum "too big" of a set?

Does the continuum allow for "too much" manipulation of the quantitative values, which, in turn, are used to identify descriptive patterns which fit data?

In the descriptions of quantum systems the wave-equation is linear, and the model for a quantum system's components is random, and modeled as a function space, where in quantum physics a function-space and its dual-space (the context of [integral] operator continuity) imply an uncertainty principle (see Powell and Crasemann "Quantum Mechanics").

Yet the adjustment (or correction) to a "solution wave-function" is based on the explicit geometry of point-particle collisions, and on a non-linear term (a connection term) in the operator structure, where this non-linear term is associated to elementary-particle collisions . . . , (for the expressed intent of particle-physics, wave-function adjustment [and which is the quantum model for material interactions]) . . . . , and is central to the wave-function adjustments (or perturbations), so this math construct (which model material interactions) is neither quantitatively consistent (ie it is non-linear) but it is also logically inconsistent {allowing both randomness (requiring an uncertainty principle) and geometry (which depends on a metric-space whose values at-least require the rational numbers to determine limits)}.

These are opposite properties: linear vs. non-linear; and random vs. geometric; yet convergences, or rather "a pair of divergences are considered equal to finding a convergence," and this has been done for two or three systems, so that these three (or so) "convergences" (which are claimed to be a valid calculation) are interpreted to mean that these opposing sets of ideas (linear vs. non-linear; and random vs. geometric) are valid, and these opposing sets of ideas are claimed to be properly related to one another within the quantum description of material systems and their material interactions.

These are the types of inconsistencies which "sets which are 'too big' allow," where the continuum is a set, which is "too big."

Modern physics

So math and science have become an endless sequence of speculating about both math patterns (not knowing how to organize them, so as to consistently identify the observed physical patterns) and interpreting observed physical patterns, but this speculation is often accompanied by careful and sometimes (or apparently) rigorous appearing discussions, and this provides the speculations with a sense of authority . . . , eg we will simply trap collision processes which result in nuclear fusion and determine how to extract the generated energy, . . . , and even though the speculations never succeed at their descriptive goal, they can lead to other speculations, which are again endowed with an apparent authority . . . , and on and on it goes . . . eg perhaps we will understand the strong-force and control how energy flows etc. . . . , so absurd math models are devised to describe forces, eg the strong force is modeled as non-linear changes of particle-states during particle-collisions, etc, where such non-linear models of particle-collisions, it is claimed, can describe "why the un-seeable quarks can bind-together to form protons" (where protons are claimed to be unstable) etc, (This is more like religion, and a scientist's absolute authority has become the model "of truth for society.")

However, this process is always consistent with the wants of big business, eg nuclear weapons engineering etc, eg collision probabilities are (remain) related to rates of reactions etc.

The oil and coal companies never get any meaningful business competition from a business supplying a clean-cheap energy source, since such a source never gets developed, and never will by using the way knowledge about quantum systems is expressed. It is in the oil and coal business interests to have the mathematicians and scientists confused about quantum properties and quantum-material interactions.

This type of "science" can only be done within a society which mistakes authority for truth, and it is clear that it is the media (the propaganda-education system) which is modeled as "a single-voice of authority," for an entire society, ie a single authoritative voice which has taught the people of society to consider truth to be authority.

Namely, the authority of the media (or propaganda-education system) which expresses the interests of the owners of society, but it is a society (and propaganda system) which, essentially, only serves the narrow needs of society's owners and is related to a process of "central planning for selfish purposes" which is done by a few very rich people (may be 5 or 10 of these barbarian [selfish, ignorant, and psychopathic] people) who own (possess the controlling shares) and operate a narrowly defined market. The so called market is a narrowly defined and highly fixed way in which to use material resources (though it is masked as "always developing," but it only develops material uses, eg electric circuits, based on 19th century science, eg it also depends on reaction-rates being related to probabilities of component collisions).

It is the owners of society who define both creativity and high-value within society, and it is defined to serve their selfish needs and wants.

It is these few owners of society whom the political-justice system serve, and the ideas and desires of these few owners of society are all expressed through the media, and these wants are backed and supported by a militarized state, a society which has been militarized since 1940.

Science and math is part of such a set of militarized management, and it is used to support the (mostly militarized) interests of the owners of society.

Perhaps, the militarization is the basis for the value of money, and a violent social-system which forces people into wage-slavery, as opposed to the natural inclination of people to know and to create things which are to be given to others as gifts (creativity and developing knowledge for selfless purposes).

Is value knowledge and creativity, or is social value all about controlling money, ie arbitrary value, and controlling the voice of authority, and (with the aid of law and propaganda [ie politics]) violently forcing people into a social structure of wage-slavery. Where the basis in law, for such a society, is property rights and minority rule, but the Constitution also had a Bill of Rights, but the government has never upheld these laws, ie the Constitution is invalid since the government has never upheld the laws which the Constitution proclaims, eg free speech is not protected and no state involvement in religion. Today the US is essentially equivalent to the Roman Empire but the emperors are the owners of society and the government serves these de facto emperors, ie the pay-masters of the wage-slaves. It is the interests and the security of the emperors to which the full power of the US state is now applied.

A new continental congress needs to re-instate the basis of US law to be "equality" and the freedoms to know, ie equal free inquiry, and to use what one knows to create, in a society where everyone is an equal creator. Only in an equal society, can there be, truly, free markets. A society must be equal since there is not a measure by which to judge all the various creative efforts, but with equality there must be self-less-ness as the motivation for creativity.

Chapter 19

Speeches

Speech 1

Use geometrization (in particular, that the most stable geometric shapes are the discrete hyperbolic shapes) to turn physical description into a geometric exercise as opposed to an exercise about indefinable randomness and (the quantitatively inconsistent) non-linearity

Should physical description be about:

1. Unstable patterns and fleeting contexts (hopefully) relatable to feedback systems, eg guidance systems, and
2. To chaotic contexts of systems briefly transitioning between relatively stable states, eg nuclear reactions,
3. That is, descriptions of marginal, unstable contexts related to the fine-tuned interests of big business, such as improving a complicated instrument's capacities (eg guiding a missile), defined within a fixed descriptive context, so as to help big business. Note: Computing is about speeding up a computers rate of switching, but a computer is limited to operating on numbers or on symbols strictly related to a process (which might be a symbolic goal). That is, new contexts can establish new ways in which creativity or new ways in which to achieve a goal (or a process). Furthermore, the increase of switching rate for a computer is approaching the obstacle of the limitation of relevant knowledge, ie traditional knowledge has great limitations, it now is being related to many failings, eg the calculations of business risks has failed because the math is failing.

Is math and science about serving commercial interests, and developing complicated instruments (for business interests) but there are limits to an instrument's performance, and/or (in regard to mathematics) does a fixed context and a (fixed) set of authoritative traditions, which are associated to business interests, also have a limit to its performance (or such a context's capabilities, such as in regard to its descriptive range)?

or

Is the range of a measurable description to be determined by a very large set of equal independent free-inquirers, all of whom are to be regarded as equal creators (not simply creators in regard to narrowly defined business interests, ie why should "big oil" or "big banks" be determining the structure of knowledge and hence the structure of creativity within society)? This allows the society . . . , ie big business who in turn has controlled big government so that now big government serves the needs of big business . . . , to identify a small set of pompous intellectual aristocrats (university academics), who are given great authority in society, but who possess a very limited and narrow range of thought, or who have limited intellectual range concerning the possibilities of precise descriptive knowledge, and its possible relation to new creative contexts.) one can only establish oneself as being correct in a fixed context, but fixed contexts are not conducive to the new "development" of precisely described knowledge, nor conducive to the developing new contexts for creativity. (big business is most profitable when the context within which it makes the most money stays fixed). Equality supports development, inequality is defined within a fixed context.

(talk)

What are stable math patterns? [They are simple math patterns]

Nuclei, atoms with more than 5-charged-components, molecules, crystals, the solar system (etc), are all stable physical systems (which implies that they are formed under controlled linear conditions) but none of these (general) systems have valid descriptions (beginning with the laws of physics and then deriving the spectral [or orbital] properties of these systems).

Is one to try to improve the descriptions of these systems by means of more complications?

or

Should one consider new contexts, and new interpretations upon which to base description?

Gödel's incompleteness theorem can be interpreted to mean "math and science are trying to do the same thing"

{Hilbert wanted to place physics on an axiomatic basis, but when it was shown that math cannot be completely developed by it having an axiomatic basis, and that both disciplines are about developing measurable descriptions of observed patterns, then this leads to this interpretation}

In this context, truth is finding "simple" patterns and processes which are:

1. Widely applicable to general systems (observed properties or observed patterns)
2. Can describe observed patterns accurately and to sufficient precision, can describe definitive, stable patterns, and
3. Strongly relates to practical creativity (measuring, fitting, selecting material or subsystem structure, building, or putting together, and controlling "physical" systems).

Geometry can be both stable and controllable, especially the geometric systems which are contained within a linear, metric-invariant, and separable (or solvable or commutative everywhere) context for their descriptions, (descriptions of physical systems (or shapes))

Geometry is more related to practical use than either randomness or non-linearity (etc).

Note: Random description of systems with only a few components have no practical value (other than perhaps to be used to identify a betting game).

The new idea:

Use the stable circle-spaces, in particular, the discrete hyperbolic shapes, to identify the stable spectral-orbital properties of observed physical systems, or to identify a finite spectral set to be used as a basis for a description's quantitative structure. Note: "Discrete Euclidean shapes" are also needed in regard to describing material interactions.

The new context:

Partition an 11-dimensional hyperbolic metric-space by means of a set of different dimension discrete hyperbolic shapes so that each subspace of each dimensional level is assigned both a discrete hyperbolic shape (of that dimension) and a multiplicative constant (equivalent to physical constants, eg c h etc). One can think of these discrete hyperbolic shapes as being "uniform" shapes, where each face of the shape's fundamental domain has the same geometric measure, ie the same spectral value. Thus the multiplicative constants are the mechanism through which the spectral sizes change between either dimensional levels or between subspaces of the same dimension.

Properties of discrete hyperbolic shapes [Coxeter]

1. The 10-dimensional discrete hyperbolic shapes are the highest dimensional discrete hyperbolic shapes which exist. Thus, the containment set contains all of the discrete hyperbolic shapes.
2. The highest dimension discrete hyperbolic shape, which is bounded, has dimension-five. Thus the 5-dimensional spectra, defined by the 5-dimensional discrete hyperbolic shapes, are the last hyperbolic metric-spaces which are related to spectra with finite values. Note: the 5-dimensional discrete hyperbolic shapes are contained within 6-dimensional discrete hyperbolic shapes, which are unbounded shapes. (Question: Are some of the faces . . . . of the fundamental domain of an unbounded 6-dimensional discrete hyperbolic shape [actually] . . . . . bounded 5-dimensional discrete hyperbolic shapes?)

The spectra of an "n-dimensional discrete hyperbolic shape" are determined by the geometric measures of the (n-1)-dimensional faces of the "n-dimensional discrete hyperbolic shape's" fundamental domain.

For example, the spectra for "3-dimensional discrete hyperbolic shapes" would be the 2-dimensional spectral set, but the existence "3-dimensional discrete hyperbolic shapes" in a "4-dimensional hyperbolic metric-space" are determined by the 3-dimensional spectra of the 3-faces of the set of "4-dimensional discrete hyperbolic shapes." Thus, in order to fit into the "4-dimensional hyperbolic metric-space" a "3-dimensional discrete hyperbolic shape" must be in resonance with a set of 3-faces which are smaller than the metric-space within which it is contained, ie it must be in resonance with the spectra defined within a different 4-dimensional subspace (which is itself also modeled as a "4-dimensional discrete hyperbolic shape," but of a "smaller size" than the "4-dimensional discrete hyperbolic shape" within which the (given) "3-dimensional discrete hyperbolic shape" is contained).

The 1-dimensional spectra are "additive" since each higher-dimensional discrete hyperbolic shape also contains a subset of 1-faces, which would also be a part of the set of 1-dimensional spectra. Thus, the set of 1-dimensional spectra is greater than the combinations, 11C1=11, and in the dimensional level and subspace partition by discrete hyperbolic shapes, wherein, if one assumes a uniform shape for each element of the partition, then one ends-up with about 1000 different-sized 1-dimensional spectra for the entire containing space.

That is, it is simplest to model this idea in relation to requiring that each discrete hyperbolic shape be a uniform shape, ie each face has the same spectral value, and then the different sizes are determined by the multiplicative constants defined between both "the different subspaces of the same dimension" and "the different dimensional levels." This simple idea of course can be adjusted.

Results:

The new context identifies a finite spectral set.

There are (can be) other "components=material" of various sizes (but contained in some one of the hyperbolic metric-space) but which must resonate with some value (with the correct dimension) of the given finite spectral-set defined by the new (descriptive or containment) context [ranging over all the various subspaces of the given dimension of the given spectral component, which is to be in resonance with the identified finite spectral set {of the 11-dimensional hyperbolic metric-space}]

Physical description of stable systems is about the set of discrete hyperbolic shapes identified by this finite spectral set in the 11-dimensional hyperbolic metric-space.

Now it is better to model functions . . . , used to describe stable physical systems and which define a function space . . . . , to be discrete hyperbolic shapes. By doing this the idea of "global commutative math structures" defined on such a function space will be more directly related to the properties of physical systems.

Sorting-out our containing 3-space

The spectral values of charges, nuclei, atoms, molecules are "small sized" and some of these shapes are, apparently, close to the "same size," and are either of dimension-2 or dimension-3, while our 3-dimensional containing (spatial) metric-space, ie our 3-dimensional discrete hyperbolic shape, is the size of the solar system, but nonetheless many "small" either 3-dimensional components or 3-dimensionally contained components resonate with the spectra of different 3-subspaces, and/or different 2-subspaces.

In such a large shape, an observer's local probing of the metric-space's shape would not detect such a large hole structure. Furthermore, the observer would "see" that which is "far away," in relation to the lattice of the fundamental domain upon which the "metric-space's shape" is identified, where the lattice is defined on an unbounded hyperbolic metric-space.

It is the hole-structure of the very stable discrete hyperbolic shapes which allows very stable spectral and stable orbital properties to be defined, and it is the stable properties of the solar system, where the orbits are filled with condensed material which takes on a spherical shape because the new structure for interaction has spherical symmetry for (free) condensed material in Euclidean 3-space, and the existence of the stable spectral-orbital properties provides evidence to show that this descriptive structure is the correct model for the containment space of stable "material" systems. These shapes solve both the stability question for the solar system, if the authorities have a better solution please present it to the world, and with the "discrete angular transformations" related to the Weyl-transformations, so that a discrete hyperbolic shape can take-on the shape of concentric orbits, thus it is a descriptive structure which also "solves" the quantum-radial equation, where charged components fit into the exact shapes of the spectral flows of the discrete hyperbolic shapes of the nuclear and atomic system's.

The real point is; that the current descriptive context fails to describe the stable properties of these many fundamental systems, and because it has failed to provide these descriptions, the new model carries more authority than does the currently accepted fixed structure for math and science.

That is, this method of both choosing and organizing a containment set is providing a measurable description for the stable spectra and orbital properties of the observed stable physical systems for the fundamental systems of existence for the physical systems of the sizes ranging from atoms to the solar system. That is, this new descriptive language is already a descriptive structure which is verified by the observed properties of stability for these systems. This is stronger evidence for the verification of this new model than that verifying evidence actually exists for the current authoritative dogmas of physics, as well as math.

The new context also allows for second order parabolic equations related to angular momentum, but the new context has more geometry within circle spaces, in regard to the set of possible angular momentum properties of the new descriptions of physical systems, where this angular momentum can link between different dimensional levels.

Note: If infinite-extent discrete hyperbolic shapes (6-dimensional and higher dimensions) have bounded faces in their fundamental domains, then the size of these bounded faces is determined by the sizes of equivalent lower-dimension discrete hyperbolic shapes which are a part of the partition. The existence of bounded faces on a 2-dimensional infinite-extent discrete hyperbolic shape would be a way of modeling electrons and neutrinos which compose an electron-cloud in an atom or a molecule.

Note:

Take Notice!

This solves the most fundamental problem in physical description, the stability of the most fundamental physical systems. This problems has been ignored since it was believed that it was too difficult that these fundamental systems are too complicated to describe, but now it is solved and its solution results in a quantitative construct based (generated) from a finite set (or stable spectral values).

In the new context the descriptive structure is "bounded" by the very stable discrete hyperbolic shapes which model both the "material component" containing hyperbolic metric-space as well as the "material components" whose dimension is one-less that the dimension of their containing metric-space. This means that the very simple but stable geometry has a dominating influence on physical descriptions (not differential equations). However, material interactions are defined between the material components . . . , though discrete [newly determined in a discrete manner every about 10^-18 sec] . . . , and these interactions are mediated in a continuous manner, in regard to space, by the discrete Euclidean shapes [or Euclidean tori] (which allow spatial continuity for the dynamic processes) so that the interactions between components are generally non-linear and they are defined in a metric-space so in 3-space there are the usual 2nd order elliptic, parabolic, and hyperbolic partial differential equations which are part of the continuous descriptions of locally measurable properties, though now there is a new geometric context for angular momentum.

Note: The solutions to the linear, metric-invariant, and separable partial differential equations related to the discrete hyperbolic shapes has been discussed by L Eisenhart in a chapter in his "about 1930's" book Riemannian Geometry.

However, for interactions between micro-components contained within, say, a thermal reservoir, it should be first noted that these micro-components are always colliding and thus the components are constantly changing from neutrally charged to being ionized, so the interactions identify a Brownian motion for each component. E Nelson (Princeton, 1967) has shown this Brownian motion for micro-components is equivalent to quantum randomness. Furthermore, the vertices of the fundamental-domain of the discrete hyperbolic shapes identify a distinguished point for each shape, about which micro-material interactions are centered. Thus, this accounts for the random motions about an apparent point-like structure for micro-material.

Concerning the media and professional math and scientists:

Succinctly-put the media makes bigger-suckers out of the "successful" professional intellects, than it makes suckers out-of the public, since it traps the professional intellectuals into a fixed context upon which the social value of these professionals depends, and it is a dogmatic context within which all valid authoritative ideas are "to be" expressed, eg peer review.

However, Gödel's incompleteness theorem implies that the professionals should also seek to consider new contexts within which to express math and physics ideas, ie professional publication should not be peer reviewed but rather the assumptions upon which ideas are expressed should be made clear and then the expression of ideas should be placed into categories

Where one category needs to be marked the context which most supports (big monopolistic) businesses.

These new ideas are expressed in a new context, in a similar way as Copernicus expressed a new context which was different from the authoritative context of Ptolemy.

It is the professionals who need to enter the new context so as to discuss the new ideas, new ideas which solve the very difficult problem of the stability of fundamental physical systems and the solution in the new context is very simple (the hallmark of a superior context (or a new paradigm)) the new ideas do not need to accept their context, even though it is their context which allows them to play the roles of wage-slave professionals who serve monopolistic business interests, the new way of expressing ideas only needs to provide an interpretation of the observed data.

Addenda:

It is strange that when confronted with such a simple solution the current authorities find the pattern interesting, but do not comprehend the significance of the true authority (its truth and great capacity for wide ranging usefulness) of this new context.

They are so caught-up in their own dogmatic authority, that they do not realize that they have been dethroned, and that their authority has already been lost. This is because they are so self-important, but it is the social structure which has created their sense of being so superior, and so fixed and traditional in their sense of possessing authority.

The overly authoritative fixed structures of containment and interpretation are used to define the "aristocracy of intellect," those chosen few, who are so keen to be intellectual aristocrats, but who, in fact, so weakly . . . . , (in relation to mental awareness about the society and about either what knowledge is, or what knowledge does (is supposed to do) within society, and how big business corrals and uses knowledge for its selfish interests) . . . . serve the interests of big business, with their business monopolies, which allow them to be such very domineering social forces, where these dominant monopolies are allowed to exist in a society, because the laws of our society are set-up and enforced, so that it is a society which values property more than both life and creativity.

Pompous, self-important math and physics professionals "believe the hype" and they believe that they are the "culturally superior people" of the society, but who are, in reality, people with personality flaws; they are manipulative, narrowly obsessive, authoritarian, and selfish people who are themselves easily manipulated by social forces (eg mainly by-means-of the media) which is why they were chosen to have these social positions, where these superior people dutifully are servile to their own image of being uppity where they promote within society the social traits of domination and fixed-ness of knowledge and its uses, due to their servile social positions, which are, in fact, anti-knowledge and anti-creativity social positions, so as to assure the rich owners of society that the façade of knowledge is all about complicating descriptions to filter the public out of social positions which would allow them greater capacity to create, and thus causing a competitive structure for products within the so called free-market.

These professional authorities are presented to the public by the media as those people who have attained the highest cultural achievements within society, but they do this by serving the few owners of society. Because they depend on their pay-master, this makes them lesser people rather than greater people. They retreat behind the big bully (the owners of society, and these bullies allies in both the government and in the justice system) who create the social images of these, so called, experts, and this is done so as to serve the interests of the owners of society.

Talk 2

The problem is the propaganda system, which is the society's sole authoritative, reliably-truthful vehicle of social expression (but which only promotes monopolistic interests) where it continuously spews-out mis-information which, when placed in the "context" of honest reporting, the public always interprets to be an absolute truth.

Thus when monopolistic, unregulated, but almost completely controlled "market-place" fails in a complete collapse, due to criminal fraud aided by the justice system and the congress the propaganda system continues to "sing the praises" of (and need for) the "magical" unregulated "market-place", where it should be clear that "unregulated", now especially, means license to steal along with a complicit justice system and political system, since these institutions have been manipulated by the propaganda system so as to simply to have become a part of the propaganda system, themselves.

The authorities (or technical experts) which serve this system by adjusting the complicated instruments for the monopolistic ruling interests (the interests of the owners of society) are also controlled by the absolute-truth espoused (in the context of "honest" reporting) by the media related to "technical-development" (but really only small adjustments to fixed traditional technologies, since the context of math and science is not allowed to seek new creative contexts within the context of stable math patterns, ie the monopolies depend on society continuing to use products and resources in a fixed way which allows the monopolies to continue to make money based on their products), where for the intellectual class, the academics, the authoritative experts, are provided with a set of "prized" problems whose context is:

Indefinable randomness
Non-linearity
Non-commutativeness, or
At best locally commutative in a context of a general metric but non-linear in regard to the containing coordinates of geodesic coordinates or set of functions

So that the propaganda system, validated by certain narrowly interested experts insists in "big bangs" particle-physics, string-theory geometry which are needed to understand the singular points of a black hole's gravitational field, and it is claimed by some possibly charges related to black holes and in regard to singularities within nuclei, so as to realize a "grand-dream" (truly a pipe-dream) of the control of worm-holes in space, or control over a "many-world" context of existence, though the math structures through which these ideas are to be described, based on probability and non-linearity, does not allow any control, . . .

since there are not any stable patterns in their descriptive context

Nonetheless these math structures do apply to the business interests:

Is math staying too traditional, fixed, formal, complicated, irrelevant?
Is math only about serving business interests in regard to:

1. Unstable patterns to be used in fleeting contexts (feedback systems),
2. Chaotic transitory systems, eg nuclear reactions, transiting between two stable states,
3. Manufacturing complications, eg formulating security codes, etc,
4. Pulling the wool over the eyes of the public, so as to provide irrelevant and inadequate descriptive structures for physical systems?

. . . , of such constructs even if they are mathematically modeled (as it is so claimed that quantum systems are correctly mathematically modeled) due to the properties of the math constructs and due to the over-whelming complexity of such models these models (of controlling a "worm-hole") cannot possibly be controlled, as the propaganda system is suggesting that they can be controlled. Consider, if the current descriptive context cannot describe the stable, definitive fundamental systems, systems so stable that it implies that these systems form in a linear controlled context, then how can their exotic models of physical systems (changing between worlds, in a many-world model) ever be realized, if they are models which are not consistent with the actual structure of the (external) world? or of these ideas

That is, these prized problems are delusional-ly based, yet the propaganda system promotes them as do the experts themselves so they become the basis for identifying an elite in-crowd of "knowledgeable" experts.

This, in-crowd, of so called superior intellectual elites, obsessed with complicated math patterns (one must note the autistic connection, the manipulation of personality types by institutional managers, used for the purpose of deceiving the public) who are led to believe, by the media itself, that they are zeroing-in on the wonderful goal of their great intellectual prowess and intellectual creativity.

Yet it is clearly a failed intellectual exercise since there are basic stable definitive physical systems which exist at all size scales but which go without valid description (based on physical law).

Nonetheless (however) when these stable systems are actually solved the intellectual community and society has a great mental inertia (almost entirely caused by the propaganda system, it unwillingness to publish the new solutions, since they define a completely new math and physics context) to realize just what they have heard, but nonetheless assured that it is they the intellectual elites who will be the ones to forge new roads into new technical landscapes and thus the elites will not (will refuse to) listen to the new ideas which must originate from an inferior mentality and thus must be wrong and surely wrong within the dogmatic authority which defines their truth for them. The elites only find these new ideas somewhat interesting but from an intellect inferior to their own since they have memorized their contexts and their always correct interpretations and their always intelligent evaluations of the state of knowledge, they are not tricked by prized problems, no not them.

That is, the propaganda system is a communication system (vehicle) which expresses a dogmatic authority (an absolute truth) which is followed by the experts which in fact determines the faith of the high-valued institutions which serve the ruling class and the experts have an absolute faith in the authoritative truth of that dogma. The propaganda system defines the true religion of society and it is a religion of personality-cult (not unlike Roman emperors or Egyptian pharaohs) and a deep belief in inequality and a manufactured property of a society dominated by selfish monopolistic interests, ie it is a society opposed to life and opposed to adapting to change

Wake-up you popes of the religion of science and math and their most revered subjects (or beliefs), eg general relativity, particle-physics, string-theory, indefinable randomness, non-linearity, non-commutativity, or only locally commutative, or commutative functions, in regard to function-space algebraic operator constructs, but whose domain spaces are related to different, non-commutative, coordinates.

Gödel's incompleteness theorem has a simple interpretation:

Because precise language has sever limitations the axiomatic basis, the containment set, the organization of this containment set, the context of the description, and the interpretations of the observed properties (of the observed measurements must be fully considered and when an alternative, well defined example of new ways in which to present axioms and contexts is provided, especially if it solves the most difficult problems in math and physics then the math physics community of experts should listen and take it seriously.

Please pay-attention (ironically) it is, me, your superior, talking to you, and it truly is, the irony is, that, in fact, we are all equal, but you have tried to achieve in the eyes of the paymaster and you have lost your way and you have made yourselves lesser (not superior), so now by artificial measures I am smart and you old experts are now stupid and disposable which is the context which your over-reaching superiority has caused the public to be, ie the public is disposable since we have "the great experts"

The crux of the problem with knowledge and education in society is its capture by corporate and private interests capture by the owners of society even though it is most often "public" education institutions, nonetheless the professors are trying to adjust the complicated instruments for the corporate and private interests and they are not concerned with descriptive knowledge in its most general and most powerful sense where new creative context get developed by new contexts for descriptive knowledge which result when assumptions, contexts, containment, organization of pattern use are considered at their most elementary levels, as Faraday developed the language of electromagnetic description while he also developed a new context in relation to the instruments related to electromagnetic properties. Though such a dramatic chain of developing events is not a necessary attribute for developing a new descriptive context at its most elementary level it means that new languages can be related to new creative contexts.

Since the professors of public universities have been captured by corporate and private interests through the mechanism of funding research and identifying prize problems in math so as to keep math traditional and under the control of peer review the talks at conferences, such as at the joint math meeting in San Diego 2013, are either about developing even more complicated theories and more complicated formal professional math language where these formal math languages have very limited if any relation to practical development they may only be marginally related to corporate interests, essentially related to bomb technology, or About applying technical complicated math so as to be able to adjust rather complicated instruments of interests to corporate and private interests, eg feedback systems, imaging systems, recognition systems and improperly defined statistical constructs which often appear to be valid, since the propaganda system is capable of making all of society to continue to use language and product in very certain narrowly defined ways so as to place an artificial stability on the statistics where such stability of the statistical context does not really exist.

Propaganda, in regard to science and math is provided in the context of great breakthroughs and, supposedly, new things but which has little bearing on the uselessness of the descriptive language except in regard to weapons technology

Science and math are often about making adjustments to systems which will reduce labor if things remain in their fixed social context in regard to the corporation's products

That is developing new knowledge in regard to new contexts and new ways in which to organize descriptive language, in regard to solving fundamental mysteries, is effectively stopped by peer review prize problems traditional authority and mostly by the process of funding which is controlled by the corporations and private businesses.

Nonetheless, there are many unsolved fundamental problems in physics which are ignored due to dogmatic authority of math and funding traditions within public educational institutions an authority which is essentially related to military development and banking investment interests

The sad thing is that now (2013) these fundamental problems have been solved but the structure of both propaganda and the "knowledge institutions" keeps-out ideas which are different from the inter-related interests of business and traditional academic authority.

Academic science was invented by the mercantile class in the 1600's (after Newton) to support their investment and productively-creative interests, in the 1500's science development based on measuring was centered around schools and about literate people, so that investment in science was centered around the schools (or universities).

However, public schools should take notice of Godel's incompleteness theorem and the logical positivists who proclaimed to limitations of precise language (or the limitations of measurable descriptions) and to consider the example of Faraday wherein he both invented a new math language to describe electric and magnetic properties but he also created the instrumental context through which these properties could be used and controlled

That is, descriptive knowledge best leads to new contexts for creativity at the elementary level of language assumption context interpretation etc not at the complicated formal level, eg no one can use the principles of particle-physics to describe the stable spectral properties of general nuclei.

That is, one wants new contexts for creativity to come from precise descriptive languages one does not simply want adjustments to complicated instruments since instruments also have limitations as to their capabilities or if there is not a better instrument to do the same thing, though digital electronic seems to be able to deal with arithmetic and math patterns well but this has led to an attempt to deal with non-linear systems but non-linear systems can only be related to feedback systems and only for limited ranges of time or distance in regard to arithmetically determined solutions.

There are many mysteries in regard to fundamental physical systems: why do there exist stable physical systems with definitive measurable properties, eg nuclei, general atoms, molecules and their shapes, crystals, life stable solar systems, there are many galaxies with planar spiral structures, etc.

These fundamental physical systems go without valid descriptions based on what is considered to be "physical law" instead one hears about all the hyped-up ideas through the propaganda system and the professional mathematicians and physicists about "big bangs," Black holes, worm-holes, Higg's particles, transforming neutrinos, all related to general relativity, particle-physics grand unification and string-theory etc, are all expressions which are wild speculations if they cannot describe the stable properties of fundamental systems whose stability implies that they come into being in a controlled context, while believing the wild speculations is mostly driven by (or caused by) the propaganda system and a traditional context of math and physics authority and a personality cult which forms around these academics who mostly interfere with the development of new contexts for knowledge and creativity but the monopolistic business interests do not want new contexts for creativity. Where it might be noted that general relativity was shown to be untrue in regard to the non-local properties of material interactions in Euclidean space since non-localness was demonstrated by A Aspect's experiments.

The correct answer as to why there are stable fundamental systems require s that the context of containment be changed in a drastic manner from 3-space and time (quantum, Newton) or space-time (electromagnetism, particle-physics (?)) and materialism where the measurable descriptions are either Classical often leading to non-linearity or Quantum indefinable randomness, spectra supposedly derived from 1/r potentials, function spaces usually non-commutative and Lie groups all used in the context of a continuum and a loose idea about convergence to this continuum, eg renormalization (something Dirac rejected) but apparently personality cult and propaganda was able to establish as an authoritative technique.

Instead consider a new context:

An 11-dimensional hyperbolic metric-space is partitioned . . . into its different dimensional levels and the different subspaces of the same dimensions . . . by discrete hyperbolic shapes which exist up to hyperbolic dimension-10. This can be used to define a finite spectra on the over-all 11-dimensional hyperbolic metric-space.

The set of all resonating discrete hyperbolic shapes which are contained within this high-dimension containing space form the bounding stable structures of the more usual physical description of material defined as one-lower dimensional shapes in each dimensional level, where the usual metric-invariant, second-order elliptic, parabolic, hyperbolic, differential equations . . . associated to material interactions or material properties . . . are defined, but the stable elliptic structures are defined on the discrete hyperbolic shapes. The elliptic case is mostly about describing condensed material contained within a (higher-dimension) discrete hyperbolic shape, or the orbital path of a material component is (becomes) resonant with the discrete hyperbolic shape upon which the component is contained (or can become "so contained," due to resonance).

The stable physical systems are the discrete hyperbolic shapes of the various dimensional levels which are in resonance with the finite spectra of the over-all 11-dimensional hyperbolic metric-space which has been so partitioned.

Each n-dimensional level "sees" the bounding geometry of the (n-1)-dimensional material "surfaces" but the open-closed topology of these shapes allows light to be observed from outside the metric-space shape's fundamental domain out to the unbounded lattice, this is especially true for metric-spaces whose fundamental domains are large eg as large as the solar system, where it should be noted that lower-dimensional shapes than (n-1)-dimension tend to condense onto the (n-1)-material's shape.

That is, the shapes imply the discontinuity of a metric-space experience between dimensional levels.

Interactions between micro-components imply Brownian motions which implies (due to E Nelson 1967) quantum randomness. Furthermore the distinguished points on discrete hyperbolic shapes implies that the interactions appear point-like, but, nonetheless, mediated by discrete shapes.

It might also be noted that this descriptive context provides a definitive spectral relation between different 11-dimensional containment sets, ie there are many different worlds where each world is well defined by a definitive spectral set, and the best instrument to realize transitions between these world might very well be a (human) life-form.

Speech 3

The observed stable, precise, patterns of physical systems are associated to finite properties, eg bounded-ness and/or the finite number of a physical system's components, eg atomic-number, and these stable physical system properties are fundamental and observed features of a reliably measurable context associated to the observers of physical patterns. This implies both the existence of stable patterns which allow reliable measuring, (or which are associated to the context of measuring (for an observer)), and the existence of stable-controllable patterns associated to a set of fundamental physical systems which possess stable features of "what may, or may-not be" "material" systems, eg nuclei, general-atoms, molecules and their shapes, crystals, solar systems, dark-matter (ie orbital properties of solar-system's in galaxies) etc, the physical patterns upon which the relatively stable aspects of our life experiences depend, and upon which our mental constructs also depend.

Thus, science and math are about identifying stable, quantitatively-consistent, math patterns which are generally applicable to these stable, measurable, and apparently controllable, physical properties so as to result in descriptions of these patterns which are accurate (to sufficient precision), and general so as to be able to describe the observed stable physical patterns of existence, so as to provide a context for practical usefulness, ie measurable and controllable, so that one can: measure, fit together (or couple), and interact with these various patterns (using the natural structures of these patterns, eg life-forms and its coordinated chemical properties (but, apparently, coordinated by an unknown structure), ie not feedback mechanisms nor carefully prepared structures so as to cause reactions), so that this descriptive knowledge can be related to "practical" creativity (as opposed to literary creativity, essentially associated to a world of illusion, ie a world without stable features).

What are (math) patterns?

Patterns are:

1. consistent relationships, or
2. operators acting on quantitative sets so as to have fixed "consistent properties" related to the application of an operator on a quantitative set, and these consistent patterns are

related to the "meaning" of the quantitative-set's elements (where quantities represent properties of: type and [measurable] size), or

3. stable shapes, etc.

   Can the current descriptive language of mathematics and physics describe stable patterns? [Apparently not.]

There are essentially the three ways in which to try to describe stable math-physical patterns . . . ,

I. stable geometry, which strongly limits both a descriptive context and the patterns it is trying to describe (the new context for physical description, the circle-spaces, or the very stable discrete hyperbolic shapes),
II. differential equations in a geometric context (unfortunately, this method most often leads to non-linear patterns),
III. differential equations in an operator context (this methods seems to only work for harmonic properties which possess actual physical attributes) . . . , so as to try to use quantitative descriptions so as to try to identify stable patterns which provide valid information, as well as control, over relatively stable (physical) system properties.

A new interaction-construct can be constructed which is general, but its stable properties are determined from a context defined by a many-dimensional set of discrete metric-space shapes, which, in turn, define existence.

The professional mathematicians and scientists in regard to descriptions of fundamental stable physical systems express symbolic nonsense, ie they provide a set of nonsense symbols which result in descriptions which are neither general, nor accurate (to sufficient precision), nor do they provide a practical context for useful creativity.

Physical systems which are very stable and definitive, but which are many-(but relatively few)-body systems, nonetheless, because these systems are so stable and definitive, it is clear that they are forming within a very controlled context, so that the descriptions (of the professionals) which are based on:

1. (vague) randomness (which is an uncontrollable description for a system which is composed of only a few components),
2. non-linearity (quantitatively inconsistent, and chaotic), and
3. non-commutative (not invertible, or equivalently, not solvable, eg non-linear or spectrally-un-resolvable), context, which is

4. contained in a continuum (a containing set which is far "too big" allowing logically inconsistent descriptive constructs to be put-together as if they belong to the same containment set), and
5. it is a description (when based on randomness) which begins from a global viewpoint (a function space) but the methods of the description focus on local spectral-particle events in space, ie it is a description which gives-up information leaving one in an inaccurate and non-useful context in regard to information.

   It is a description which "in general" is not accurate, yet it also is a description which is "intent on" losing information about the stable definitive properties of the [assumed to be random] system.

That is the descriptive structure of the "dogmatically pure" set of experts of math and science is simply a bunch of nonsense.

Yet one must list the places and contexts within which it is a valid descriptive context:

1. It is a description which is relatable to a system whose initial conditions, and initial properties are carefully put-together so as to be a system which is easily broke-apart, so as to form a transitioning system which is chaotic, so that the rates of reactions (in this context, based on component-collision probabilities) are determined by cross-sections of the broken-apart components, where these cross-sections determine the rates of certain aspects of the (a) reaction, and
2. They are descriptive contexts which relate a limited set of metrically measurable (observable) properties to a feedback structure, which is mostly associated to the critical-points and limit-cycles of a non-linear (usually classical) partial differential equation, where the range of relevance of the differential equation is difficult to determine or to control. Furthermore, the initial or boundary conditions of this type of a system relate to the properties of the descriptive context (or properties associated to the solution) of the system's differential equation in a chaotic manner.

That is, difficult math methods . . . , which are related to fleeting, unstable math patterns . . . , are descriptive constructs which have no content (and possess no useful information), they are patterns which apply only to unstable contexts, where control emanates from a higher abstract, and manipulative, context imposed on properties which are only definable in a metric-space, and which requires a lot of preparation (in regard to sensing and reacting in the desired way to the detected properties), a context which is at-odds with the system's natural properties, ie rather than controlling a system by simple adjustments to affect the system's properties in regard to affecting the properties of several system-components being coupled together.

So we have the tradition of "western hypocrisy," where failure is rewarded if those who perpetuate it, are in the high social classes.

What is wanted, by the owners of society, is that the social structures through which the powerful derive their power are kept in place.

That is, it is a social structure which is opposed to new, creative changes and thus it is also opposed to equality and the creativity associated to equality. However, the traditional social structure which upholds dominant interests so violently, and it expresses its interest in "lyrical creativity" in regard to the science and math experts . . . . , where these authoritative experts define the "literary" creative development of science and math, which is authoritative, but unrelated to practical creative-development, and the owners of society support the "creativity" of the elite artists, those who also compete in a "narrow context of authoritative cultural value," as well as those journalists and intellects whose ideas are judged (by the owners of society) to possess "cultural value," so that the ideas expressed are consistent with the ideas of (or can be used by) the owners of society, so as to be distributed by the material-instruments of the media, which are owned and controlled by the owners of society . . . , then even the failures of the experts can become part of the social structure which allows the powerful to remain powerful.

The top-intellects and top-artists are defined as a social class, along with artists and journalists, so that the intellectuals can dogmatically dominate those many-others who question the authority of assumptions, or who have different ideas.

The main tool used to maintain the power of the owners of society is the single voice of authority which the media has become (most clearly controlled by ownership, or by a set of funding processes).

That is, it is violence and domination (intellectual domination) which is fundamental to social power, not knowledge.

Knowledge is relevant, within today's social structure, only in regard to the creativity which is a part of the organization of society (ie business productivity) which, in turn, maintains the power of the few. However, the organization of society, and the use of resources and the ownership of technology within society, essentially, remains fixed and traditional.

For example, the many-purpose phone, eg an i-phone, is about developing 19th century ideas of electromagnetism, and the micro-chip circuit boards in these devices depend on 19th century optics.

Whereas identifying stability "as a needed property" in both math and physics, in regard to the useful descriptions of controlled (or controllable) physical systems, is a focus (in regard to the valid descriptions of math patterns) which the math professionals, apparently, have not considered.

Furthermore, very simple math patterns can be used to create new math patterns, which can be used to describe the stable material properties, so that these descriptions are based on a finite quantitative set, within which the descriptive containment of physical properties depends, ie the containment set is not a continuum and the derivative and its integral-inverse function-operators become discrete operators (the continuum can, instead, be the set of rational numbers).

In fact, the math patterns of stability are very simple, and relating these simple structures (which are best characterized by the stable discrete shapes, or circle-spaces) to many-dimensions, can be done by a simple process of partitioning the dimensional-levels of a hyperbolic 11-dimensional containment metric-space (base-space) by means of stable shapes, ie partitioning the dimensional levels by means of the discrete hyperbolic shapes (or circle-spaces), so as to form a finite spectral-orbital set, where the sequences of spectral-size are defined (either increasing or decreasing) as the dimensional level increases, so that these size-sequences of spectra are fundamental, in regard to how the description is organized, so that a finite spectral set is the basis for physical descriptions of the observed spectral-orbital-material order which the stable (material and containing metric-space) structures of existence possess.

Speech 4

Stability

In order to describe stable, "measurably consistent," and precise patterns of material systems one needs stable, "quantitatively consistent" math patterns.

In math such patterns are (quite often related to) linear, metric-invariant, separable partial differential equations, ie non-linearity does not work. Stable and quantitatively consistent math patterns also deal with the geometric models used for measurable quantities of the line (or line segment) and the circle, where these two geometric (or quantitative) structures can be easily organized so as to be quantitatively consistent with one another, eg the real-line and the complex-number-plane.

Circle-spaces

The circle-spaces fit both categories which identify reliable and stable quantitative descriptive (or measurable) constructs. The circle-spaces are the tori (the dough-nut shape) and those shapes with toral components, these include the discrete Euclidean shapes (single torus) and the discrete hyperbolic shapes (composed of toral components). The patterns of the circle-spaces are stable and quantitatively consistent. The circle-spaces are characterized by the properties of being related to non-positive constant curvature metric-spaces where the metric-functions have constant coefficients. The geometric properties of circle-spaces fit into the geometric structure of analytic complex-functions (where an analytic function is supposed to be consistent with the algebraic structure of quantitative sets, but the series must be put into the context of finitely defined polynomials to ensure quantitative consistency)

Lie groups

These are also the discrete isometry subgroups of the classical Lie groups (associated to metric-invariance). These spaces (shapes) include the discrete Euclidean shapes of R(n,0), the discrete hyperbolic shapes of hyperbolic n-space, which in turn, is associated to the general (n+1)-space-time spaces of R(n,1), as well as more general R(s,t) spaces and their associated discrete shapes related to circle-spaces.

Properties associated to metric-spaces

These metric-spaces, which may be modeled as discrete shapes, have associated to themselves both math and physical properties, so their discrete shapes can identify both metric-spaces and stable material components within the metric-spaces.

That is:

1. The property of position in space (in relation to the distant stars), these are related to the Euclidean spaces, which includes the property of action-at-a-distance, when the shapes have distinguished points and the context is the very rigid properties of hyperbolic spaces filled with sets of very rigid discrete hyperbolic shapes.
2. A stable well-defined pattern or shape, ie properties which are continuous in time, or conserved properties (or conserved patterns, or shapes), this property is related to the hyperbolic spaces [and the R(s,t)-spaces]. Time states are defined by the properties of the opposite flow of time advanced and retarded potentials, as well as the opposite pair of wave-equation solution functions.

Opposite time states and wave-propagation

How are these properties related to the propagation of wave-functions in either odd-spatial dimensions of a metric-space (distinct directions of time associated to wave-propagation [surface]), or even-spatial dimensions (always a mixture of time states, waves fill space after propagation "surface" distinguishes a wave property of a solution wave-function)?

Spin and unitary fiber groups

The assignment of properties to metric-spaces leads to the existence of pairs of opposite metric-space states and this, in turn, leads to the two: real and pure-imaginary subsets, of the complex coordinates, and the unitary fiber groups, as well as the spin-groups, where the spin-groups spin-rotate between the pairs of opposite metric-space states, where pairs of opposite time-states (associated to hyperbolic-space) are a part of the local dynamic process, and they are a part of the stable spectral-flow structure, which exists on the discrete hyperbolic shapes, and is related to the sub-face structure of a discrete hyperbolic shape's fundamental ["cubical"] domain.

Note: For new (3-dimensional) material associated to (or contained in) R(4,2), there are two time-dimensions so that each direction of time can be associated to a different domain space for the spatial positions of distinguished points (of the discrete hyperbolic shape models of material components) being transformed by SO(4)=SO(3) x SO(3) in (x,y,z,w)-space.

Size (Cardinality) of quantitative sets

[problems in description which are introduced by using sets which are too big]

There is another problem in regard to the size (cardinality) of a containing set, which is used in descriptions. The continuum is "too big of a set" allowing the inconsistent properties of both stable geometry (of particle-collisions) to be defined by means of convergence to the domain (or containing) space so that this explicit geometry (of particle-collisions) is defined in a context of assumed fundamental randomness, in regard to quantum physics, and (non-linear) particle-physic's random math structures do not allow microscopic material geometries (such as the geometry of a particle-collision). The continuum is "too big of a set" since it allows inconsistent constructs to be defined through convergence into the same containing (or domain) space, which is a continuum (containing both particle-collisions and fundamental random math structures).

To create a finite quantitative set

Thus, one needs to base quantitative descriptions, especially of physical systems, on a quantitative set built upon a finite spectral set. That is, one needs a stable, quantitatively consistent, and finite based quantitative set upon which to base a measurable description of physical systems.

Properties of discrete hyperbolic shapes

By following (or using) the patterns associated to the stable discrete hyperbolic shapes:

1. the last type of discrete hyperbolic shapes which is compact is 5-dimensional,
2. the dimension of the last type of discrete hyperbolic shapes to exist is 10-dimensional (hyperbolic-dimension) and it is an unbounded shape.

Partitioning a many-dimensional containing-space with shapes, so that the partition depends on dimension and the set of separate subspaces (of the same dimension, as well as of different dimension) which are defined on a many-dimensional space.

The number of n-dimensional subspaces "for n less that 11" is given by the "combinations" 11Cn, eg 11C2=55.

The properties of the dimensional partition

The idea is to partition the different dimensional levels . . . , and the various subspaces of each dimensional level, of the 11-dimensional hyperbolic metric-space . . . , by a (finite) set of (bounded when possible) discrete hyperbolic shapes, ie circle-spaces. Thus an 11-dimensional hyperbolic containing metric-space which is so partitioned is only continuous within each dimension, and subspace shape, of n-dimensions, and the continuity defined in such a n-dimension subspace is mostly defined (identified) in regard to the set of (n-1)-dimension boundaries of the (n-1)-dimension material component shapes contained in the containing n-metric-space, which, in turn, is contained in an 11-dimensional space. Inside a discrete hyperbolic shape, one does not see holes in the shape, but instead sees an open-closed topological space, ie one sees the space within which the lattice [("cubical" partition) of the discrete hyperbolic shape] is defined.

The nature of the spectral sequence based on the partition

That is, one is defining (on most of the "many sets of subspaces," defined within the 11-dimensional space) an increasing spectral sequence (as the dimension increases), and where the spectra are defined on the bounded, discrete hyperbolic shapes, which are a part of the (finite) partition, though on some subset of "the set of subspaces" there may be some decreasing spectral sequence defined (as the dimension increases).

This can be used to define a finite spectral set upon which the stable shapes and components must be resonant in order to be contained within such a containing set (or metric-space) structure. The sets of subspaces within which an increasing spectral set is defined (as the dimension increases) (can) have material-components contained within themselves (within the discrete hyperbolic shapes which model hyperbolic metric-spaces) so the contained material components (within the partitioning shapes) will have stable geometric-spectral properties which resonate with the finite spectral set of the partitioned 11-dimensional over-all containment space. Furthermore, condensed matter . . . . , ie material components whose size is too small, in regard to both the dimension and the subspace (of that same dimension) within which the material component is contained . . . . , will still be contained within a discrete hyperbolic shape, within which the condensed material can be in an orbit (orbital structure defined by the discrete hyperbolic shapes, within which the condensed material is contained). That is, the stable solar system can be interpreted to be evidence which proves that these ideas are correct.

Thus, the idea that "geometry dominates (or is more important than) the traditional authority of the partial differential equation of a physical system, defined in a context of materialism, non-linearity, and (undefined) randomness."

The decreasing spectral sequence, as dimension increases (the current, fixed, overly authoritative viewpoint)

On-the-other-hand a decreasing spectral sequence (as dimension increases) will not have any stable properties "to speak of," except for the material components themselves, and in such a decreasing spectral sequence (as the dimension increases), the high-dimension containing metric-space should be continuous for all the material components, but since this is not observed (that is, we do not experience the fact that we are in an 11-dimensional space, when we confine our observations to a low-dimension subspace of this 11-dimensional containing space) so these small spectra are curled-up, eg string-theory, so that the idea of materialism is maintained . . . ,

where the existence of material, ie materialism, is related to the fact that both the 1-dimensional discrete hyperbolic shapes and 2-dimensional discrete hyperbolic shapes are close to the same size, while the size of 3-material-components, in our metric-space, are the size of the solar system, thus, in our containing metric-space, the atoms and molecules (which are 3-dimensional discrete hyperbolic shapes) are condensed material, which have relatively smaller energy ranges of stability, than do nuclei,

. . . , and with a decreasing spectral sequence (as dimension increases) (in a metric-space whose subspace structure of a higher-dimension containing space is not detectable), one is left with descriptive structures which are unstable, indefinably random, and quantitatively inconsistent, where logic becomes irrelevant, as is now the "current way" in which quantitative language is now being organized to describe the observed material properties. Furthermore, this organization of precise language is held onto with stifling authority, and this is because the probabilities of particle-collisions are used in the randomly-directed transitioning system (of a nuclear reaction), so as to identify properties of rate and energy release of such a reaction (where a reaction is randomly transitioning system, wherein material interactions are modeled as collisions).

That is, the current descriptive context assumes that the spectral sequence of the containing structures decreases as the dimension increases, so all of the "many-component systems" (but still containing relatively few components), become "too complicated to describe," if one is trying to describe their stable patterns based on material interactions, which, in turn, are based on random, non-linear, (but nonetheless geometric) particle-collisions.

Other new attributes of description

In the new description there are new sets of operators, or properties, but the geometry of the description becomes the dominant attribute of the description, often this is because the new geometry describes (identifies) the new context, in a most dramatic way.

1. There exist conformal factors defined between dimensional levels (ie physical constants), as well as between different subspaces of the same dimension,

2. (perhaps) conformal factors can be defined between toral components of a discrete hyperbolic shape (though this might simply be relatable to the existence of particular varied discrete hyperbolic shapes which can exist based on the reflection group structure of the lattices at vertices which can also be related to the various possible sizes of the faces of the fundamental doamin)
3. Discrete Weyl-transformations of angles can be defined between toral components of discrete hyperbolic shapes (these Weyl-transformations define "allowable folds on the lattice" of a discrete hyperbolic shape).

The derivative becomes a discrete operator in regard to:

1. Time intervals defined by the periods of the spin-rotations of opposite metric-space states
2. Dimensional levels
3. (possibly) Between toral components of a discrete hyperbolic shape

This new descriptive structure defines material interactions using many aspects of Newton's law of inertia. Thus, there are also the "usual types" of 2nd order material interaction math (or equation) patterns:

1. elliptic (or orbital)
2. parabolic (or free, or angular momentum)
3. hyperbolic (waves with physical properties, or collisions of material components)
   But these types of interactions depend on the spectral values, and they apply in a more restricted geometric context, but now it is within a more diverse many-dimensional construct where quantitative descriptions are to be guided by the geometry (both shape and size) of any of the dimensional levels.

Note:

It should be noted that with both materialism and the belief that the spectra of higher-dimensions need to identify a decreasing spectral sequence as the dimension increases, means that conventional science tries to sort-out the spectral properties, of quantum systems composed of "five, or more," charged components, by attaching a 1/r potential (also associated with spherical symmetry), for each charged component, where the assumed spherical symmetry, of each 1/r term, is "deduced" from the assured-ness that the random-particle model of material interactions is to be spherically symmetric in each dimension (if the dimensions are not curled into small shapes), because of fundamental randomness, so if particles are being emitted from a "force-field source," then the field-particles will emanate in any random direction, and then this will define a spherically symmetric force-field.

But such a model . . . , (or such an assumption that a "physical description of material interaction is to be based on fundamental, indefinable-randomness, associated to random particle-collisions," so that particles are emitted in random directions from a "force-field source" so as to cause a spherically-symmetric force-field, in all dimensions) . . . , is far from the truth.

In fact, material interactions are mediated by a toral shape, associated to action-at-a-distance toral shape defined for each (small) time interval (~10^-18 sec), where this time interval is defined by the period of the spin-rotation of metric-space states, so that the tangent structure of each interaction-torus (defined for each time interval) is related to a 2-form force-field, which, in turn, is related to the geometry of the fiber group (of the containing space of the interaction torus), and this geometry (of material interaction) only results in a spherically-symmetric force-field for SO(3), ie the interaction torus which is contained in R(3,0).

Quantum randomness of point-particles

The new model of material interactions, defined for small components whose properties of "being neutral" or "being charged" changes rapidly, results in these small material-components defining Brownian motions, where these Brownian motions, in turn, determine an appearance of quantum randomness. Furthermore, the vertices of the fundamental domains of the circle-space shapes, define a distinguished point on the circle-space shapes, about which material interactions are defined. This, in turn, creates the illusion that material-interactions are interactions between point-particles.

Chapter 20

Dimensions, shape (holes, stability), size, measurable description, and spectra

The dimensions of the set of 2-forms defined on an n-metric-space is also the dimension of SO(n), where dim(SO(n))=dim(spin group of SO(n)).

In the new descriptive construct the geometry of the 2-forms in an n-space is related to the local (tangent) geometry of an (n-1)-dimension "discrete Euclidean shape," ie an (n-1)-torus, which in turn are related to the geometry of the SO(n) fiber group of the n-base-space, in regard to the local coordinate changes of the positions of the interacting materials (determined by both the 2-form force-fields and the local coordinate transformations) where these local coordinate changes are associated to each discrete time interval, in turn, defined by the spin-rotation of metric-space states (~10^-18 sec).

Thus dimensional relations can be found between the various possible spaces related to a material interaction and the associated spectral properties (ie properties of spectral-size) of the different containment levels.

The list of the dimensions of the 2-form spaces of the different dimensions up to Euclidean 6-space . . . , since the spectral sizes of discrete hyperbolic shapes of dimension 6 and up to dimension-10 are infinite extent, so that shape loses its intuitive sense of bounded-ness, thus the geometry of the interaction-shapes are difficult to identify . . . , are as follows:

2C2=1, 3C2=3, 4C2=6, 5C2=10, 6C2=15

From this list . . . . , and along with some information about the geometric structure of the SO(n) fiber group, such as SO(4) = SO(3) x SO(3) . . . . , one can make some determinations (or guesses) about the nature of force-field interaction geometry. The relation of the fiber group geometry to the 2-form geometry to the geometry of the containing n-dimensional metric-space, or possibly the geometry of an (n-1)-dimensional metric-space associated to the: subspace, material, and dimensional structure of our containment set.

That the 2-form on n-space has the same dimension as SO(n) means that there needs to be a geometric-vector relation to force-fields acting on (n-1)-shapes contained in n-space. The "charges" on (n-1)-shapes are the (n-2)-flows (or (n-2)-faces of an (n-1)-shape) so the geometry of SO(n) is related to the geometry of (n-2)-flows on (n-1)-shapes in turn, contained in n-space.

1. For 3-shapes contained in 4-space it is useful to identify the normal to the 3-shape with time instead of the 4th spatial-dimension, so the geometry of the 2-forms is related to 3-space, while
2. for the 4-shape contained in 5-space so the geometry of the 3-flows (or the 2-forms) is related to 5-space, but
3. for 5-shapes contained in 6-space the 4-flow (or 2-form) geometry is again dimensionally more convenient to treat is being related to the toral 5-shapes again letting the normal-direction of the 5-shape to be associated to a time-direction, etc.

In regard to odd-dimensional spatial subspaces, which are the material-component containment spaces, the dimensional properties of the 2-forms are related directly to the local tangent geometric properties of the containment space.

In regard to even-dimensional spatial subspaces, which are the material-component containment spaces, the dimensional properties of the 2-forms are related (directly) to the local tangent geometric properties of the interaction toral shape which is 1-dimension less than the dimension of the containment space, with the normal direction to the toral shapes being associated to a time-dimension (instead of the extra dimension of the containment space).

The most important geometric relation in regard to our "3-space experience" is that the geometries of the 3-space and the 4-space can have a fairly complicated relation with one another, especially since (or if) the spectral-size of 4-space seems to be the size of the solar system, thus the geometry of the 3-tori do not manifest, in regard to our human size-scale on earth, in 4-space, but rather are related to the 3-space in our experience (due to their spectral and subsequent resonance relation with another 4-dimensional subspace, a subspace which possesses "smaller spectral values" than does the 4-space within which we are contained, where our planetary-orbits are defined by the large 4-spectra of our 4-subspace containment). That is, our 4-space material geometries are in fact 2-surfaces contained in a 3-space, due to issues of spectral size in our subspace for 3-shapes, thus the 2-forms defined on a 3-torus (contained in 4-space) which define the interactions of 2-surfaces, which model material components in 3-space, must be related to 3-space. This is possible since SO(4) = SO(3) x SO(3) so for 4-space, (x,y,z,w) can be separated into a pair of

3-spaces (x,y,z) and (x,y,w) subspaces of the 4-space, and thus relatable to 3-space and to a 2-plane in 3-space.

Thus, there is a geometric relation of the condensed material of our 4-subspace to the spectra of our 3-space containment, allowing a 6-dimensional electromagnetic-field, etc.

Whereas for a 2-form defined on a 4-torus contained in 5-space, the 5-space relates to either a 4-field (or a 4-vector) and a 2-form of 4-space structure, or a pair of 5-fields (or a pair of 5-vectors).

Whereas a 5-torus defined in a 6-space can be related to three 5-vectors, or a 5-vector and a 2-form of 5-space, etc. or a pair 2-forms defined on 4-space and a 3-vector defined on 3-space (or a 2-form defined on 3-space), or a 2-vector-field defined on 2-space, the intersection space of a pair of 4-subspaces defined on 6-space.

The geometry of the 2-tori in 3-space, ie the vector-fields defined on the 2-torus which is contained in 3-space is, thus, associated to SO(3) (which has the geometry of a 3-sphere), is the geometry which causes inertial interactions to be spherically symmetric in 3-space, while the geometry of 4-space and SO(4), along with the spectral sizes of 3-dimensional discrete hyperbolic shapes in 4-space seems to also allow for some aspects of spherical symmetry for the inertial properties of material interactions being related to 3-tori, which are contained in 4-space.

The partition

Partitioning the dimensional levels by defining a discrete hyperbolic shape for each subspace of each dimensional level.

On the other hand, the new viewpoint requires that a catalog of spectral values be found for the different dimensional levels, which are related to (or modeled by) bounded "discrete hyperbolic shapes" of which the 5-dimension "discrete hyperbolic shape" is the last such "discrete hyperbolic shape" which can be a bounded shape.

This spectral-catalog would be similar to the periodic table of the elements.

Thus there is:

The spectra related to subspaces as follows:

[11C1 + 11C2 + 11C3 + 11C4 + 11C5] = [11 + 55 + 165 + 330 + 462] (respectively) = 1023,

. . . , where 1023 is the number of subspaces of the various dimensions in regard to the 11-dimensional containment set, where each subspace of each dimensional level can be associated to "discrete hyperbolic shapes," whose shapes might be bounded, so as to define a specific (well-defined) spectral set for each subspace.

This number, 1023, may also be interpreted to be the number of 1-spectra which compose the finite spectral set upon which all material systems (or bounded discrete hyperbolic shapes) depend for their existence by means of resonance with this finite spectral set.

This spectral set will define sequences of spectral size where the sequence is defined as the dimension increases, so that these spectral sequences may be increasing (allowing for stable material structures) or decreasing which would imply that order to material systems would be much more limited, where an increasing spectral sequence allows the lower dimension shapes to be contained in the upper-dimensional shapes.

The time subspaces in higher-dimensions

Consider the metric-spaces whose metric-functions have constant coefficients, R(s,t), such as R(3,0) [Euclidean 3-space] and R(3,1) [space-time], where s is the dimension of the spatial subspace, and t is the dimension of the time subspace of R(s,t).

In such spaces the time-dimension changes when a new material is added to the structure of existence.

For example, one can hypothesize that for R(2,0) there is only the material property of inertia, which exists as a 1-dimensional discrete shape, when charge is added into the descriptive context, it is a 2-dimensional discrete hyperbolic shape, and then it is contained in R(3,1). Thus, one might hypothesize that there is a new type of material to be contained in R(4,2), namely, the odd-dimension 3-shapes which possess an odd-number of holes in their shapes, ie an odd-genus, but nonetheless, the inertial properties of the material interactions, ie changes in spatial position, of the material contained in R(4,2) would be in R(4,0) space, so "in general" inertial changes of interacting materials' positions in an R(s,t) space would be identified in R(s,0) Euclidean space.

Sizes of discrete hyperbolic shapes

The spectral size sequence, defined as the dimensional levels increase, can be increasing, decreasing or neither increasing nor decreasing, but the sequence of increasing spectral sizes allows the associated material systems to be ordered and stable, while decreasing spectral sequences do not allow for the order of material systems, or decreasing spectral-orbital sequences only allow for a limited (stable) spectral-orbital order to exist.

There is also the issue of infinite-extent "discrete hyperbolic shapes," where the last existence of bounded discrete hyperbolic shapes existing in a 5-dimensional discrete hyperbolic shapes contained in a 6-dimensional hyperbolic metric-space, where the idea that lower-dimensional shapes are contained in a larger higher-dimensional metric-space which (also) posses shapes would only exist (or be possible) for an increasing spectral sequence.

However, there are discrete hyperbolic shapes which have the property of being "infinite-extent" (or unbounded shapes) within (or for) all hyperbolic metric-spaces.

The neutrino is best modeled as an (semi) infinite-extent discrete hyperbolic shapes, due to its, apparent, zero-mass, where its property of being semi-infinite-extent allows a neutrino to be both infinite-extent and to possess a spatial position (associated to the atom which the neutrino is a neutrally charged component), but such an infinite-extent model of a discrete hyperbolic shape (at low dimensions) allows the "infinite-extent property" to, subsequently, be contained in a higher-dimensional metric-space shape, which is bounded, and this allows the low-dimension material systems, wherein these low-dimension systems possess components like neutrinos, whose unbounded geometric property comes to be contained in a bounded metric-space shape.

The infinite-extent properties of neutrinos can be contained in a bounded metric-space. Thus, the systems which contain neutrinos can still be contained in a bounded spectral (or metric-space) set, thus metric-spaces of higher-dimensions can carry within themselves the finite spectral-set which defines a "world of experience."

In turn, this would allow other higher-dimensional infinite-extent discrete hyperbolic shapes to determine (by containment) an "arbitrary" bounded, (relatively) low-dimension spectral set, so that this spectral set can include atomic-type systems, whose components are neutrinos.

However, such arbitrary bounded spectral sets require a higher dimensional experience, ie higher-dimensional containment.

Chapter 21

Empty of content (Apparently, No stable patterns exist)

The content, or focus, or motivation of today's science and math languages used in professional (peer reviewed) journals are the elaborate and complicated techniques that are either unrelated to the observed patterns, ie the observed stable patterns of material systems, or these techniques are unrelated to reliable descriptions of stable, well-defined, measurable patterns, (where well-defined patterns are: shapes, reliable quantitative relationships, observed stable and precise properties, laws which are truly applicable to a wide range of different contexts so as to provide relatively accurate solution functions (or spectral sets), and the math-physical conditions which allow for reliable measurements). The dogmatically authoritative literature of the science and math communities has become devoid of content. It has become elaborate complicated descriptions of a world of illusions. Its main social function is to define an authority which identifies inequality, yet its descriptive context is primarily formulated to develop weapons and to allow ever more control over communication (channels).

Today's professional math-scientists are not describing stable patterns, since the context of the authoritative descriptive precise language, and its associated techniques, are unrelated to describing stable identifiable patterns, rather the intellectual content of their descriptive focus is only about describing complicated elaborate calculating techniques which are unrelated to the observable order of the world.

The context of the professional dogmatists is defined by:

1. indefinable randomness (the elementary random events are not stable and they are not well-defined),

2. non-linearity (quantitatively inconsistent), and
3. contained within a set of measurable properties (or measurable coordinates) which have the properties of a continuum (a very large set, high cardinality), and either

3a geometries or

3b. functions spaces
both of which whose properties are non-commutative, where

 3b1. the functions-space spectral techniques associated to random descriptions are focused on local spectral properties, as opposed to

 3b2. solution functions which provide global system information, and

4. the description has the property of being about random descriptions of the relatively few components which compose a stable system, the logic of this construct is inconsistent.
 Furthermore, it is a description which is logically inconsistent where convergences . . . of geometric properties which are based on random descriptive structures . . . where such geometry and randomness are both defined upon (into) the domain space's continuum structure.

Furthermore, there is an improper focus on "as to what constitutes" a valid frame of descriptive reference for a physical system:

I. A valid "frame" for the containment of a physical system's properties is not about coordinate frames associated to motions as in general relativity, this unduly narrows the context within which a physical system obtains its ordered-form, but rather

II. Physical systems are determined by, and contained within, the shapes (both macroscopic and microscopic shapes, which are defined for all dimensional levels) of metric-spaces, where these discrete (isometric) shapes are models of metric-spaces, and where these metric-space shapes are needed in relation to a finite spectral-orbital set observed for the observed material systems at all size scales, where a finite spectral-orbital set is to be defined on "an over-all" high-dimension containment set (an 11-dimensional hyperbolic metric-space), in which, the partition of the dimensional levels by shapes, defines, either "increasing or decreasing sequences of spectral-sizes as the dimension increases" within the high-dimension containment set . . . ,

. . , so that, within the partition of the dimensional levels by spaces with stable shapes, there is an associated, and prevalent, stable set of holes on these dimension-partitioning shapes.

Note: The genus of such a shape is the number of holes defined on the "discrete isometric shape," upon which the "holes in space" are prevalent, but these holes are not seen by the observer who is contained within a (particular) dimensional level.

Furthermore, there is an inability (of a learner) [in a society dominated by monopolistic businesses] to question the traditions and authoritative structure of "what has come to be thought of as a discipline," but dogmatic authority, ie religion, is a fallacious mental context "in which to develop both knowledge and a descriptive language," where the descriptive language should not be fixed, but rather a change should (always) be considered in regard to a precise descriptive language. That is changes in a precise descriptive language "should be the main focus" in regard to the intellectual context (or condition) of causing changes in descriptive knowledge for the better.

The both overly-authoritative and fixed intellectual state (of our US society) is a result of the experts "need to be subservient to the process of 'peer review,'" which protects, or ensures that, the knowledge fits into the interests of the monopolistic businesses which dominate the society.

That is, one needs to question:

1. the continuum,
2. indefinable randomness,
3. non-linearity,
4. materialism,
5. what are the fundamentals of a differential equation, and
6. the context (of knowledge associated to business interests) of newly forming systems

. . . . , after a stable system has become unstable and has broken apart so as to transition to a new stable state by a series of many-component collisions (where business concerns are interested in the probabilities of collisions and its relation to rates of reactions).

These dogmas need to be questioned if one wishes knowledge to develop and change.

Yet one must list the places and contexts within which it is a valid descriptive context:
(see below for more details)

1. It is a description which is relatable to a system whose initial conditions, and initial properties are carefully put-together so as to be a system which is easily broke-apart, so as to form a transitioning system which is chaotic, so that the rates of reactions (in this context, based on component-collision probabilities) are determined by cross-sections of the broken-apart components, where these cross-sections determine the rates of certain aspects of the (a) reaction, and

2. They are descriptive contexts which relate a limited set of metrically measurable (observable) properties to a feedback structure, which is mostly associated to the critical-points and limit-cycles of a non-linear (usually classical) partial differential equation, where the range of relevance of the differential equation is difficult to determine or to control. Furthermore, the initial or boundary conditions of this type of a system relate to the properties of the descriptive context (or properties associated to the solution) of the system's differential equation in a chaotic manner.

These contexts identify structure related to (1) nuclear weapons and (2) guiding missiles and drones.

Furthermore, there are the overly general contexts, wherein the experts consider holes in shapes, but they view holes in shapes as arbitrary structures, which most often, the experts, relate to complicated shapes and distortions of very general, but unstable, geometries, and these overly general contexts also need o be questioned, though seeking generality can have great value, it should not be a dogmatic command, since it is the limitations which allow useful information about patterns to be found.

Though there is a great imagination, by the experts, for great generality, but nonetheless there is an unimaginative viewpoint about "how holes in shapes" can be (might be) related to physical descriptions.

The nature of shapes, and shapes with holes in themselves, identifying valid descriptive frames (or frames of containment) through which the spectral-orbital properties which characterize physical systems can be modeled.

Holes in space affect (or interfere) with single-valued-ness of values determined from integral operators, yet the relation that holes have with stable spectral values is seldom considered. Namely, the discrete hyperbolic shapes, placed into a new construct, which is to be used for a new precise descriptive language.

One might note that:

The observed stable, precise, patterns of physical systems are associated to finite properties, eg bounded-ness and/or the finite number of a physical system's components, eg atomic-number, and these stable physical system properties are fundamental and observed features of a reliably measurable context associated to the observers of physical patterns. This implies both the existence of stable patterns which allow reliable measuring, (or which are associated to the context of measuring (for an observer)), and the existence of stable-controllable patterns associated to a set of fundamental physical systems which possess stable features of "what may, or may-not be" "material" systems, eg nuclei, general-atoms, molecules and their shapes, crystals, solar systems, dark-matter (ie orbital properties of solar-system's in galaxies) etc, the physical patterns upon which the relatively stable aspects of our life experiences depend, and upon which our mental constructs also depend.

Thus, science and math are about identifying stable, quantitatively-consistent, math patterns which are generally applicable to these stable, measurable, and apparently controllable, physical properties so as to result in descriptions of these patterns which are accurate (to sufficient precision), and general so as to be able to describe the observed stable physical patterns of existence, so as to provide a context for practical usefulness, ie measurable and controllable, so that one can: measure, fit together (or couple), and interact with these various patterns (using the natural structures of these patterns, eg life-forms and its coordinated chemical properties (but, apparently, coordinated by an unknown structure), ie not feedback mechanisms nor carefully prepared structures so as to cause reactions), so that this descriptive knowledge can be related to "practical" creativity (as opposed to literary creativity, essentially associated to a world of illusion, ie a world without stable features).

The violent nature of today's society

The entry into science and math of a set of overly authoritative dogmas (essentially, defined by the authority of the peer-review process), which are overly protected (dogmas), where the protection is accomplished by means of extreme mental, social violence, and the extremely violent-intellectual demands of the dogma, which are required for a person to be admitted into the realm of being a "valid authoritative person within society." It is authoritative dogmas which also define the image of "true science and math" and this adherence to narrow dogma turns the wage-slave scientists and mathematician into a protector of a fixed viewpoint of high-valued knowledge. However, this overly demanding authority comes to be knowledge which serves only the narrow, monopolistic, dominating, business interests (the interests of the owners of society) within society.

The social-instrumental structure with defines both high-value and an "authoritative truth" for all of society is the media, and the ideas which are authoritative are the ideas which are to be expressed on the media, all of the media (including the alternative media).

Most of what the professional (peer-reviewed) science and math communities do, is marginal at best, and it is essentially irrelevant, where its focus is on creating (in a literary sense) [not in a practical sense of creativity] elaborate methods which are associated to unstable contexts which possess only fleeting patterns (measurably distinguishable, but in an unstable context) which are only useful within a larger global context (than the structure of observed stable measurable patterns) of measuring abstracted components of an (unstable) arbitrary context (a context within which stable measurable patterns [associated to the partitioning-components of the measurable attributes] do not exist).

That is, distinguishing features in an arbitrary and abstract context is not evidence that there exist stable patterns within such a context.

An idea appears to be about a context, but one also needs to identify properties, and one needs to possess an ability to measure these properties in a reliable context, and furthermore, these properties also need to be associated to (valid) measurements which are described within a context, wherein, the properties and the contexts are (should be) related to the capacities of human capabilities to create in a practical manner, to measure and to control the patterns. This is about the relation that a measurable descriptive knowledge has to creativity.

The unstable contexts, which nonetheless possess short-lived, identifiable (or measurable) [but unstable] patterns can be used . . . . either in a context of feedback-systems where the "relevance of the measurable context" is difficult to define (determine) or in the random context of a transitioning system, [from a broken-apart system, so as to transition to its final (relatively) stable state], where this transition takes place under a context of component-collisions, and where the probability of these collisions is related to the rate of the reaction, ie the context of nuclear reactions . . . , this is one of the main business interests associated to our society's fixed way of organizing society, so that this fixed structure is upheld and maintained by extreme violence, in which the violence and coercion is needed to in order for the society to remain fixed, so that the monopolistic business structure can continue to exist.

In the context of feedback of locally measurable (and non-linear) properties the observed properties (observed in the context of a metric-space) seem to be "relatively stable," but in fact, are fleeting and unstable patterns, which are built upon abstract interpretations of contexts, [in turn, built upon an underlying set of fundamental stable material properties].

The social context of the descriptive structure is built upon an overly fixed social-structure, and the symbolic structure, which the owners of society impose on material constructs which define the products of businesses within society. If knowledge is structured primarily to be used to build weapons and to control knowledge and information then this is what people will be best suited to create.

However, the context of un-identifiable spectra, or the context of unstable events placed in a context of physical attributes which are measurable in a metric-space, is a descriptive context which possesses virtually no relation to:

1. the stable patterns [of "stable spectral-orbital material systems" which exist at all size-scales], nor to
2. valid models of chemistry,
3. models of life, or
4. mind, and even at
5. the higher abstract level of human experience, often labeled "religion," where one can perceive the "world as it really is," or as it could possibly "appear to be" in regard to the "true nature" of a living observer (What is the complete context of life?).

The dogmas of science and math are expressed in the social context of "intellectual exclusion," and it is formalized (or defined) by peer-review, where these authoritative dogmas are used to protect business interests.

This is possible because of the fact that the dogmas and the "elite structure of science and math" exclude the development of knowledge, in regard to creativity, which is not under the control of the monopolistic business interests (which the justice system so violently upholds), where business can control science by, controlling (1) an authoritative media (2) the laboratories, and (3) educational institutions, which are owned and/or controlled by the business interests, and thus these institutions serve these (same) business interests.

It is these institutions, and the associated authoritative dogma so presented by the media, which is used by the media to express both the identity (and absolute authority) of science and the business interests of those few people who dominate the society and its organization.

What are (math) patterns?

Patterns are:

1. consistent relationships, or
2. operators acting on quantitative sets so as to have fixed "consistent properties" related to the application of an operator on a quantitative set, and these consistent patterns are related to the "meaning" of the quantitative-set's elements (where quantities represent properties of: type and [measurable] size), or
3. stable shapes, etc.

Can the current descriptive language of mathematics and physics describe stable patterns?

Consider:

I. A new context which is identified by the special shapes (circle-spaces) which relate the local quantitative operators more directly to a more stable (and fixed) set of separable solution-functions (which might exist for a system's set of partial differential equations), where the process has a more significant, and more restrictive, geometric dependence (than does the notion of materialism), each dimensional level is given a more restrictive context, but in this new construct there is a many-dimensional context which is highly relevant to the set of observed properties, and how math constructs can be related to these observed stable, definitive patterns of existence.

II. Partial differential equation (a process for finding formulas for measurable properties of a system by relating local measures of the system's measurable properties, to the local measures of the containing space's coordinates), ie sets of locally linear operators which relate function values to domain values, seems to be a construct which is "less important in the new math construct" than are the importance of shapes.

III. But, in the current "descriptive authority," derivatives are being used to identify "local" spectral values "associated to random local particle-spectral events," so that operator-types are believed to be related to various spectral-types. Thus, the descriptive context is about finding (complete) sets of commuting Hermitian operators, which identify (in a unitary-invariant (or energy-invariant) context) a system's set of identifying local particle-spectral set which, supposedly, can be used to identify the system's set of spectral-measurable properties.

    That is, the containment set is defined in a context of measurable, local, random spectral-events, so functions represent the randomness of the system's components, and the spectral sets represent the containment set of the system's identifying measurable properties.

    After, nearly 100-years this idea has not been successful at a level of generality which is needed to make such an idea valid.

That is,

There are essentially the three ways in which to try to describe stable math-physical patterns . . . ,

I. stable geometry, which strongly limits both a descriptive context and the patterns it is trying to describe (the new context for physical description, the circle-spaces, or the very stable discrete hyperbolic shapes),

II. differential equations in a geometric context (unfortunately, this method most often leads to non-linear patterns),

III. differential equations in an operator context (this methods seems to only work for harmonic properties which possess actual physical attributes)

. . . , so as to try to use quantitative descriptions so as to try to identify stable patterns which provide valid information, as well as control, over relatively stable (physical) system properties.

If a measurable descriptive language is without the properties of stability (ie stable properties of the description do not exist) and the descriptions are also without the property of quantitative consistency, but the descriptions are still associated to relatively distinguishable patterns, which, unfortunately, are unstable and fleeting patterns, then one's math methods end-up being only complicated exercises, which possess no content (or, at best, unstable patterns may allow for control by feedback in a fleeting pattern whose range of stability is even more difficult to identify than is the unstable pattern).

If a descriptive structure is associated to many elaborate techniques for "framing and describing" observed physical phenomenon, but if actual, generally accurate descriptions are not forthcoming from such techniques, and the context has virtually no practical purposes, then such a descriptive structure is without content, and is only the basis for elaborate, but irrelevant, techniques, which are devoid of any content, and such techniques are without the capacity to identify (in an accurate manner) stable patterns.

That is, the descriptive patterns of the current (overly authoritative) beliefs (dogmas) of math and science are (have become) irrelevant, in regard to using these (such) patterns to describe the observed, "relatively stable" and definitive, and (often) discrete, patterns of material systems. Furthermore, they are "descriptive" patterns which have no relation to practical creative developments. That is differential equations based on geometry work for some classical systems, and the operator viewpoint acting on function spaces only works for waves (harmonic functions) whose attributes have physical properties, but not for general, stable, precise quantum systems (whose defining property has been assumed to be the randomness of their spectra-carrying particle-components), so it seems that only the context of a very limiting geometric context which also defines a new context for the derivative as a discrete operator on discrete geometries is available for a valid way in which to represent and quantify descriptions of stable observed patterns.

The question is open: What other ideas are there?

When (If) math procedures are: non-commutative (in regard to both geometry and when used in the context of function spaces), non-linear, and indefinably random . . . . (where indefinably random events are events which are neither stable nor calculable), ... then the math patterns, which these procedures are trying to describe, are fleeting, unstable patterns, which are neither "generally accurate," nor do such unreliable-patterns have any practical use.

That is, difficult math methods . . . , which are related to fleeting, unstable math patterns . . . , are descriptive constructs which have no content (and possess no useful information), in regard to the stable observed patterns which do exist.

That is, we are equal creators, and this needs to be expressed in a context of equal free-inquiry where a precise description needs to be related to some type (preferably new types) of practical creativity, creativity intrinsic to the intent of life, not creativity which identifies and maintains inequality and its associated violence.

Knowledge is not fixed, and knowledge does not need to only be related to some intricate instrument's further development, or to some intricate authoritative viewpoint about "how knowledge should be developed." That is, knowledge is based on elementary properties of language and the patterns realized which are related to these beginning elementary language structures are sets of assumptions, and contexts of a set of descriptive patterns. Furthermore, a particular set of assumptions, contexts, and interpretations is not moving toward an absolute knowledge, where this is because descriptive knowledge is limited as to the patterns which it (the language based on assumptions) can describe, and the limits of the patterns and/or the practical usefulness of a set of patterns (of a precise descriptive language) can be reached, so that further development leads to irrelevance and illusion. This is analogous to the idea that there are limits to the capabilities of instruments (precise descriptive languages built upon sets of assumptions), and other contexts might very-well be related to a better way to do the same (functioning aspects to which the patterns of language are related) capacities of the fixed set of (complicated) instruments.

The set of assumptions about which the current overly narrowly defined authority depends are both "far too general" and also "far too restrictive" based on an (almost) arbitrary narrowness emanating from (social) authority. There is the very narrow idea of materialism (which assumes that no-holes exist in the material-containing coordinate-space modeled as a continuum), continuity of dimension (or the holes of spaces which define higher-dimensions, and the spectral-lengths which fit-into these higher-dimensional constructs, get smaller), and randomly based spherically symmetric force-field geometry (often an inverse square field), which emanate from material-particle components, where a quantitative structure (a measurable pattern) is imposed on a blank-canvass (but consistent with materialism) either by means of a (geometric) solution to a differential equation, or by a set of operators acting on a function-space (which model the randomness of harmonic local point-particle-events), both of these structure-imposing constructs are defined upon what is believed to be a blank structure, wherein a prevalent spherically-symmetric inverse-square force-field is always assumed to be that which imposes a spectral-orbital structure, but wherein holes, twists, and cuts-points (in all generality) are believed to be relevant (valid), apparently for shapes imposed by material properties, or properties of high-dimension space whose spatial regions are assumed to diminish in size, while [and analogously] the high-energy spectra (observed in particle-colliders), whose origin is assumed to be in higher-dimensions, the values of the spectra must descend in size (implying increased energy). The idea is that if one finds either the force-field or the energy structure which applies to the operator structure (associated to a material system) then one can identify, by calculation on a blank canvass, the observed order of the material system.

The construct of "a blank canvass which hosts the "material" of random points of spherically symmetric force-fields, modeled as non-linear random relationships," has no relation to any (general) precise stable pattern which is re-construct-able from the laws of this context. This description has no relation to a stable pattern.

It is a descriptive context which is neither general, nor accurate (so as to have sufficient precision), nor practically useful. It is relatable to a state of free-material components transitioning between stable states, ie the reaction rates (or collision probabilities) of transitioning systems, where the original system has broke-apart.

Instead (alternatively), holes are prevalent at all dimensional levels (and in all subspaces), the dimensional structures are partitioned by an increasing set of spectral-orbital values as the dimension increases (at least on most [or some] subspaces), so that the spatial structure of the high-dimensional containment set is not "continuous between dimensional levels," ie furthermore continuity in n-space is defined by the continuity of (n-1)-faces of the lower-dimension metric-spaces, (or equivalently "the material-components," which the n-space contains, or a system's spectral-functions (or the functions in the function spaces), are tied to the geometric structure of the coordinates, so that: orbits, angular momentum, the state of being free-material-components, and component-collisions, as well as the properties of physical waves are all "closely tied" to a limited set of stable shapes and the usual second-order differential equations, whose context is now (in the new context) limited to the prevalent shapes (of existence), on which both spectra and orbits are analogous constructs. These fundamental spectral-orbital properties are related to a hole structure of the shapes, but the shapes are placed in a many-dimensional context, which allows material-components to be contained on (or to exist on) "linear shapes," which, nonetheless, guide the material to an orbital structure based on the material trying to adhere to the geodesics of the (linear) shape (defining envelopes of orbital stability).

The spectral-orbital properties which determine the organization of material structures are defined by the geometric-measures of the faces of the (difficult to perceive) fundamental domains which determine the shapes of the metric-spaces and the material components which determine existence (though there can also exist condensed material).

The new interaction construct is general, but its stable properties are determined from a context of the metric-space shapes of existence.

Re-iterating

A new interaction-construct can be constructed which is general, but its stable properties are determined from a context defined by a many-dimensional set of discrete metric-space shapes, which, in turn, define existence.

The professional mathematicians and scientists in regard to descriptions of fundamental stable physical systems express symbolic nonsense, ie they provide a set of nonsense symbols which result in descriptions which are neither general, nor accurate (to sufficient precision), nor do they provide a practical context for useful creativity.

Physical systems which are very stable and definitive, but which are many-(but relatively few)-body systems, nonetheless, because these systems are so stable and definitive, it is clear that they are forming within a very controlled context, so that the descriptions (of the professionals) which are based on:

1. (vague) randomness (which is an uncontrollable description for a system which is composed of only a few components),
2. non-linearity (quantitatively inconsistent, and chaotic), and
3. non-commutative (not invertible, or equivalently, not solvable, eg non-linear or spectrally-un-resolvable), context, which is
4. contained in a continuum (a containing set which is far "too big" allowing logically inconsistent descriptive constructs to be put-together as if they belong to the same containment set), and
5. it is a description (when based on randomness) which begins from a global viewpoint (a function space) but the methods of the description focus on local spectral-particle events in space, ie it is a description which gives-up information leaving one in an inaccurate and non-useful context in regard to information.

   It is a description which "in general" is not accurate, yet it also is a description which is "intent on" losing information about the stable definitive properties of the [assumed to be random] system.

That is the descriptive structure of the "dogmatically pure" set of experts of math and science is simply a bunch of nonsense.

Yet one must list the places and contexts within which it is a valid descriptive context:

1. It is a description which is relatable to a system whose initial conditions, and initial properties are carefully put-together so as to be a system which is easily broke-apart, so as to form a transitioning system which is chaotic, so that the rates of reactions (in this context, based on component-collision probabilities) are determined by cross-sections of the broken-apart components, where these cross-sections determine the rates of certain aspects of the (a) reaction, and
2. They are descriptive contexts which relate a limited set of metrically measurable (observable) properties to a feedback structure, which is mostly associated to the critical-points and limit-cycles of a non-linear (usually classical) partial differential equation, where the range of relevance of the differential equation is difficult to determine or to control. Furthermore, the initial or boundary conditions of this type of a system relate to the properties of the descriptive context (or properties associated to the solution) of the system's differential equation in a chaotic manner.

These contexts identify structure related to (1) nuclear weapons and (2) guiding missiles and drones.

That is, difficult math methods . . . , which are related to fleeting, unstable math patterns . . . , are descriptive constructs which have no content (and possess no useful information), they are patterns which apply only to unstable contexts, where control emanates from a higher abstract and manipulative context imposed on properties which are only definable in a metric-space, and which requires a lot of preparation (in regard to sensing and reacting in the desired way to the detected properties), a context which is at-odds with the system's natural properties, rather than controlling a system by simple adjustments to affect the system's properties in regard to affecting the properties of several system-components being coupled together.

These professionals are deemed, by the media, to be the intellectual top-experts of the society.

Yet their failed descriptive context is claimed to be the best descriptive range that they can offer. Namely, a descriptive structure which essentially destroys the context of creative development, by the experts providing a failed descriptive structure.

Nonetheless, these experts proclaim that only "the dogmatically pure" can join in on the discussion.

That is, the professionals are getting high-marks (big salaries) [by the owners of society, ie those few who assign value within society] for playing a role of top-intellect in society. Yet their true goal, which they seem to not be aware of, (which is to develop knowledge, which, in turn, is useful and applicable over a wide range, in regard to developing practically useful physical systems).

All that these, so called, top experts do is to develop contexts which are hopelessly narrow in their application, but which demonstrate elaborate and complicated methods, but they are methods which do not describe stable patterns, they do not (one cannot use the laws of quantum or particle physics to) generally and accurately describe the observed stable patterns of existence (of general but fundamental quantum systems), and these descriptive structures have virtually no relation to practical creative development. That is, the experts can provide patterns which have literary interest to other experts, but these, essentially, unstable patterns . . . , (which are contained in an illusionary world) upon which the experts dwell . . . , have no physical interest (or they have no relation to the stable patterns of the physical world).

Like most aspects of the current society, those on the top tiers of society are held in high social esteem for being total and complete failures (the media and the corruption of institutions allows this).

This is the result of the justice system of the US society, where according to the Declaration of Independence US law is supposed to be based on equality . . . ,
(the point of "freedom based on equality" is about each person having the right to develop knowledge as they want (by the process of equal free-inquiry), so as to be able to create what they envision, and then give as a gift, which the individual can give to society in a selfless manner, where the society cannot judge their value, and the society is committed to giving everyone the material needs to live, prosper, and, subsequently, to create in a selfless manner [note: only in this context can a truly free-market exist, but the profits of the most successful products should be well below 1%]), ... , but the elites, who opportunistically began administrating "the independent US nation," instead of instituting equality within the law, so that selfishness was to get punished, acted in a selfish manner, so as to base law on property rights and minority rule, [which is the essential law of the emperor of the Holy-Roman-Empire]. That is, the US governance began as a total failure, so as to be run by opportunistic elitists, who instead of instituting equality and "free-inquiry based on equality" so that knowledge and creativity were to be developed by the culture, instead the elitists in charge used to law to steal, coerce, and destroy, those in the lower social classes, in the name of their own selfish advantage, the selfish advantage of the few, ie power and production were based on social domination of the many by the few.

So we have the tradition of "western hypocrisy," where failure is rewarded if those who perpetuate it, are in the high social classes.

What is wanted, by the owners of society, is that the social structures through which the powerful derive their power are kept in place, ie it is a social structure which is opposed to new, creative changes and thus is is also opposed to equality and the creativity associated to equality. However, the traditional social structure which upholds dominant interests so violently, and it expresses its interest in lyrical creativity of the science and math experts . . . . , where these authoritative experts define the "literary" creative development of science and math, which is authoritative, but unrelated to practical creative development, and the owners of society support the "creativity" of the elite artists, those who also competed in a "narrow context of authoritative cultural value," and those journalists and intellects whose ideas are judged to possess cultural value, so that the ideas expressed are consistent with the ideas of (or can be used by) the owners of society, so as to be distributed by the material-instruments of the media which are owned and controlled by the owners of society . . . , then even the failures of the experts can become part of the social structure which allows the powerful to remain powerful. The top-intellects and top-artists are defined as a social class, along with artists and journalists, so that the intellectuals can dogmatically dominate those many-others who question the authority of assumptions, or who have different ideas. The main tool used to maintain the power of the owners of society is the single voice of authority which the media has become (most clearly controlled by ownership, or by a set of funding processes). That is, it is violence and domination (intellectual domination) which is fundamental to social power, not knowledge.

Knowledge is relevant, within today's social structure, only in regard to the creativity which is a part of the organization of society (ie business productivity) which, in turn, maintains the power of the few. However, the organization of society, and the use of resources and the ownership of technology within society, essentially, remains fixed and traditional.

For example, the many-purpose phone, eg an i-phone, is about developing 19th century ideas of electromagnetism, and the micro-chip circuit boards in these devices depend on 19th century optics.

Whereas identifying stability "as a needed property" in both math and physics, in regard to the useful descriptions of controlled (or controllable) physical systems, is a focus (in regard to the valid descriptions of math patterns) which the math professionals, apparently, have not considered.

Furthermore, very simple math patterns can be used to create new math patterns, which can be used to describe the stable material properties, so that these descriptions are based on a finite quantitative set, within which descriptive containment of physical properties depends, ie the containment set is not a continuum and the derivative and its integral-inverse become discrete operators (the continuum can, instead, be the set of rational numbers).

In fact, the math patterns of stability are very simple, and relating these simple structures (which are best characterized by the stable discrete shapes, or circle-spaces) to many-dimensions, can be done by a simple process of partitioning the dimensional levels of a hyperbolic 11-dimensional containment metric-space (base-space) by means of stable shapes, ie discrete hyperbolic shapes (or circle-spaces), so as to form a finite spectral-orbital set, where the sequences of spectral-size is defined (either increasing or decreasing) as the dimensional level increases, so that these size-sequences of spectra are fundamental in regard to how the description is organized, so that a finite spectral set is the basis for physical descriptions of the observed order which the stable (material and containing metric-space) structures of existence possess.

Furthermore, these simple ideas seem to be much better ideas than are the ideas which the experts possess, [ie than are the ideas that the professional "dogmatically pure" intellectual-army of experts (who work for the owners of society) possess], where the "top-intellects" of society (as proclaimed by the media, where the media is the single authoritative voice of the society) allow the ownership (the management) to bend the minds of these so called experts, ie the pay-masters bend the minds of the salaried-help, but those who possess the best resume's get the best jobs (as everyone competes to help develop the high-value of society, but the high-value of society are those ideas which are proclaimed [or expressed] by the media) the minds of the experts who serve the high-value (defined by the media) have their minds bent in any way which the management wants to bend their minds.

That is, demonstrating high-value in regard to an external model of high-value compromises the internal value (and thus the real creative value) of a person, and it destroys (or greatly limits) knowledge and creativity.

That is, the commercial world is related to a fixed stationary way of behaving or acting, a commercial structure is a very narrow context, based on a limited range of creativity and a fixed way in which to use material resources. The power of business monopolies depend on society not changing how it uses the material resources a business monopoly supplies to a society. The law is supporting this type of narrowness, essentially based on property rights and minority rule (creditor vs. debtor, smart vs. stupid, etc), and it supports such selfish actions with great violence. In fact, the economy is tied to a fixed narrow way in which to live and create, and this model of monopolistic economies is being used as a means to conquer ever larger populations, but it is being put into-place by means of extreme violence and coercion (often an economic coercion).

Does one want a society to be based on a fixed way to use material, and a fixed way in which one is to serve the material based, and fixed structure of society, and a fixed overly authoritative organization of descriptive knowledge, so that this type of power, and associated narrowly defined knowledge, depends on expansion in the form of an ever greater exploitation of particular types of material (usage)?

Part III
Newer Material

Chapter 22

The nucleus

The nucleus is composed primarily of one type of positive-charge, as well as (independently unstable) neutral components (called neutrons), yet the nucleus is relatively stable, and it has stable identifiable spectral properties.

Why this is possible?

The observed stable spectral-properties of the nuclei, have (or possess) no valid description, where such a "valid description" is to be based on the, so called, (currently accepted) laws of physics.

According to the new descriptive context, the nucleus, which is a small positively charged material-component, apparently composed of protons and neutrons, which, together, form into a relatively stable, and correspondingly small, discrete hyperbolic shape . . . . , so as to be a small positively-charged material component which is thought of as being "in the center" of: atoms, molecules and crystals . . . , so that this small, centralized, positively-charged material component is in resonance with the finite spectral-set.

The, so called, "needed" binding-energy for the nucleus is (would be) an energy which is intrinsic to the spectral-values of the discrete hyperbolic shape, where the stable spectral of the nuclear discrete hyperbolic shape is in resonance with the finite spectral-set, which is defined (in the context of real numbers and in the context of a fixed metric-space state) for the over-all 11-dimensional hyperbolic metric-space containment set.

Within such a context the "natural" questions are:

"What is the dimension of this relatively stable discrete hyperbolic shape, which is assumed to model the nucleus?" and

> "How do the two-electric charges (protons and electrons, both of which, apparently, identify a point-position in a metric-space) get organized in this stable shape?"

These are the main questions, since the only particle-like material components (or waves with apparent particle properties) which are stable, are:

the proton
the electron
the photon, and
the neutrino,

while the neutron has a relatively-short ½-life of about 900 seconds (a measure of the particle's decay rate). (while most ½-lives (ie the phrase "½-life" is a phrase indicating the rate-of-decay for unstable systems) of the elementary-particles are very-small fractions-of-seconds, so as to "put into question" whether these "brief events, which are interpreted to be, particle-properties" even exist as (real) particles (or valid entities) at-all, but rather their brief properties can be interpreted to indicate, either "the breaking-apart of some high-dimensional shape," into other shapes, which are the result of a high-energy particle-collision (with the originally stable shape), and because many of the resulting newly emerging shapes, one of which could be a neutron-shape, are often not stable shapes (as the neutron is not a stable shape), ie the spectral properties of many of these newly emerging shapes do not resonate with the finite spectral-set of existence's containment-set, ie there is a cascade of decaying higher-dimensional shapes whose (apparent) existences have been caused by a high-energy particle-collision which breaks apart the original, stable, 3-dimensional discrete hyperbolic shape, ie and creates new shapes which either do not fit into the metric-spaces [within which our laboratory metric-space-frames are defined], or "they do not resonate" with the over-all containment space's finite spectral-set, and thus, disintegrate.)

The relative stability of these few relatively-charged material-components (electrons and protons) means that these charged structures are related to stable shapes, and are to be thought-about as the stable system-substructures which are a part of a (valid) description of the properties of the nucleus.

To re-iterate

The fact that "the neutron is unstable" can be interpreted to mean that it is composed of an electron-proton charged-pair, which, in turn, form into a relatively stable metric-space shape, but this stable metric-space shape is more directly related to the relatively stable nucleus's shape, than it being related to a pair of opposite-charges forming a system (or forming an independent stable shape), so that the unstable neutron is built from an electron and a proton, ie when the neutron shape is independent it is unstable (ie when the neutron is by itself "it is unstable").

However, a metric-space shape which does "form into a nucleus" would be hypothesized to be a shape of the "same type of shape" as that of an atom, but of a smaller size-scale, ie a set of concentric toral shapes, identifying the stable spectral-orbital properties of the nuclei (but there could also be other orbit properties which perturb the nuclei's spectral properties).

Thus, one needs to consider "what is the form of the relatively stable electron and the proton material-components within this orbital structure?"

Is the form of either the electron material-component or the proton material-component, within the shape of the nucleus (or within the shape of the neutron), that of a lower-dimensional shape of localized condensed material, eg point-charges, which has a somewhat (or a relatively) independent orbit?

Where this orbit would be within the metric-space shape of the nucleus, or is the form of either the electron material-component or the proton material-component within the shape of the nucleus (or within the shape of the neutron), a model in which the system's charged-components are independently and individually contained within (or modeled as) lower-dimensional shapes (than the dimension of the nuclear shape) where, in turn, these charged components occupy the orbital toral-components, so that both the charge-shapes and the nuclear-shape possess relative sizes which naturally fit together, ie so the charged-components form rings-of-charge which naturally fit into the "natural" spectral-flows (or as faces of the shape's "cubical"-simplex structure) within (or on) the nucleus's concentric toral components? That is, the size of the charged components would be similar to the geometric measures of the faces of the nuclear shape's faces (when the nuclear shape is thought of as a "cubical"-simplex).

The small-size of the nuclear shape would require that such a shape be of high-energy. The, so called, "needed" binding-energy for the nucleus is (would be) an energy which is intrinsic to the spectral-values of the discrete hyperbolic shape.

In order to envision the geometry of the nuclear shape (which is really a 3-dimensional shape), one can consider a 2-dimensional model of the orbital shapes of the concentric toral-components, where there is a central electron-charge, where if one assumes that each of the toral components, which are occupied by electrons, are about of the same size, but the different toral components are bent (or folded) at different Weyl-angles to one another, and then there is also an outer set of positively charged toral-components perhaps with a similar folded structure.

Thus, one might hypothesize that the neutron is to be modeled as a 2-dimensional discrete hyperbolic shape, which is occupied by both electron and proton components, where a simple pair of toral-components (of the neutron's (genus-2) discrete hyperbolic shape) have been bent (or folded), so that the charge-sizes (of the low-dimension circles) which compose the neutron are resonating with the finite spectral set (which is defined by the high-dimension containing space) but, most likely "the 2-dimensional, discrete hyperbolic shape (used to model the neutron) . . . , into which the two-charges (of the neutron) form, or have come to occupy . . . , is not in resonance, at the dimensional-level of 3-dimensional spectra, with the finite spectral set," so as to cause only a partial resonance (ie where the electron and proton shapes are in resonance), which cannot sustain the stability of the 2-dimensional shape "model of a neutron."

However, the properties of the nuclear components (ie the so called, unstable elementary-particles), which emerge from particle-collision experiments, define a SU(3) pattern of particle-decay, which one would expect to be related to a corresponding (real-number) 3-dimensional shape. A 3-dimensional discrete hyperbolic shape (which is the new model of the nucleus), naturally, fits into a 4-dimensional metric-space, where the dynamics of such a 4-dimensional space is related to SO(4), where, in turn, SO(4) is naturally, divide into two-parts (ie SO(4) = SO(3) x SO(3)), where one-part is directly related to the 3-space which we observe, and the other-part would define dynamics which would (mostly, except possibly for 2-plane motion) "not" fit into the 3-space of our laboratory frames.

The model currently (2013) used (in peer-review literature) for the nucleus . . . , as provided by wikipedia, . . . , is based on material interactions, which are modeled as (or as a result of) random particle-collisions.

Note: Wikipedia reports the narrowly defined intellectual-dogmas which most support the interests of the investor-class.

Note: the elitist-banker structure of society is also expressed within the context of intellectual-elitism (and intellectual-domination). Thus, there are dogmas which form the basis for intellectual-competitions ie the competitive model of the US education-system, where the, so called, "winners" of these competitions provide a small set of well-indoctrinated intellects from which the investors can chose the personnel (or some of these top-students) to manage their technical projects into which they have invested, ie the dogmas of the intellectual-educational-contest also happen to be the narrow intellectual viewpoints which best serve the investor-class.

One sees the way in which the social position of the intellect has changed by the history of Socrates and Plato. Socrates identified equal free-inquiry as the best pathway to gain ever more useful and accurate knowledge, but the idea of equality put-forth by Socrates was enough for the patricians to have Socrates be put-to-death.

Then there followed Plato, who expressed the idea of elitism, ie the philosopher-king was the type of person who should (or who is to) lead society. Plato's elitist ideas were acceptable to the ruling-class, so he "got published."

And this has been the relation between the intellectual and the ruling-class ever since in western society, ie so it has ever been in western civilization.

The socially-weak intellectual has been forced into narrow viewpoints, which support the ruling-class, and the ruling-class became ever more powerful, and the intellectuals have become ever more dogmatic and authoritarian, and intellectually-domineering.

The ideas of Socrates have been echoed in modern times by Godel's incompleteness theorem, ie that there exist many patterns which a precise language, based on fixed assumptions, cannot describe. The conclusion is, that one wants quantitative language to be developed at an elementary level where there are "no" well-defined experts, ie no intellectual-elites, ie where there is equal free-inquiry.)

Return to the dogmatic descriptive language of the accepted theory of the nucleus

This old descriptive context (of the, so called, strong-force) primarily describes the, so called, quark-interactions (where quarks are claimed to be unobservable) so as to form protons and neutrons (from quarks). That is, particle-physics provides a detailed description of an unobservable, but (nonetheless) a, supposedly, material context.

But this type of a model (based on quarks and the strong-force and random-collisions) identifies an unstable proton (which is a property which is not observed).

Note: Since a description, which is based on the property of randomness can be used to interpret data in a wide variety of ways, it is likely that eventually some unstable event will be found which can be interpreted to support the unstable proton idea, especially, since (1) the quarks are unobservable and (2) the randomness of this descriptive context is based on unstable events. In fact, the so called verifications of particle-physics seem to all be about the interpretation of non-existent properties, or briefly disintegrating patterns, as representing verification of particle-physics patterns, but these patterns are most-likely a unitary representation of a relatively stable 3-dimensional (real) shape, which exists in 4-space (and has been broken-apart by particle-collisions), and where the metric-space states cause it to be in complex coordinates, and thus the observed patterns of disintegration are related to a unitary fiber group.

(note: random events need to be stable-events if a reliable probability is to be fashioned from these events.

Thus, randomness which is based on unstable events is a description which possesses no [rational, or quantitatively valid] content).

In the current model, the nuclear force, ie the (so called) residual force (associated to quark-interactions), which, supposedly, holds the nucleus together, has no definite model for either the potential-energy term or the interaction structure, and combinations of "harmonic-oscillator potentials" and box-potentials (where box potentials, and the height of their potential-walls, are convenient for fitting data), which are used to model this residual nuclear-force, are expressions of ad hoc data-fitting, wherein there is also defined a context in which both

(1) various ways of pairing-up protons and neutrons so as to form Bosons (within the nucleus), and
(2) various different constants,
are used as models of component interactions based on the residual nuclear-force so as to use such a varied model (ie pick the model which best fits) to fit the observed data of the relatively-stable nuclear spectra.

This is clearly a data-fitting exercise, and it is not a valid-way in which to consider these relatively very stable spectral properties of general nuclei.

This data-fitting is being done in an institution which, supposedly, believes-in the dogma of representing physical-law as partial differential equations, and believes-in an assumption about using a deductive-method for the precise descriptions of physical systems, where physical-law is, apparently, to be represented as a form of an absolute-truth (apparently it is assumed, from which the truth may be deduced).

Consider

It should be noted that the atomic number is almost exactly ½ of the atomic weight in the first 3 rows of the periodic table (of the elements), while at row 4 (of the periodic table) and beyond (the atomic weight) > (2 x "the atomic-number"), ie there are more neutrons than protons in the nucleus, from row 4 and beyond.

A neutron decays into an electron and a proton and an anti-neutrino, or changes from a proton to a neutron and a positron and a neutrino, etc.

The new descriptive language

Perhaps (the above mentioned, in regard to the old descriptive structures of) "the pairing-of-nucleons so as to form Bosons" is better modeled (in the new descriptive language) as toral-components (of a discrete hyperbolic shape) which can contain both protons and "the fewer number of" electrons, where the different toral-components possess various Weyl-angle relations with respect to one another, within a 3-dimensional shape, whose geometric properties are contained in a 4-dimensional hyperbolic metric-space.

Thus, also giving rise to the idea of a set of unseen (undetectable) components within the nuclear system, ie unseen because the system's geometry is actually in 4-space (whereas our labs are in 3-space).

This, 3-shape in a 4-dimensional context, can be of much importance in regard to modeling van der Waals forces (also called London forces) which are an important aspect of both the vaguely identified nuclear-force, and the much needed, and equally mysterious, van der Waals forces, which, apparently, are the main contributing forces to the formation of crystals (or condensed matter), which form out of neutrally-charged atomic and molecular material-components.

Note: The various Weyl-angle relations can (might) be thought of as the various latitude-angles, which are defined by the integer, m, in the spherical harmonics of the H-atom (a 3-shape contained in 4-space), where the natural splitting of the dynamic groups SO(4) into SO(3) x SO(3) can also be used in understanding this geometric-dimensional structure (note: where one would expect angular-momentum to be considered to be a dynamic property).

There are many geometric models of a nucleus, which can be based on the geometry of toral-components of a discrete hyperbolic shape, where the toral-components can have different sizes, so that the toral-components are bent (by Weyl-angles) in relation to their adjacent toral-components, though one assumes that "the beginning discrete hyperbolic shape" is a linear-string of toral-components, where the toral-components have various sizes. Then various ways of arranging charges and anti-charges, etc, which can be modeled as the natural (spectral) flow-structures on the (variously folded) toral-components of a discrete hyperbolic shape, so there can be modeled many different charge distributions (for a model of a nucleus).

The reason that one wants such a geometric model of stable quantum systems (such as the nucleus) is that: (1) this allows for such systems to be stable, (2) such geometries can be related to both other geometries which exist in the metric-space, and to the (same) finite spectral-set (which is an intrinsic part of existence's over-all high-dimension containing metric-space), (3) the higher-dimensional context of existence allows for greater range of creative possibility, and (4) the idea of practical creativity . . . , which a probability based description . . . , is only about the creativity of "fitting-data," but this type of "practical-creativity" is unrelated to any form of practical creative development.

Note: The main point of a probability based description of a quantum system, which is composed of a relatively few material-components, is that these systems cannot be used for the practical development in regard to using them in relation to other systems and to still be within a descriptive context of control, yet the stable and regular properties of these systems implies that they actually "do" form, within a context of control.

That is, apparently, quantum systems are formed in an interactive-context of control
So
"What would such a 'context of control' be?"

Is it "stable shapes contained in a many-dimensional context," where existence is determined by "a stable shape being in-resonance with a finite spectral-set"? [Yes!]

Or

Is the correct descriptive context:

indefinable randomness (eg basing probability on unstable events, etc),
non-linearity,
non-commutative patterns for operators,
quantitative inconsistency,
logical inconsistency,
eg placing the description in a containing (domain) space which is "too big,"

so that the descriptive structures, within the domain space, are (or can be) logically inconsistent, and so that all descriptions are to be based on a set of partial differential equations (the context through which physical laws are, supposedly, to be represented) and they are partial differential equations which cannot be solved? [No!]

The new descriptive context

The neutrino is similar to the neutron in that it is (hypothesized, in the new descriptive context) also composed of an electron and a proton (but not in the context of hadrons but rather in the context of leptons), which are bound together so as to form a, seemingly, unbounded discrete hyperbolic shape, but it is a stable, "unbounded," discrete hyperbolic shape. This is to be interpreted that in the context of stable leptons the electron-proton pair either defines a stable geometry of an H-atom, or the stable context of the entire metric-space, which, in turn, can contain material-components which possess geometrically measurable properties. This is the "point-position" vs. "an entire metric-space" (where the metric-space possesses a stable shape, which might be a bounded shape) wherein a material-component's position can change, so that spatial-displacement invariance implies momentum invariance. The hadron (or neutron) shifts the electron-proton system into a component which has the property of possessing a (point) position in a metric-space, while the lepton electron-proton-pair (or neutrino) shifts this system into an unbounded representation of a metric-space's subspace (in a many-dimensional context of a containment-set), or perhaps the neutrino is bounded by its containing metric-space, if the metric-space within which it is contained is also bounded. Note: The unbounded geometric model of the neutrino implies a non-local structure, ie an action-at-a-distance capacity in regard to the descriptions of material-systems in metric-spaces (especially within Euclidean space).

Such a geometric distinction between hadrons and leptons is a significant idea in its own-right.

Note: Quarks are most likely the result of 3-dimensional geometric shapes which exist in 4-space, where SU(3) is related to these shapes in complex-coordinates.

That is, quarks are the intellectually-abstract results of stable patterns of 3-shapes being broken-apart by particle-collisions, so the patterns of disintegration (after being broken-apart) exist in 4-space, but these patterns are projected into 3-space, where such a projection into 3-space is "naturally" contained in Euclidean 4-space, whose fiber group is SO(4) = SO(3) x SO(3). Thus quarks and leptons are not the natural pairing, rather one needs to look at the stable components, ie electrons and protons.

An important question is:

Why is the neutrino stable, while the neutron unstable?

Can this be related to the fact that the neutron has both positive and negative ½-spin properties and this is also true for both the neutron and the anti-neutron, while the neutrino has a ½-spin property, but the anti-neutrino has a (-½)-spin property?

That is, the neutron is a local, bounded material-component (with a shape upon which two-opposite metric-space states are defined), whose fundamental property is to define a point-like position within space, but it is a shape which does not quite resonate (or only half-resonates, ie the charges which it contains are in resonance) with the finite spectral-set (where the instability of the neutron was also described above), while

The neutrino is an unbounded shape (or perhaps bounded by the large metric-space shape within which the neutrino is contained), where this, seemingly, unbounded shape is identifying the subspace of the metric-space within which the laboratory frame is defined, where this is being done in the many-dimensional containment context, so that the anti-neutrino is identifying a subspace, but that subspace is in the opposite metric-space state, which is related to the same, but opposite-state, laboratory metric-space frame.

That is, the neutrino defines both a dimensional level and a subspace of that dimensional level, in the real-number context, while the anti-neutrino defines the opposite metric-space state in the same dimension and in the same subspace, and where the two opposite metric-space states define a complex-coordinate containing space, which, in turn, is related to a unitary fiber group, and the, subsequent, mixing of opposite metric-space states in these complex-coordinates, where locally the dynamics (of the two opposite metric-space states) define inverse spatial-displacement relations.

That is, the neutron identifies a position in space, where this position is related to a (bounded) material component, while the neutrino identifies the metric-space within which the material-components (associated with the neutrino) are contained, so that mathematical processes can be defined in regard to these material-components and their spatial-positions.

The electron-proton-neutrino component structure of the nucleus could have (1) a central-electron toral-structure (as expressed above), and (2) an orbital-proton component toral-structure, and then since the neutrino is also being contained within the shape of the this nuclear-component, (3) the reach of the neutrino may carry, the central negatively-charged property, out to the boundary of this shape (since the neutrino is assumed to be composed of both a proton and an electron, which is to fit into the neutrino's unbounded shape, unless bounded by its containing metric-space shape, thus, the electron-part can be at the far-reaches of the shape of the neutrino-component's shape), and where the nuclear-shape is contained-in a higher-dimensional metric-space, ie the neutrino effectively extends the negative center of the nucleus to the boundary, where this boundary is seen in the nucleus-containing metric-space, so that from the outside of the nuclear metric-space-shape the boundary of the nucleus appears negative. Furthermore, the nucleus could be composed of protons and neutrons and neutrinos, so that the number of neutrinos could also be similar "in number" to the atomic-number. That is, the neutrino carries the main (or central, ie the point-position vs. metric-space-boundary dichotomy [boundary of a metric-space subspace]) charged property of a shape, so that this property is (or could be) observed on the boundary of the shape when the shape is viewed from its higher-dimensional containing metric-space.

The great weakness of the quantum picture is that it is a probability based model, and probability based models are all about fitting data.

Thus, there seems to be no-end to the ways in which the quantitative models, in a context of random (and often unstable) events, can be adjusted to fit data, so as to subsequently, claim that the descriptive context is a valid description. For a math construct based on randomness, which is applied to stable systems which are composed of only a few components, means that the math properties identified in this descriptive method, for this, assumed to be random, system cannot be used in any practical process so as to control such stable systems, yet the stability of these systems implies that they form in a controlled process. Only idiots would insist on continually following this course for a quantitative descriptive language.

Furthermore, quantum physics is also based on doing math in an arbitrary manner, such as the so called solution to the H-atom, where the diverging series-solution function is arbitrarily cut-off so as to fit data . . . , etc etc, etc. Nonetheless, the illusion provided to the public by the propaganda-education system is that these descriptive efforts are circumspect, careful, and (delusional-ly) rigorous.

But this (arbitrary data-fitting) is, essentially, the same claim made by the Ptolemaic epicycle model of the planetary-motions . . . , where epicycles are also designed to fit data.

But neither model (quantum or epicycle) is relatable to practical creative efforts, ie it does not lead to a practical creative context, other than the types of "creative efforts" which are needed to fit data, yet the measurable properties of these systems (which quantum physics is trying to describe) are stable, and thus not random.

Chapter 23

DNA as a blueprint for life

DNA as life's blueprint?

DNA is, supposed to, contain all the information which is needed for a life-form to live, but it needs to be activated (so as to live) by doing (performing) chemical processes (which, apparently, have no descriptive relation to science) within an intermediary cell, before it possesses the attributes needed to actually possess the capacity to live, when this DNA molecule is placed into, yet another, living cell.

Introduction

There is an alternative descriptive language which provides a descriptive context for these intermediary chemical processes in the other living-cells. This shows the versatility and superiority of this new alternative measurable descriptive language. But it extends the descriptive context into many "new sets of possibilities" which transcend the constraints of materialism.

The endless failings of the current scientific paradigms of both material and living systems, which are essentially based on the failed ideas of: materialism, (partial) differential equations, algebraic equations based on average measurable values, (indefinable) randomness, and DNA; yet these obvious failures seem to not be enough for people to challenge these failed paradigms.

Instead the public are enamored by the results of the failed paradigms, which: money, propaganda, and intellectual-bullying can bring about so as to appear to be promising results, but which exist within the very limited context of solvability and controllability (within which precise descriptions of the properties of material systems are framed), but the entire contexts' (the entire paradigm's) obvious relation to failure.

note: without a valid model of both chemistry and a chemical model of living systems the idea about DNA as a "complete information construct" for living systems is sure to remain a failure; and thus another partial-truth which modern-man is turning (these many partial-truths) into a source for man's own extermination (where this extermination will result from the improper use of: nuclear "wastes," chemical toxins, biological imbalance, and burning of carbon fuels).

This inability by people within society, to challenge failed belief structures, has to do with the terror in which the public is held, and the mis-placed intellectual arrogance, which is based on ignorance, to which the public agrees, in their relation to the propaganda-education system which is associated to social domination of the public by the ruling-class, to which the public forms a collective-effort wherein the members of this collective society both blindly and in terror agree to the failing beliefs of this collective support for the ruling-class.

This domination is leading to the earth's destruction, and the intellectuals and their intellectually-bullying role within the propaganda-system, has the result of their being some of the main contributors to the propaganda-education system, and they . . . . , who are other than and different-from the ruling-class . . . , are the main reasons, for this path to disaster for this social system's failure.

They stay fixed in their attention on the failed systems of: materialism, (partial) differential equations, algebraic equations based on average measurable-values, (indefinable) randomness, and DNA, and spend their time considering irrelevancies which are applied to this failed intellectual construct.

On CSPAN-TV, on Book-TV, on 11-16-13, in AZ, C Venter presented his claim, or he expressed his belief, that he had proved that DNA is exactly life's blueprint.

Note: Venter exemplifies how commercial investment interests confine and control the categories or contexts of allowed intellectual dogmas (eg use the chemical model of DNA for commercial interests of developing medical drugs), ie the model of DNA used by Venter appears to be too narrow.

Though there is much information about the chemistry of proteins in DNA, there may be a much better way in which to understand (or precisely describe) the structure and organization in shape and space and in regard to properties and functioning of how this "chemical" information in DNA is related to the cell than simply relating it to a statistical model of (inert material) chemical reactions in time and space within a thermal context of (statistical) randomness and average thermal-values.

There might be a much better way in which to understand its context of existence and its functioning and/or its information context than the statistical randomness of inert chemical reactions.

Introduction

There must be a change in point of focus of the main process by which the justice system violently upholds the ruling-class within society (a society with an oligarchical social structure).

This focal point is that law is "being based on materialism and property rights and its associated relation to violence," and, subsequently, we live in a society in which morality is based on materialism, lying, stealing, and murder, and this is being upheld with arbitrary violence applied by the justice system; whereas those who are not in the ruling-class who engage in: lying, stealing, and murder are (most often) severely punished by the justice system.

However, this barbarous behavior of the ruling-class results in similar behavior by the public and their local domains where they express their limited domination by violence, rather than justice being based on defending the equal creative efforts of every individual.

Why are we destroying the earth?

Since our collective social hierarchy wants to do this, since it is in the selfish interest of the ruling-class, ie they have invested in this destructive pathway, and the justice system, ie the national security state, protects the interests of the ruling-class, and the media protects the intellectual beliefs upon which the ruling-class depends for their (productive) power.

However, human creativity, mostly, exists outside the idea of materialism, it exists in a many-dimensional, stable geometric context, where there is a constructability to this higher-dimensional and geometric context to existence (wherein human-life participates in the creation of a manifestly different existence, ie life can change the properties of existence, itself), where this constructability might be similar to Venter's construction of DNA, but more encompassing, ie more thoroughly expressed within a wider range of applicability.

The chemical-constructability of DNA

Venter's proof that DNA is exactly life's blueprint, was (within the TV presentation) that his lab (apparently within a lab he owns) was able to create DNA molecules, which can have (possess) exactly a DNA's "letter" sequence of an arbitrary species (in this case the DNA sequence of a particular Bacteria) and that when this manufactured DNA molecule is

1. first placed into a yeast cell where a molecular change in the manufactured DNA is realized, and
2. then this DNA is subsequently placed in a different particular (or other) bacteria,
3. then the new DNA destroys the DNA of the "other bacteria," and
4. the cell with the destroyed old DNA and the new "manufactured" and "molecularly processed within yeast," and after the new cell's original DNA is destroyed by the new "manufactured" DNA, then the new DNA begins to function, in this cell, as the organism which is defined by the "manufactured DNA."

Though this is quite remarkable, it does not prove what Venter is claiming that it does prove.

This belief of C Venter's, ie that DNA is exactly life's blueprint, has been expressed as an authoritative dogma, almost, ever since Crick and Watson identified the structure of the molecule DNA (around 1952), and thus, it has been the focus of a great amount of investment.

And

Then it was believed to be determined that the sequence of "letters" within the "double-helix twisted" molecular structure of DNA, can in turn, be related to the processes of "the making of proteins" which are associated (by composition) to the sequences of "letters" on the DNA molecule, and the relationship of these "letter sequences" to both RNA and to "the molecular structures of proteins," and their "letter-related building-block-molecules," in the "letter" sequences (of both proteins and DNA).

That is, the information in DNA is supposedly determining "the protein context" within the cell, in which the DNA is contained, and "the timing" in the, supposed, sequence in which protein is being manufactured Is also supposed to also be information which is "hidden," but existing within the DNA.

The proteins in living-cells play the dual role of being both catalysts, as well as being the structural building-blocks of an organism's structural properties, where these structural properties of cells are put-together by various other chemical reactions but with the help of catalysts, so

The obvious question:

"when does a protein play one role (catalyst) and when does it play the other role (chemical building-blocks)?"

These are all complicated chemical processes which need to be organized.

How is this done?

Chemical reactions are modeled to be determined by random-collision events. Thus, it is difficult to define an ordered structure from a random structure. This type of statistical-order is possible only in regard to quantitative properties which are averages over many-components in closed-systems, whereas molecular activity in living-systems is based on relatively few components which fit into geometric structures (protein-building), and, apparently, the geometric-components come into being at just the right time, and it is not so much about random-collisions of molecular-components, but rather the precise orientation of stable shapes and a geometric fitting together.

However, an energy-generating (odd-genus, discrete hyperbolic shape), stable, organized geometric structure (with a memory) can provide a whole new context within which to organize both the:

(1) chemical enzyme, and
(2) structure-building of the shaped-molecular processes involving proteins in living cells (see appendix).

The claim that DNA is exactly life's blueprint can only be proved if

1. all DNA sequences can be manufactured, and
2. a molecular-cell can also be manufactured, by using only laboratory molecular processes, and
3. without recourse to intermediate processing stages, wherein the molecules are placed in other living cells, that is, so that the molecules are not ever placed in other living cells, so that instead, the DNA can be placed into the "laboratory manufactured cell," and
4. all forms of life (as well as new forms of living-organisms) can be replicated by this process, (where all molecular processes are done exactly in a laboratory and put together in the lab so as to function as a living system).

That is, it is not sufficient to show this process in regard to, "the true building of life within the laboratory," for only single-cell life-forms.

It is not clear what condition of chemical organization . . . , beyond molecular processes, which are taking place within living cells, (where cells exist in very complicated molecular contexts), and chemistry associated with living properties, apparently, can only be created (or caused) by doing chemical-living-cell processes within living cells (but what are considered to be only molecular processes) . . . . , within the context of "pure" laboratory chemistry, can actually cause life-forming molecular processes to occur (and be organized) so as to build cells which are functionally living.

To have a model of a living-cell as a thermal-chemical system of many different molecules colliding in a statistically-individualistic context, may not be a correct model for a living-cell, and so that in this model there are only a few ways in which molecules can be detected, and the processes of the cell can only be related to a (or to these) few detectable molecular-types, so that a detection-method along with "life-process identification," where the life-process is occurring in the cell, is information which is used (put-together) in a statistical construct of "correlations of events," so as to become a model of a living-cell as a set of correlations, which exist between molecular-detection and living-process-occurrences within the cell,

Why follow this highly constrained and very limited viewpoint, a viewpoint, which is very limited by its dogmatic assumptions, and rather than this old form of failed thought, perhaps, a more unified (new) crystal-molecule shape structure.

Furthermore:

If there is not a valid model of the chemical processes which are related to the chemical actions of catalysts (and/or enzymes), then all chemical-models of living systems can only be deeply flawed models.

It is admirable that Venter, apparently, motivated by his narrow dogma, has constructed what he claims to be a very precise model of DNA of a particular bacteria.

Of course the obvious question "Is his lab construction of DNA as precise as he claims?

The dogmas of commercially responsive science have driven the determination of scientific truth "back to the standards of truth" used by the Ptolemaic models, but with the added proviso of "if there exists (or even theoretically exist) a couple of measurements" which display precise agreement with an, essentially, empirical model, then the scientific truth has been verified.

Of course, by this standard, Copernicus's model of the solar-system would have been professionally dismissed, since, in many instances it was not nearly as precise as the "calculating" methods of the Ptolemaic model of planetary motions.

But the Ptolemaic model was not true, and most of the modern dogmas in science are also obviously not true, eg quantum physics, general relativity, particle physics, the models of chemical reactions being primarily related to rates of random molecular-collisions is a very limited picture of chemistry, ie the theory of chemistry is very limited, and the idea that "living systems are exactly the result of material based chemistry," are all obviously not true, where this judgment can be based on the results which science has so far provided, since the observed stable properties of: nuclei, atoms, molecules, crystals, internal-control of living systems, have no valid descriptions which exist to sufficient precision, and which apply to general types of these listed systems, based on the so called laws of physics. That is, modern science has clearly failed in a most fundamental way.

. . . , though Venter has pushed the "DNA molecular model of life" to a vaguely, almost, slightly believe-able level, if his DNA constructs are as precise as he claims. But, nonetheless, molecular manipulation in a system context of the chemical reactions being within a living-cell before the manufactured DNA becomes functional as a "living" molecule. Thus, his statement seems to be at least three-quarters wrong.

This set of intermediate steps associated to chemical-living systems should lead to fundamental questions about chemistry and life (see appendix about life as a 3-shape contained in 4-space).

But Venter seems to ignore (or dodge) this question, by apparently, upholding the (accepted) dogma of a statistical molecular-collision model of chemical reactions, and subsequently, a belief that in a context in which only a few chemical-molecular properties can be detected, with a [detached (or intrusive)] set of lab-chemical tests, that one can (then) identify a chemical-life-process relation by a statistical correlation, wherein life-processes have occurred and certain chemicals have been detected, (where the detected chemicals exist on a short list, ie the point-is that only a few chemical can be detected in this way) so these detected chemicals are assumed to be central to the life-process, and

Thus, there is a subsequent molecular-life model which is built upon a "statistical correlation determined model" which depends on chemical-detection and life-process model of correlations, in what is essentially an arbitrary statistical context, ie a context which is highly dependent on the arbitrary nature of chemical-molecular detection's limited methods and limited capabilities, ie there exist only a limited number of detectable molecules. It is unknown what chemical processes exist in living-systems.

Yet, the main property of life is its precise control over its own very chemically complicated living system, and this type of control will never emerge from descriptive models based on randomness, ie chemical processes are based on random molecular-collisions.

Living-systems must possess some connection to an actual controllable-system context, ie a linear and solvable context, is evident (see appendix about life as a 3-shape contained in 4-space).

It should be noted that in our science models of material systems only statistical constructs based on large numbers of components for a system's model of composition, eg large numbers of atoms or molecules, which are measurably related to physical properties, and which are average-values, which exist in closed systems, can be related to controllable inter-relations of these measurable properties.

But DNA's relation to life is local-molecular control of individual molecules and their relation to the direct building of structures within the (local) cells of living systems, and to other life processes.

Appendix:

Life can be modeled as a 3-shape contained in 4-space and a simple stable geometric connection to a 4-dimensional geometric-shape (these are linear solvable shapes), which splits between inert material and "living" material where the natural spilt is caused by both (1) the two types of shapes one with an even-genus and one with an odd-genus, and (2) the property of the fiber isometry group of SO(4) = SO(3) x SO(3), so that both subspaces upon which such a fiber group would naturally act in (x,y,z,w)-4-space are the usual R^3, ie (x,y,z), and an R^3, ie (x,y,w), which is separated from (x,y,z)-space.

The 3-shapes in (x,y,w)-space (do) have an odd-genus, and thus, when their spectral-flows are occupied by charges these odd-genus 3-shapes will naturally oscillate, so as to generate their own energy, these shapes will also possess a natural "memory structure" which is attached to their unitary fiber group SU(4), ie associated to real metric-spaces and to SO(4).

The SO(3) x SO(3) structure of the SO(4) fiber-group both separates and mixes the shapes in relation to the common 2-dimensional space of (x,y). Thus, the two types of shapes (even- and odd-genus shapes) can work on both (1) the, so called, inert physical properties of chemical shapes and (2) the energy-generating properties which are 3-dimensional and only physical in an (x,y,w) 3-space, which we are taught to ignore.

That is, this loss-of-perception of these energy-generating shapes, supposedly related to our own life-forces, might be related to our own true relation to our own life-form shape. If we are truly related to the properties of the sun, so that the sun is a part of our life-form shape, then this would mean that this part of our shape (and of our being) is very large, very large beyond our expectations of a material world with what we consider to be normal sized shapes on the earth's surface.

However, this large-size of a part of our life-form shape would also explain the so called (religious) myths from which religion descends, of heavenly-beings coming to earth and giant-beings from the heavens.

Nonetheless, there could also be stable circle-space shapes in the (x,y,w)-space which are a part of our being which are also the same size as the molecules which make-up the cell-structure of our bodies. These are questions about both (1) resonance with the over-all high-dimension containing space's finite spectral-set and (2) the discrete and discontinuous set of properties, which can exist between dimensional levels, where the size-scale can discontinuously change between dimensional levels, but resonances can allow this type of a system to be composed of many size-scaled circle-space components.

This is a simple model of life, and it depends on the existence of a (stable circle-space) shape, which is similar to the shape of a molecule which exist in (x,y,z)-space, and where their oscillations can be associated to a rhythmic-flow within the living system, eg or a heart-beat which generates such a circulatory flow.

Stable 3-shapes in (x,y,z,w)-space can exist as either (x,y,z)-shapes or as (x,y,w)-shapes, which are in resonance with the finite spectral set of the high-dimension containing space (of "all" existence). However, these 3-shapes exist within a 4-space, within which exists a 3-shape which identifies the solar-system, these are 3-shapes within 4-space (and may, or) if small enough would also be within the solar-system's 3-flow. Thus, the 3-shapes which materially interact with the 3-shape, which is the solar-system, in a significantly strong (magnitude) level, are at least the same order of size as the solar-system, otherwise the interactions between these small 3-shapes (within the 3-shape of the solar-system) are the small (in magnitude) van der Waal type of forces, as well as the collision interactions.

That is, within a cell's chemistry, which is "an interaction relationship" between 3-shapes, (ie van der Waal forces and collisions) an inert-material molecule can interact so as to form (or to connect with) a new 3-shape (which resonates with the finite spectral set, so that it can exist as a relatively stable shape) which oscillates, and thus would then have (possess) the required properties for a living energy-generating system. Apparently, this is the mysterious chemical-living reaction which takes place within the living cells and which can endow chemistry (chemicals) with energy-generating life properties in 4-space, but related to 3-shapes.

However, it may be that these 3-shapes exist within a 4-shape whose 3-flows identify the solar-system, and these are 3-shapes bound within the solar-system's 3-flow.

An energy-generating, stable, organized geometry, with memory, can provide a whole new context within which to organize the enzyme (chemical) and structure-building molecular processes involving proteins in a living-system. It also seems to identify a constructible context, but connecting to the higher-dimensions seems to exist (only) within the cell, although a model of the radioactive nucleus seems to also connect to the higher-dimensions, while the stability of the solar-system can only be explained in a context of higher-dimensions.

Summary

It is very admirable that Venter has proceeded to act on his beliefs and to progress so far as he has in an attempt at DNA building, though his claims about the precision of his DNA models should be challenged, [but who would fund such a challenge?] and to relate these molecular constructions to "chemical processes" in living-cells by means of intervening in the cellular-molecular processes, so as to get results which are consistent with life, is truly remarkable.

Yet, in the context in which the physical models of both

1. "chemistry" and "chemical-thermal" systems are very limited, and
2. when catalysts are also not well-modeled (not modeled at-all), then interesting models of life are non-existent, so that only a very questionable molecular model of DNA as the basis for life is allowed

Thus, other models should be considered, especially, if the catalysts can be modeled in such a similar geometric-constructive manner, as the constructive techniques which are being used by Venter, for his model of DNA, and such a model is available now, then why not challenge the accepted dogmas?

Why does Venter, as a scientist, simply accept the dogmas about chemical-thermal-living systems?

If there was an attempt to use other constructive-geometric models, especially, since such models are available, of for these chemical-thermal-living systems, then life-science (as well as physical science) could have the same constructive nature as Venter's constructive approach to DNA.

But as it now is, within the domineering dogmatic intellectual models of the commercial category determined science it is difficult to separate his results from mysteries about structure and process of the so called chemical-thermal-cellular systems.

Chapter 24

Ruling-class's creativity

The ruling-class and their limited creative intent

The key properties of our society which need to be changed in order to over-come the domination by a failed system of both knowledge and, more to the point, a failed society, ie a failed system of social organization and social-manipulation, where this society is well on its way to the destruction of all of earth by poisoning (petroleum-coal pollutants and nuclear and DNA pollutants) and an appalling disregard for the earth and all of its life.

The key social properties are the three categories of social activity and discourse which shape and determine western society so that it has ever-remained a barbarous society:

1. Property-rights and minority rule (the essential law of the Roman-Emperors, where the model of the Roman society was to militarily-conquer other societies so as to steal the riches of these other cultures, and then to do "brick-laying" so as to create engineering structures, which, in turn, formed the basis for establishing colonies and taxing these distant people)
2. The propaganda-education systems which both categorizes the society's educational interests so as to conform to commercial interests, and (news [related to institutional social structures]) focuses on inter-personal domination roles, so as to help sustain the national security state, a state of violent domination of the many by the few, by focusing on local personal roles of domination, eg categories of academic experts and roles of male-domination etc.
3. The supposition that the economic system is about quantitative laws of economic-flow and economic-value, however, the only economic flow, which the ruling-few are

interested-in-measuring, is the flow of "all the worlds money and wealth" flowing into their monopolistic businesses, ie the banks-oilmen-military industries, ie the economic system is a measure of resource-containment and monopolistic domination.

There are two main issues, in regard to the relation between the economy and the justice system and propaganda system and these main issues are (1) expansion (the point of the an empire), and (2) the regimentation of the society into which the economy is expanded, so as to facilitate a narrowly defined (or the relatively small set of categories which define the society's) expansion process (within the empire).

The main issue in US society (or within the propaganda-education system) is concern about an expression of either domination or superiority, where this has meaning in regard to the collective nature of the western society, which exists as a collective society, which supports the few (apparently the 10 dominant banks, ie the central-planning committee which narrowly confines knowledge and creativity), where these 10 banks effectively compose the ruling-class, and the investments of these few banks determine both the class-structure and the judgment of what arbitrary category possesses high-social-value, eg these investments determine the academic categories which exist at public universities.

This is not a community (society) in which there is a set of moral principles through which either individual actions, or the actions between two (or more) parties, are uniformly judged, in regard to whether they are an allowed-set of "right or wrong" actions.

Rather, in the US society, there is "first" a judgment concerning "the superior or dominant social position of one of the parties over the other (eg military superiority)," as to whether the action of the party possessing the superior social position is allowed and supported. Overwhelmingly any actions of the superior party over an inferior party is to be allowed (this is how property-rights and minority rule determine a context for a, so called, moral judgment within the US society, and within US law). It is based on a violent insistence which allows the context of judgment (concerning what possesses arbitrary high-social-value) to be pre-determined.

Thus, the idea of domination and/or superiority in regard to personal inter-actions, where the two parties are supposed to be equal, ie members of the same social class, is the (a) fundamental concern in the society, in regard to judgments concerning right or wrong, or (equivalently) in judgments about dominant social positions, so that there is constant attention paid in the media to personal issues concerning the judgments of superiority and domination in regard to people's two-or-more-party actions in the lower social classes: guns, sex-roles, birth-control, educational-level, sports, categories-of-intellectual-pursuits, race, wealth, possessing material things, business-ownership, etc etc (the activities, which are used to identify social hierarchy, need to identify an illusionary context for all-of society, all of these activities need to be organized in such a way so as to form the illusion of their (it) being an all-inclusive social context, the activities which are chosen need to be both limited and then organized in a way in which opposition is central to the activity, but rather than this propaganda-built illusion of society, in which social activities are both narrowly defined and always competitive, (instead) people can be inventing other categories, and not identifying "the acquisition of money" as a requirement for survival, etc)

In order to realize a different society whose focus is on knowledge and practical creative-works, consider the following simple replacements:

1. From law based on property-rights and materialism; to law based on equality and knowledge and creativity (we are all equal creators) [exceptions can be dealt-with]. That is, the law opposes selfishness, especially, selfishness based on acquisitive material concerns.
2. Instead of a propaganda-education system, use the inter-net and broadcasting (or any other communication system) not as an instrument for narrowing control, and to disseminate failed expert knowledge, but rather use it as an instrument which can easily separate the categories of thought and expression, so that in the different categories (including many new categories) the thoughts can be distinguished in regard to the assumptions contexts interpretations organizations upon which they are based, so that new ideas are easily identified and considered in regard to their relation to practical and non-destructive (and not risky to social survival) creative efforts.
3. Instead of an economy (or set of trade relations) of domination, let each equal creator have equal access to the market for their creations, and protect this equal access. The particularly good ideas related to production can be run as cooperatives and controlled where that control is based on resources and the environment, and they are regulated, so that they do not dominate the inter-relationships of a, truly, free-trade-market.

BUT
As it now is:

1. Law is based on property rights and minority rule, and this requires an amoral application of violence to enforce this social structure, since all the land of the US was all stolen from the native peoples, it is thus, a social structure based on materialism, and subsequently based on lying, stealing, and murdering, and coercive terrorism as well as a counterinsurgency policing action against the public, it is the use of violence in order to protect property-rights and minority rule which defines the "national security state," and within the propaganda system there are the most profound expressions (in the society)

concerning claims about inequality, and a false claim to superior knowledge, and the expression of the Calvinist vision (of God favors the rich) of superior morality is a "just" (superior moral) reason, in order to gain property (ie religion means nothing in such a context, since this idea is relating the non-material world [where social relations in the material world are formed by lying, stealing, and murdering] exactly to the material world)

(though it has been claimed that, in the old testament Bible, the Jews were given property-rights by God, but these property-rights were not granted in the middle of the Egyptian civilization, rather they were granted in a place where the Jews could: lie, steal, and murder so as to "acquire" the land, ie to acquire the land by military-force. Remember God ordered Abraham to murder his son, and Abraham was following these orders, but murder is against the ten-commandments, so either one or the other is amoral, so the Jews God and His ten-commandments are all arbitrary, and the "God of the Jews" is all about identifying "the Jews" as the superior people within the culture (or the literature) of the Jews) [note: where the ten-commandments are, supposedly, intrinsic to God, so they are intrinsic to both God and Abraham])

2. The propaganda-education system, and a fixed set of arbitrary dogmas related to commercial categories, eg physics departments of public universities are primarily bomb engineering departments, as well as a propaganda system which both defines arbitrary categories of social-value, and provides an arbitrary presentation of a sexist male-dominated set of social roles, to be applied to society, etc, etc, ie all aspects of culture are focused on the domination of the many by a few, so for a small set of fixed narrowly defined categories, there is a competition for domination, eg intellectual categories are primarily associated to business or commercial interests, and violence is integrated into the national security state where violence is directed to protect the few in the ruling-class, where all of these categories (all of which) are used to identify a, so called, top-performer in that category, so these categories identify inequality within the society.

3. A false economic model of the so called free-markets, supposedly, wherein products are supposed to exist within a supply and demand mode of barter (or moneyed economy), whereas the western economy is really a command-market driven by fixed traditions and ways-of-living, or life-styles, built, or maintained, by a "propaganda system form of marketing," whereas politics is all about the business community selecting personal whom fit into (as the ruling class wants them to fit) this propaganda system, which commands markets, where one of the markets, also happens to be, political elections, ie most usually

the politicians are selected by the business community so as to fit into their propaganda system.

The, so called, free-market system is not a quantitative structure (based on measuring amounts of money, or debt), for freedom and creativity, rather it is a quantitative structure used for: (1) social domination, (2) control, (3) social exclusion, and (4) capture of the world's wealth.

The quantities of money are measured to determine where the money flows so that the monopolies want some large percentage [90%-95%-99% ?] of that flow to be within their market domains.

The, so called, free-market economy is not a quantitative structure which can be used to develop new knowledge and creativity as well as new creative contexts, rather the, so called, quantitative economic structure is used for: 1. Domination, 2. Control, 3. Exclusion, and 4. The capture of the world's flow of wealth into the economy that the (very) few-investors control.

For example, the, so called, high-intellectual value, which is expressed in the propaganda-education system is an expression of the limitations on knowledge and creativity which are imposed on all of society by the banker-investments.

That is, the, so called, high-level educational institutions, such as MIT, are divided into intellectual categories associated to the technical needs of (or technical dogmas associated to) commercial-investments, and these high-level educational institutions are not about free-inquiry and challenging these narrow limited dogmas, rather they are organized as to be contests in which the winners are expected to extend these dogmas.

Violence and barbarity

Despite the basic barbarity of this social system, the public is fairly docile.

This, apparently, is caused by their fear, by the public, of the justice system.

On the other hand, there is an attentiveness by the justice system (or spy system) to the edges of the violently-dominant types of people within society (controlled, in part, by the language of category, ie the limited set of categories associated to of personal domination expressed through the propaganda-education system), where the combination of random societal violence . . . , (which is certainly (or "one would expect" to be) a main part of the US social structure, since violence is the basis by which the state is upheld) . . . , and the controlled violence of the national security state's terrorist activities (ie activities of the spy state) which are carried-out against the public, and which are an intrinsic part of the control-by-violence justice system (or national security state system).

Note: Categories of social activity can be used to identify "types of (eg game-like) behavior within these categories," wherein sets of statistical samples which measure average behaviors can, in turn, be used to define normal curves, and thus, there is also identified marginal behaviors.

The role of the media in this violently based model of justice, is that the media regularly reports local violence to the public, so that the public will demand ever more police protection . . . , (so as to protect their (the public's) small [petty] set of owned properties, wherein they usually do not even possess the mining rights of the land they might own), the protection of the public is not really the purpose of the local police, and putting many-more police into the community, when the policing is getting ever more militarized, means that the police can form into a bigger local army to use in the war against the lower-classes, . . . , but there is also the (deep-state) terrorist violence (which has often been an important part of the European colonies), which, mostly, goes unchecked (or [in this spy-based-society] goes unstopped until after the fact, and even then not prosecuted [even-though the law has been broken]), eg the shooting-up of local political-headquarters (by forever unknown assailants), but most often these terrorizing acts go unsolved (even though we live in a totally-surveilled society), which identifies a justice system involved in terrorism, just as, in the "old" south, eg the KKK was likely composed of the people within the justice system (of those communities), and many of the prominent people in the enforcement-arm of the justice system, such as J E Hoover, who was a racist, knew well the police procedures which can be used to terrorize people with violence, (eg the violent disruption of the 2000 Florida presidential election re-count, which is not dealt with by the justice system in an effective manner, ie the election re-count results were violently disrupted, yet the justice system did not stop this violent disruption), or the police il-legally suppress public-descent, so as to terrorize people from entering (or participating) within the political system, etc, etc.

The narrowly defined collective society of the US society, which so dutifully supports the few, as an amoral extremely violent justice system requires the public to act to support the few, who compose the ruling-class; that is, the ruling-few who possess the set of controlling-shares of all of the monopolistic corporations, as well as having the total support and help of the society's justice system.

Failures in academia

However, it is the narrowness of this social system which is leading to substantial external failure (eg expert knowledge is practically useless) and its own internal failure, ie economic stagnation.

People can clamor (within the propaganda-system) to remain a collective which supports ever narrower pillars of social power, But it is not enough to be given, so called, "equal opportunity," to be a part of an education system which only teaches the narrow dogmas used by commercial interests.

Nor do the, so called, attempts within the media (alternative and main-stream) . . . , to realize equality for the people considered to be marginal (or of low value) . . . , have any far reaching affect.

Unfortunately, the education system only teaches the narrow dogmas used by commercial interests.

This is, exactly, what the MIT's and Princeton's and CIT's and UCa's (or the entire educational system) . . . , ie and most particularly the highest-level educational institutions . . . , all provide and support (the narrowly categorized commercial interests).

"In these highest educational institutions," the idea (or the institutional effort) is to "extend the (too narrow) dogmas," (where it is within a dogma that one is identified as an expert) and where these dogmas are provided to these educational institutions by the narrow set of categories of commercial interests, so as to only provide an education which is related to corporate interests and supporting corporate interests.

[Yet, both alternative media and corporate media do not identify this.

In fact, corporate media is more likely to identify this issue, of delusional academic endeavors (as L Stahl once did on 60-minutes, concerning useless math research), than is alternative media. * However, it is the duty of the right-press to marginalize the, so called, intellectual-left.

This is because of the main message of alternative "intellectual" media is that the academic authorities are opposed to the militaristic behaviors of the governing institutions.

But this is an empty (or meaningless) concept, since the academic authorities are "all about" supporting the categories of intellectual endeavor which support the narrow interests of the investor class. The category of thought most often mis-represented is economics. The quantitative under-pinnings of economics are not well defined (and it is not clear that they can be defined), as, also, is the nature of (the, so called, precise representation of), so called, economic "knowledge." For example, arbitrary economic contexts are claimed to be determined by quantitative cause and effect relations, but these claims are seldom, if ever, true.]

Or

There can be an education system which seeks knowledge and knowledge's, subsequent, relation to creativity, and to new creative contexts which exist outside of the corporate dogmas.

That is, physics is not (necessarily) about bomb engineering, yet the fundamentals of bomb engineering, ie particle-collisions, is the main focus of physic's commercial-military determined category of academic interest within the current education system, and math is not at its peek, in regard to both accuracy and reliability, when it primarily is considering:

1. the quantitatively inconsistent non-linear and non-commutative sets of operators (or a partial differential equation model for a system which is non-commutative), and
2. logically inconsistent indefinably random set of considerations, and
3. in regard to all of these mathematical un-patterns . . . , and in such a non-quantitatively-measurable context, which possesses no stable patterns, . . . , one (in a university math department) does not have to consider math-patterns, which actually connect to the world, but rather only need to be related to a word-dogma, which is related to a delusional world.

Although there is always present, in academia, math categories which have commercial interests, due to funding.

Yet, as often as not, (even in these commercial categories) the math considered in math departments is about either non-linearity and feedback systems, or indefinably random systems, where both of these descriptive contexts (especially the indefinably random contexts) have very great limitations in regard to the validity of the information they provide, and/or the reliability of the feedback product.

Rather (in regard to both accuracy and reliability)

Instead one should consider: linear, metric-invariant, and continuously commutative (almost) everywhere descriptive context for the partial differential equation models used to identify the measurable properties of physical systems (in both the geometric and the random descriptive contexts).

But this is too limited of a context for partial differential equations.

Partial differential equation models do identify a non-linear, and, apparently, random, context, which is associated to the measurable properties of physical systems, which are modeled in regard to partial differential equation models, the domain space of the so called system-containing set, is not modeled correctly for the, so called, solution functions of partial differential equation models (but where such solution functions are never found [or only rarely found in the limited number of linear solvable contexts]).

That new avenues for creative work can be uncovered through the process of equal free-inquiry is easily identified in a few paragraphs which follow, wherein are identified a set of math patterns (based on stable geometric patterns which exist in a many-dimensional context), which better relate the observed properties of both physical systems and living systems to a useful math construct, a new math construct which is a map (a guide) into a whole new range of creative ideas.

Instead

The descriptive construct needs to also transcend the context of the "partial differential equation" as providing the main basis for determining measurable properties for systems existing within containment-sets (it also must transcend the idea of materialism). The models associated to partial differential equations can be useful in regard to providing a perturbing, ie lesser, set of influences, which are somewhat independent of the primary stable geometric and spectral properties (which most fundamentally determine the properties of existence).

Thus, one needs to consider new set-containment math (quantitative pattern) contexts which are consistent with the math properties of these patterns being linear, metric-invariant, and continuously commutative (almost) everywhere as defining the correct descriptive context for describing math patterns, ie this is a stable and controllable context, but it is a more expansive context than the context of partial differential equation formulations (for physical systems), where in the context of physical description there is the included assumptions of both materialism and, subsequently, too many limitations on the dimension of the descriptive containment set.

The natural category for this new containment context is (the Thurston-Perlman context of) geometrization, with its focus on the discrete isometry shapes, in particular, with a focus on both discrete Euclidean shapes and the (very stable) discrete hyperbolic shapes, [This context can be easily related to complex-coordinates and the unitary context.] and the idea of geometrization is naturally placed in a many-dimensional containment space.

This places the description in a new math context of a descriptive language needing to be both (1) measurably reliable and such that (2) there exist stable patterns, where the stable patterns are used both as a part of the descriptive construct and which are the type of (stable) patterns which one is trying to describe.

Furthermore, though this many-dimensional structure transcends the idea of materialism, it also accounts for all the observed properties of the material world, including the stable properties of the wide set of the most fundamental systems such as: nuclei, atoms, molecules, crystals, living-systems, the stable solar-system, etc, as a proper subset of this new context of containment. This is something which the context of modeling physical systems by partial differential equations cannot do.

Education,

Education must be about each person being an equal creator and all ideas getting challenged and most ideas being about building new precise languages based on new assumption interpretations, new contexts, and new organizations of ideas

We must live as equal creators not as a barbarous crowd, which only seeks each individual's petty material-wants, and petty roles of domination, to be realized by violence.

We are not the obsessive group of people who can be manipulated by a desire for domination

Rather

We want to know, and we want to create, and we can be selfless in these actions, if the common welfare of the cultural community of knowledge and creativity is taken care of and if this is done in a manner which is in harmony with the earth and its resources the resources which nourish all of the lives of the earth.

The economy does not possess a cause and effect quantitative model, there exist players, in a game-theory model of economics, who are such big players that their money (their control over property) can be used to dominate and control all of the, so called, free-markets, so the quantitative structure is all about determining and maintaining domination, in particular the domination, in regard to how and where money flows, where this control is realized through the propaganda-education system ie through the control of language and thought, and by terror, and a subsequent lack of confidence.

Here education is all about identifying commercial categories, and then associating to these categories an intellectual dogma or high-social-value, and a subsequent intellectual competition, which appears to identify a truly superior-set of intellectually-superior people. But, in fact, these dogmas identify a failed intellectual structure which is too complicated, and full of incomprehensible details, that this intellectual structure is both failing and confusing, but which nonetheless commands a sense, in the public, of high-value (due to the propaganda-education system), but which when followed, it is a language structure only leads further into irrelevant constructs, so as to increase the complication and confusion, and this, so called, construct of knowledge's incomprehensibility.

This is an "authoritative-dogma-and-investment" inter-relationship, which expresses social domination, and thus, there is also the behavior of a few people who want to be intellectually-dominant (violence and/or intellectual-domination within a too narrowly defined set of educational categories).

The key point . . . , where violence is used to determine and maintain a hierarchical system (society) of domination and social-class . . . , is the justice system, where law is based on property rights

Whereas

In fact, in the US law is, by historic fact, supposed to be based on equality, ie the Declaration of Independence, and a government is to be "by and for the people," and where the justice system is supposed to enforce social equality, in the context of equal creators and equal free-inquiry, (the point of freedom of belief, and freedom of expression, as expressed in the first amendment) while the material side of life is supposed to be about the government ensuring the common welfare of the public (Preamble of the Constitution).

But those who have possessed power within the organizational context of the constitution, ie the so called separation of powers, never enforced the Bill of Rights, so the constitution has nullified itself due to its own incompetence.

That is, instead of a collective based on materialism and mindless accumulation, in turn, based on monopolistic control, and subsequently defining a collective society which supports the ruling-class of property owners (or the ruling-class, who own the controlling share of corporations, eg the investment network which is controlled by the 10-big-banks) the society is supposed to be a set of individual equal creators, which live in harmony with the resources and with the ecosystems . . . , where it is the natural systems which have allowed the nourishing of life on earth, . . . , where the individuals contribute in a selfless manner to the collective of individual creators, where the development of ideas (which are relatable to creative efforts) can now easily be organized around the communication channels of the inter-net and where "word searches" which can be designed to distinguish between ideas which have a basis in a particular set of assumptions

Thus, there can be categories of new ideas and categories of fixed dogmas, which are to be easily distinguishable.

In particular in regard to measurable and stable descriptive patterns which can be related to practical building the assumption-basis of the descriptive language is key to distinguishing in a fairly elementary manner, the differences in assumptions, and thus the differences in creative contexts of any particular practically creative descriptive language.

The normal curve and defining the margins and subsequently, using and manipulating the margins of a too-narrowly-categorized society

The use of the propaganda-education system to identify and establish fixed categories and institutional social structures so as to define a normal curve for social behavior and then to use the extreme ends of this normal curve so as to identify obsessive behaviors which are used and manipulated for social purposes

The main category is dominant behavior which is defined in regard to a deep belief in inequality, where particular but arbitrary categories of behaviors or of intellectual or economic interest are identified and then competitions . . . . within a fixed social structure which maintains the definition of these categories . . . , are defined . . . or high-social-value is defined and where issues of an assumed (or natural) type (or context) of behavior exists is focused-on by the propaganda system, such as the value of property in a community, or an ability to provide so called low prices for a people who are being economically oppressed or gay-rights, abortion, marriage, the social roles involved in relations between the sexes, and undefined categories of freedom, or undefined but seemingly defined categories of social behavior such as women's rights and where freedom is incorrectly put in opposition to free-market economies, and in foreign policy, ie war, freedom vs. oppressive regimes

How much of Tiger Wood's "social-behavior" problems were, actually, caused by spying and resulting interventions of agents?

Spying: How often do spy-agents interfere with the lives of the public?

Apparently, they are allowed to interfere with anyone until that person proves that they are "valuable enough" within their (the ruling-class's) own limited definition of value, and then if such a valuable person becomes one of the 10-bankers, then such a person will be sure to get protection by the national security state.

Who are these agents attached to the spying and interfering-agents of the justice system?

One can be sure that many of them were members of the KKK, and were recruited in J E Hoover's "law"-enforcement arm of the justice-system.

This is a result of a social hierarchy, which is not based on rational decisions, but rather is based on arbitrary judgments and upheld by violence.

That is, the aristocracy of intellect is as responsible as are the other failed social-institutional structures of this social hierarchy.

Does the "aristocracy of intellect" have the right to put Socrates to death?

One cannot look to the expert scientists and then to proclaim that the public must "follow their expert knowledge," since the experts are mostly about:

(1) failing in their purpose, but nonetheless,
(2) building and maintaining the products and institutions which best support the ruling-class, and which
(3) are causing the wide-spread "poisoning of the earth," and
(4) the destruction and "exploitation of its people's capabilities,"

eg exploiting people's intellectual capabilities where (all) intellects are focused on failed authoritative-dogmas, though the existing hierarchy does seem to promote greater population growth, where this is because of its narrow regimented social-structure, making it easier to organize and, subsequently, sustain populations, but Knowing, exactly, what the nature of the human population is to earth's capacity to nourish the earth's life . . . , ie the life which is on the earth . . . , is an open question, ie knowledge (which is wide ranging (instead of narrow and regimented)) when used in new ways may be able to sustain a wide variety of life at high population levels.

Note: That both the KKK and the "aristocracy of intellect" are similar social structures, which exist and have their function of excluding other different ideas or different groups of people in an arbitrary hierarchical society.

Both of these organizations (clubs, gangs) are arbitrary and irrational.

One, the KKK, deals with the violent and arbitrary enforcement in a "deep state," ie a "hidden" state which is based on spying, and the subsequent actions of agents in regard to personal interference with the public, which, it is claimed, must be required for such a (hidden) society, and its hidden rulers, while the other, "the aristocracy of intellect," demands that a narrow dogma be the only intellectual viewpoint which can be considered by all of the public (within public educational institutions) and this viewpoint is consistent with the intellectual dogmas associated to narrow commercial categories, and arbitrarily excludes valid free-inquiry, even though the authoritative dogmas of the intellectual aristocracy are failing in their purpose of describing the observed properties of the wide ranging set of general physical systems to a sufficient level of precision, eg nuclei, atoms, molecules, life, crystals, the stable solar system, etc.

Since human life is, primarily, about knowledge and creativity, this means that the "aristocracy of intellect" is also excluding life-energy from society in order to serve the commercial interests about which they have been formed.

That is, "to be rational" then one must evaluate the purpose of a descriptive structure, and then one must determine if that purpose is being realized (or being met) by the set of assumptions interpretations etc of one's dogma.

If the purpose is not being met, then new sets of assumptions and new types of free-inquiry need to be considered.

Is this any different from claiming racial, or cultural superiority over others?

It should also be noted that Godel's incompleteness theorem implies that free-inquiry is always valid, and should never be excluded.

Furthermore, one should not blindly compete in a dogmatic context for the petty purpose of trying to claim intellectual superiority over others.

Is this any different from claiming racial, or cultural superiority over others?

Widening the range of the creative context

If someone has something to say (that is new) then they will not be published, and all new ideas come from people who are defined to be "not valuable," ie not a member of the ruling class or not a person who supports the ruling class gain ever more social power.

For example, the so called "sensitive artists, or brilliant artists" are being used to validate the theft of the common culture, ie the copyright laws.

That is, creativity is not for petty personal gain, rather it is to enhance others further creativity.

Creativity is not for personal aggrandizement . . . , through a market controlled by means of a propaganda-education system, a propaganda based structure of (social) domination, and an expression of inequality, and the subsequent theft from culture . . . , rather creativity is about its further relation to other creativity.

Propaganda is both blatantly ridiculous and subtly clever, eg N Chomsky [who most often mis-represents knowledge and creativity, and this is probably since he is a part of the aristocracy of intellect, if one looks at Chomsky's professional contributions one sees an attempt to relate language to DNA, a look for the "gene of language," and there are other quantitative considerations (related to describing the properties of language) about using indefinable randomness, so that his "professional" work is similar to the usual science considerations, of the other scientists at MIT, by trying to use indefinable randomness, a context and method, which cannot describe a stable pattern, at least it cannot "do this" for the vast majority of stable and fundamental quantum systems, which make-up our experience, and if a stable pattern is not describable with a method then the discussion has no content.

eg he implies (suggests) that MIT is developing technical knowledge based on free-inquiry (but this is far from being true),

ie the picture is painted by the propaganda-education system that they (the superior intellects) are on the road to an absolute truth about science, but their claim about the vast set of fundamental stable quantum systems, is that, these quantum systems are too complicated to describe, but these systems are stable, and more than anything else this means that they are linear and solvable,

These, so called, public institutions are "not" closing-in on any form of valid descriptive structure, for either physical or living systems, but instead the, so called, technical knowledge at MIT is based on the narrow dogma which serve business interests, and "what is developed at MIT" is "what business wants developed at MIT,"

ie MIT intellectual efforts are all about competitions to extend the dogmas, and this extension is into categories which of the greatest use by business interests].

Furthermore, he (Chomsky) identifies the basic context of the so called, "conspiracy-theory" ie "the privileged few determine what happens in the US society," but he refuses to consider the unknown contexts of "deep state" actions, [as the "deep state, (or the secretive state)" is defined by P D Scott], in a spy-infiltrated and propaganda-driven society.

Consistently, one hears Chomsky be an apologist for the state,

He believes truth is determined by consensus, ie by the statistics of an arbitrary-event frequency-curve, but if the intellectuals do not judge truth within themselves then their intellectual work is all about manipulating symbols so as to fit into their, so called, high-valued-group, eg about spying he says that people should not be surprised, but wide-ranging spying implies fundamental personal interference of a free citizenry by the hidden-state. Thus, there are many reasons that awareness of this type of continuously prevalent and personally invasive spying by amoral agents, should lead to strong complaints; that the people in control of this operation, and just as importantly, the people carrying-out the spying and the social-interfering, are amoral . . . , as well as being some of the most destructive, violently domineering, and selfishly motivated people in society, ie lying, thieving, murderers,

Where a nod-of-acceptance is sort-of implied in Chomsky's, apparent, apology for state-spying, ie spying is to be expected.

Yet this spying is all about "fixing" the few competitive categories in which the people of the society are allowed to participate, intellectual categories associated to fixed and failed dogmas, not to mention the fundamental, amoral, personal interference of a free citizen, by the spy-state, and it also fails to grasp the relation that such spying has to the development of knowledge, ie a hidden process of excluding ideas which are, thus, never considered by institutions.

A few narrowly defined competitions are set-up within society, and then the investors, or their justice-system agents, cheat in these competitions, so the society is even more narrowly defined than it appears, and yet, it appears that the society is failing since it is far too narrowly defined, while the few categories of creative actions have no bounds defined (or no regulations which are enforced) in regard to disposing their toxic and destructive wastes.

The free-market is ever-more narrowed by the actions of the justice system, and this is a result of the psychologically impaired people who compose the justice system, the public is subjected to ever-more terrorism and selfish destructive actions applied to the public by the justice system (the spirit of the terrorist justice structures, such as (was) the KKK, are alive and well in the US society, and the ruling-class believe that this is needed to maintain wage-slavery, or to maintain "total social domination").

But where free-markets are really controlled-markets, and "the quantitative structure of an economy" is not truly related to cause and effect markets, which are supposed to be determined by supply and demand, and a related idea about price, but rather the quantitative structure is about measuring and implementing monopolistic domination through the market.

The apparent quantitative properties of an economy within a hierarchical society, where the social hierarchy is defined by control of money (or control of products and trade) and control of product and control of language and thought, ie the propaganda-education system, ie market-control where the quantitative properties of the economy, and its "main" players (or stakeholders) are such that "there exist individual players" in this economy, whom can totally change markets, by how they position of their huge reserves of capital (in new ways) within the economy, and

Thus, these players can control the flow of money within the economy, so that the quantitative structure of an economy is really about measuring the amount of domination the few banks have over a wage-slave society, such as (and) the control of markets and the control of language and thought within society, and where wage-slavery is violently imposed on society by the justice system, so that the public is forced to channel its flow of wealth into the narrow monopolistic channels of the market.

The confinement (or narrowing) of intellectual endeavors within society by commercial interests shows (or identifies) how and/or why intellectual dogmas are rigidly held by society . . . , and by the so called intellectual community, ie the public is not capable of being intellectual (only the academic experts are intellectuals) . . . , by confining the range of what ideas are allowed to be expressed so that this has the affect of keeping investment from facing (other) risks, such as the risk of the formation of new creative-contexts, and a, subsequent, need for new investment in laboratory instruments, as well as the opening-up of explicitly-new markets, whereas truly free-markets are difficult for the structure of investment to control, etc, etc.

But truly free markets can only be related to societies in which everyone is considered to be an equal creator, and where equal free-inquiry exists, so that an "aristocracy of intellect" . . . (where such a small highly valued community of intellectuals, who focus on narrow dogmas, is part of the propaganda-education system, which has been put in place by bankers) . . . , does not dominate "who has the right to express (technical or precise) ideas."

That is, the "aristocracy of intellect" is authoritatively domineering, and it is a creation of the bankers, so as to make their investments less risky, so that society and its intellectual foundation will stay within the narrow social-intellectual-creative context into which the bankers have invested, and thus the investment bankers, have guided society (apparently, since around 1400, when the bankers started determining who was pope, in the "church dominated" western-European Holy-Roman-Empire, which was a continuous extension of the Roman-Empire, and where Constantinople fell about 1450).

That is, European culture has been a collective society, where the collective intent is that the public upholds a small ruling-class, apparently, in exchange for brick laying, and roads, and the crafts of gadgets, which markedly changed in their amazing properties with the development of combustible fuel-engines, and knowledge of thermal physics and electromagnetism, electric-engines and circuits, where electromagnetism was principally developed by Faraday and Tesla (which though they each had a primary position in the creative efforts centered around using the properties of electromagnetism, the Tesla's and Faraday's (ie the original creators) are never the ones who "get rich," nonetheless the claim of the patent and copyright lawyers, is that "intellectual property allows, or motivates, greater creativity in society."

This is quite false, and these laws are about stealing the knowledge of the culture, so that only the bankers gain economically from their, so called, investment in creativity, But, it should be clear to most people, that engines based on burning carbon-fuels should have been abandoned by 1900. This idea was expressed even then.

That is, limitations on production should be imposed, based on limited resources, and based on the toxicity of the wastes which result from production.

Another technical change occurred with the development of the atomic bomb, but the development of this technology is even worse than the development of carbon-burning engines. The accumulation of large amounts of this deadly material implies eventually, that this toxicity will result in the demise of mankind by its toxic effects.

Supposedly, another change occurred with the finding of DNA, but though this idea is simple and seemingly useable, but blindly manipulating the chemistry of DNA, without a valid knowledge of this chemistry, can lead to a catastrophe for life on earth, the real fact is; for both nuclear and genetic knowledge these subjects are not understood, yet they are being used in a very irresponsible manner, as is the use of carbon based fuels.

The mis-management of: gadgetry and energy and knowledge; is more than a sufficient basis for ending the oligarchic social structure.

Yet, one of the main issues concerning implementing such a shift in social organization is the social-inertia which can be associated to the "aristocracy of intellect," which harms all aspects of communication, and where this community of "aristocratic-intellects" is controlled by the ruling-class.

The best example of such a relatively new creative context is the development of the micro-chip computer-components, (though it is somewhat minor in its newly defined range of application, ie it is mostly an extension of the electric circuit, this seems to also be the range of application which best suites the interests of the ruling-class), where this new context for creativity, in regard to electrical engineering, was 1st filled with the traditional business model such as Microsoft (micro-computer programming), but this new context also allowed some real "product competition," ie a new product (aimed at a more democratic use of technology by the public) came from Apple, which in turn, resulted in the new type of business manager, whose business model was not about the domination in regard to a limited range of product, where this limited range is caused by the social domination of monopolistic economic forces, but rather

The new-type of manager was about new product and new creative contexts, and an apparent attempt to relate "the computer" directly to interests of the public, ie art and computer-aided publishing (apparently, originally aimed at the public).

However, because all of these aspirations for public creativity were dependent on a managing-person (who tries to do them) having accessible to a-lot-of money, they (these new-product managers) have, subsequently, been integrated into the dominant investment structures, (as well as the capability of "the economic system" to manipulate S Jobs . . . , though Jobs seemed to have more "creative integrity" than most money-managers for the ruling-class, ie the investor-class . . . . , he learned to fit into the violent materialistic domination and thoughtlessness upon which the "management social-class" is based,

Apple came into being because the micro-computer-chip recently came into being, and it "needed" to be commercially developed. This opened the door, but the finance system would soon re-shape the behaviors of its new-type of managers, Nonetheless, but he also continued to expand the access of the public to technical capability, possibly, more than any other modern "titan of industry," where one of the main goals of the money-managers is to hide from the public this public access to technical capacities, ie to limit market-competitions), eg Apple's operating system (which was originally developed by Xerox) was taken over by Microsoft, and computer animation was taken over by G Lucas, Pixar, and Disney (where Disney's role was, originally, its dominant relation to the propaganda (or marketing) system).

However, this individual creative capability, associated to individual computers, was seen by the elite rulers as a way to "gain even greater social control," in particular, when the internet was formed, and this resulted in a new way in which to dominate communication channels, and subsequently, new communication-channels for spying, and ever more control over individual activity within the, so called, public communication channels.

Within the communication channels, the reach of property-rights, ie copyright law (the right to steal from the culture, where this "right to steal" is, essentially, based on arbitrary violence), is expanded so as to try to control the free expression of ideas, whose real creative origin (of most products) is the culture of the society, but all aspects of that culture are to be organized to serve the power of the ruling-class, as this ruling-class leads the earth to its destruction.

There is also the blacklisting of an individual, would result in there being great social limitations for an individual, and blacklisting is about the petty judgment of the ruling-class that such an individual (who they blacklist) is not going to support them, ie not going to support the ruling elite.

That is, the psychopathic domination by the ruling-class over society is the type of behavior which the justice system (as well as the economic system) demands, and this results in ever less creative range for society.

Thus, one can hypothesize that a successful businessman will become a self-centered, selfish monster, since this is the type of behavior which the justice system demands of the ruling-class.

In turn, this results in ever less creative expansion.

Whereas, the, so called, economic model for society provided to society by the ruling-class (a system which is controlled by the ruling-class), demands ever more economic expansion, ie building new things which investment allows. Thus, since commercial interests stay narrow, and, subsequently, ever more destructive, as they focus on fewer resources, which are associated to the narrow and fixed "creative" context upon which the, so called, market (of the ruling class) depends for its social power, eg oil usage.

The understanding of these types of destructive implications, associated to economic and social domination and its relation to narrow development, which is modeled by narrow dogmas, where social dominance and its maintenance are the main functions, especially, in the structure of the justice system, are totally lost on the, so called, "articulate class."

This is because the intellectuals come from the universities, ie the propaganda-education system of intellectual competition based on very narrow dogmas, and so, these indoctrinated people believe the nonsense about the determination of the, so called, "top-intellects," and that truth "needs to be ever more complicated" since "we already know, essentially, what is true," our high-intellects possess an absolute truth, which, according to the propaganda-education system, the public is too stupid to understand.

However, this "authoritative truth" is based-on the hidden relation of intellectual expression, both within universities and in the media, to (its) a very narrow categorical-and-overly-authoritative relation to commercial investment interests, where monopolistic investment interests confine and control the categories, and contexts, of the allowed intellectual dogmas which can be expressed, and to still get paid.

But the justice system wants to exclude any new creative context (which is either not controlled by the ruling-class or not controlled by the intellectual-servants of the propaganda-education system, ie those in the aristocracy of intellect) from consideration by society, or at-least until the ruling-class identifies a legal route for social domination in relation to the new ideas, ie finding a way to steal and control the new ideas, eg copyright laws.

[However, the easiest way to exclude an idea, is to not let the idea be expressed within the media.]

(note: the justice system most manifests at the point at which the social-power of the ruling-class intersects with public institutions, and this point of intersection is mediated by its relation to the justice system (the crux of the national security state) where the social function of the justice system is to uphold the social position of the ruling-class, where this is done by means of the extreme violence and by the arbitrary nature of the justice system (through which the ruling-class is maintained) . . . , ie the relation that the legal process has to upholding the property-rights of the ruling-class . . . , where this process includes the lying, stealing, and murdering which is done by the justice system for the ruling-class, and this amoral legal support by the justice system in regard to upholding the social position of the ruling-class is the main means through which the ruling-class gains control of society.)

One could say that such a social structure is based on an "absolute truth" about materialism, so that the law is to be based upon the morality of Calvinism, which is the religious belief that the rich are favored by God.

[whereas it is generally agreed that the main teaching of religion is: equality, and a belief "not in materialism."]

The relation of . . . "the ideas which are considered, and the ideas which are creatively implemented by society" . . . to the ruling-class's controlling interests, is central to the relation of the ruling-class's to social power.

Any particular narrow intellectual-social condition desired by the ruling-class is imposed on society with violence by the justice system, eg the justice system requires that the public be wage-slaves.

Furthermore, intellectual-property is defined as: "forming a small variation of an idea" which is a part "of the common knowledge of the culture."

It is defined so vaguely, so that the common knowledge, which is a part of the culture, can then be stolen by corporate interests. Where when the idea is within the corporate entity, then the stolen ideas are protected by the justice system, so that personal-corporate advantage is given to the corporation which provides a variation (in an idea from common knowledge) which fits into a commercial context, ie a context into which the ruling-class invests.

However, personal, or corporate, advantage from a variation of an idea (which emerged out-of common knowledge) can only occur in the commercial context of the propaganda-education system.

That is, any intellectual-context which possesses economic-value (and is controlled by the ruling-class) is defined in the context of the propaganda-education system, ie and no investment occurs outside of this context, ie no single person (who is not in the ruling-class) can earn all that much money from a patented or copyrighted idea without depending on the propaganda-system helping the person market the idea (or work).

This destructive relationship between knowledge and creativity which the justice system upholds in order to support the ruling-class represents (or demands) can be rectified by; instead of basing US law on property rights, rather base US law on equality,

The original claim in the Declaration of Independence was that US law was supposed to be based on equality . . . . , where equality is defined in the context of knowledge and creativity.

Thus, "equality" would not be about (what is considered to be an absolute truth, namely) the idea of materialism.

Subsequently, equality to know and create, would allow a truly free-market to exist, instead of a market based on the arbitrary violence imposed on society by the justice system.

In order to protect a truly free-market this society "big (economic) operations" would not be allowed to be dominant.

Thus, these big operations can be changed if they become ever more destructive, in their narrow intent.

This is about individual equality as opposed to the collective nature of our society, where now (2013) the justice system is so intent on "maintaining a collective society," which upholds the ruling-class, ie the social model of the Roman empire.

There must be a change in regard to the process by which the justice system violently upholds the ruling-class within society (a society with an oligarchical social structure) and the implications of this process.

The implications concerning the violent process concerning both the law which is "based on materialism and property rights" and the process of violence by which the law is being enforced where the law focuses on the idea of materialism as being the main focus of human beings but materialism implies a contest for acquiring material which must be associated to violence, and, subsequently, we live in a society in which morality is based on materialism, lying, stealing, and murder, and this is being upheld with arbitrary violence applied by the justice system to (against) those who are not in the ruling-class, rather than justice being based on defending the equal creative efforts of every individual.

but this results in similar behavior by the public and their local domains where they express their limited domination by violence,

Why are we destroying the earth?

Since our collective social hierarchy wants to do this, since it is in the selfish interest of the ruling-class, ie they have invested in this destructive pathway, and the justice system, ie the national security state, protects the interests of the ruling-class, and the media protects the intellectual beliefs upon which the ruling-class depends for their power.

However, human creativity, mostly, exists outside the idea of materialism, it exists in a many-dimensional, stable geometric context, where there is a constructability to this higher-dimensional and geometric context to existence (wherein human-life participates in the creation of a manifestly different existence, ie life can change the properties of existence, itself), where this constructability might be similar . . . , but more encompassing, ie more thoroughly expressed within a wider range of applicability . . . , to Venter's construction of DNA.

Summary

The oligarchy of the bankers-oilmen-military is quite similar to the oligarchy of the Roman-emperors.

A violently installed justice-military system allows the ruling class to lie, steal and murder with impunity.

The education-propaganda system is of paramount importance so as to define both narrow dogmas which the domineering intellectual class follows and then the intellectual class expresses the justice-system's violent viewpoint concerning social domination and social control, so the dogma-justice system effectively controls both how one is required to act (or behave) and what is thought (or said) and what is created within society. The intellectual-class acts within the illusion created by the propaganda-system that they are intellectual superior to others.

The highly controlled economic structure relating knowledge to commercial creativity, thus, leading to the dogmatic commercial-intellectual categories which confine the intellectual range of consideration and

The social behavior of getting ahead in a narrow, unfair social system of domination, ie be a manager.

Economics is all about expanding into ever greater regions and then dominating all human activity everywhere.

The oligarchic society within which the people within the US live is an arbitrary social hierarchy based on violence, lying, stealing, and murder (or terrorism), ie barbarism, which are techniques used to acquire ever more control over material resources. It is the same model of society as the Roman-Empire (as has been the case for all of European history, since Rome, but it is the banker-oilmen-military who now play the role of the emperors the 150 or so main corporations may be effectively controlled by as few as 10 banks and possibly fewer number of individuals.
The justice and military system is a moralistic-intellectual and arbitrary system of expression for the ruling class which allows extermination. The justice system upholds a social hierarchy put into place by violence it is a war against the public and against "inferiority," where being superior is defined by the ruling-class.
All of the social institutions in this social hierarchy are also hierarchical and all the institutions are narrow dogmatic and fixed and of course exclusive

The image of the empire is that it brings high-value, superior knowledge, superior technical constructions, and superior institutional organizations to society but its functioning depends on the public being forced into both being wage-slaves and into being a collective-society which supports the ruling-class, so that all of the people's action within society are forced, by the violence of the justice system, to be directed, so as to support the interests of the ruling-class, ie the society is a collective which supports the few in the ruling-class.

Chapter 25

Preface 2

Knowledge and investment are more deeply connected than the image of a highly-valued intellectual within society suggests, where a deceptive image of "independent intellectual activity" within our society is concocted by the propaganda-education system, where this image is based on a deception about judging-merit. Simply-put, the bankers determine merit not the intellectuals. That is, intellectuals compete in a narrow (highly-valued) context determined by investment interests.

Introduction

San Diego Joint AMS and MAA meeting 2013

These are new ideas about math, which I submitted to the math community, so as to be expressed at one of their meetings. I was accepted by them for a presentation at the San Diego Joint AMS and MAA meeting on 1-12-13, where I presented a talk on these new math ideas, which use geometrization to create a new containment-set for either physics or math.

However, these ideas cannot be expressed in peer-review publications, since they are truly new ideas, and thus, there is no one to peer-review them.

Nonetheless, they are both carefully expressed and valid math ideas, which the math community, to their great credit, has been willing to have expressed at one of their professional-math meetings.

But professional math is not about ideas, but rather it is about funding and working on (very important, at least according to "the investors in the technical projects") traditional technical projects, into which the ruling-class invests, ie narrowly defined traditional technical projects which (occasionally) use the "overly complicated, too abstract, and far too narrow in their viewpoint" dogmatically-authoritative math ideas which are expressed in peer-review publications.

However, since these new ideas were allowed to be expressed at a very-big professional meeting this should be interpreted by the public to mean that these "new ideas" are "officially legitimate" math ideas, and can be considered seriously by anyone. (Though they may be ideas which are not of interest to the ruling-class, since they do not fit into any of their commercial categories, about which technical authority is used to help build, and designed to obscure, so as to be narrowly used).

Just as children are instructed (or led) by the propaganda system to insist on brand-name-products, so too, have adults been led by the propaganda system to insist on only considering ideas which have been "officially sanctioned" by the propaganda system, ie officially sanctioned by the ruling-class.

Note: This is the main-function of the intellectual-left within the propaganda system. They are in the media to claim that truth is only determinable by an expert, and, furthermore, only an expert who is allowed to have a voice within the propaganda system, eg one who is peer-reviewed.

Note about peer-review, those who "question global-warming" have access to "being peer-reviewed," but carefully considered, and carefully expressed ideas by an expert which question the authoritative-dogmas of science, ie questioning the dogmas which serve big-business interests, are not allowed in the peer-review publications.

That is, both children and adults have been taught by the propaganda-education system to not believe their own ability to understand and discern truth, so that the public has come to believe that they are not capable of deciding anything for themselves, whereas the dogmatic intellectual-left believe that only an expert, usually on the intellectual-left, is capable of discerning truth. The intellectual-left are true-believers in the idea of inequality, but for some reason they seem to believe that the high-truths of our culture, actually, come from intellectuals, and not from the bankers, oilmen, and military ruling-few, but in regard to this belief they are very wrong.

The most unfortunate thing about the narrow confinement of knowledge . . . , and its use, . . . , is related to the use of partial-knowledge, which is acquired mostly by observation, . . . , but there is an intent to use the observed patterns within the context in which full knowledge is needed . . . , so that without a vision about how the partial-knowledge is truly connected to a system's structure . . . , eg how molecules within life-forms are related to the living-system as a whole (or within the complete context within which the molecules function in the living-system), or if unstable "particle-states" are related to the properties of material-systems, . . . , this partial-knowledge is used in a destructive manner, and/or it is used to decieve.

For example, the molecular properties of DNA etc, which are related to forming proteins, where proteins function as both building-blocks for tissue and as enzymes for other molecular-system-processes within the life-form, but nonetheless, this limited molecular knowledge of a semi-determined cause-and-effect (essentially, determined by random correlations), is being used to alter the DNA structure within life-forms and then these life-forms are introduced into a large scale ecosystem so as to interfere with the DNA of the entire species, eg GMO (ie genetically modified organisms) corn being produced on a commercial scale where the benefit is "a petty economic benefit," where the limited benefit may quickly "molecularly dissipate" (or be circumvented by the life-forms which are related to the economic benefit, eg chemical toxins in corn (due to DNA manipulation) used to attack "crop destructive" insect "pests" where the destructive-insects develop immunity to the toxins) so as to leave behind the genetic-destruction of a species, as well as the destruction of ecosystems. Not to mention the grave danger to which a life-form is subjected by ingesting a GMO, since the chemical relation of the GMO to the digestive system is, apparently, now (2013) known to be damaging, and, apparently, creating new destructive bacteria within the gut.

Another example, is the use of radioactive elements for both nuclear weapons and nuclear energy, though the particle-collision model works for a system transitioning in a random manner between two relatively stable states, (the partial-knowledge of) the particle-collision model of quantum-material-intreractions is of no-value when trying to model the observed (relatively) stable spectral properties of nuclei (the larger scheme concerning the nature of a material existence), thus isolating radioactive substances and accumulating them in large quantities, has basically been an exercise of creating extremely toxic material accumulations, which no civilization can assure life-on-earth that this accumulation will not eventually exterminate life. That is partial-knowledge put to petty selfish usage is a grave threat to all of life.

This destructive-end resulting from mis-using knowledge, is being represented in the propaganda-system in the context of a superstitious religious viewpoint . . . central to Christian church (as devised by Paul), and instituted on a large social-scale by the Roman-emperor Constantine, as the propaganda structure for empire, . . . concerning "the end-of-days,"

or

In the pseudo-scientific vision of E von Daniken's story of "ancient-astronauts from distant galaxies, who possess superior-intelligence and superior-technology, and who genetically created mankind," where that superior-role in our society is being played by the ruling-class, and their right . . . , due to their moral-intellectual-superiority . . . , to interfere with the balances of the living-systems by using partial-knowledge, for their petty selfish-interests. That is, von Daniken is presenting a vision of oligarchy.

Though the mysteries to which von Daniken refers, are mysteries concerning the ancient world which do require consideration in regard to a wider-range of thought, than is presented in the propaganda-system. They seem to be issues about the true nature of human-life, and the true relation of human-life to the properties of existence.

An alternative society to the "collective-society supporting the petty-rich," is a society of equal-creators with a will to use equal free-inquiry to gain knowledge, so as to form new creative contexts, and then to create in a selfless manner.

The great-value of being human is the ability to know and to create. [note: In fact, human-life may be directly related to the creation of existence itself.]

As opposed to basing social-power and human-value on the material-world, so as to base law on property-rights and minority-rule.

Instead base law on "equality to be a creator."

The point of being equal and free, is to gain knowledge (not necessarily related to the dogmas promoted by the media) and use it to create new things . . . , humans try to gain knowledge and they are creative . . . , rather than,

Being a member of a collective, a wage-slave society (full of high-paid intellectuals) wherein this collective society the public only wants to know-things and to help create-things, which the ruling-class wants "the public to know and create." That is, a collective-society whose, seemingly, only (or main) concern is to uphold the relatively "few numbers" of people who compose the ruling-class, where this ruling-class also determines those categories of intellectual-efforts which the intellectual-left seem to believe (that these categories of intellectual-efforts) exclusively define the (only) important intellectual-efforts (for all considerations) for all of society.

Note: The alternative intellectual-left propaganda-outlets express this idea of an "exclusive, high-valued, narrow, superior intellectual viewpoint" of the intellectual-class, ie but this narrow intellectual viewpoint is really the narrow dogmas which most support the ruling-class.

A collective-society rather than a society of equal-individual-thinkers and equal-creators for whom the government promotes the common welfare of these equal individual creators.

However, as it now is (2013):

The public must "politely ask for jobs," and if they do not "get a job" then the collective-society, effectively, tries to exterminate them, since without money they cannot buy the necessities of life.

Whereas there are no bounds on the behavior of the ruling-class, and their (carefully selected) automaton servants (though the servants may be high-paid).

Science-math and the creative context within society, is as much about how the communications within society are organized, and the limited range of creative efforts to which knowledge is developed and applied (it is [effectively, only] applied to those creative-projects into which the ruling-class invests), as it is about math-science ideas

These are not simply issues about math and physics ideas, but they are about the control of all communication channels in society, so that all thought is directed at supporting the ruling-class.

That is, math and physics ideas are also dependent on social issues, and the exclusion of a person . . . "who is expressing ideas" from being allowed within a rational arena, such an exclusion of ideas . . . , is not based on the validity of the ideas, but rather these judgments (about exclusion) are about the organization of social-power within western society. This is exactly analogous to the church excluding the ideas of Copernicus, but who is now (2013) doing the excluding, is not the pope, but rather it is the ruling-investor-class. Thus, the understanding math and physics is also about understanding how social-power within the western society is organized and functions, and how it uses knowledge, exclusively, to serve the interests of the ruling-class.

Thus, there is a double (quadruple) edged sword about expressing truly new creative technical ideas

(1) is the nature of the intellectual-class (professional wage-slaves, who have competed for high-value within a dogmatic intellectual competitive structure, ie the "true believers" in (or the religious faithful to) the dogma provided to them by the ruling-class), and
(2) the actual structure of power within society (the investor-class) and their investments in technical creativity, and
(3) that "within society" ideas are not about independent equal free-inquiry, but rather they about dogmas which the highest-level social forces create, and thus new ideas are effectively excluded, since the implication of not being peer-reviewed is that the ideas are not valid, the ideas are not valid in regard to the technical interests of the ruling-class, and then there is
(4) the fact that the ideas, which are now be expressed in peer-review publications, are completely failing to describe the observed basic stable properties of the vast majority of the fundamental physical systems which exist, yet a lot of effort is used-up describing the original instants of the (very speculative) and "fundamentally irrelevant" ideas concerning the big-bang, or concerning the Higg's particle, or forming a geometric model of the Higg's particle in string-theory etc . . . , but these ideas are derived from particle-physics, and the laws of particle-physics cannot be applied, so as to accurately describe the stable spectral-properties of a general nuclei . . . , so this line of inquiry (or line-of-thought) is irrelevant . . . . , except in regard to the particle-collision model of nuclear weapons . . . and thus it is a dogma which is associated to the military-industry.

Yet the public is not attuned to rational-thought, but rather they are attuned to institutional-authority and social-domination, the public pays attention to the social-cues denoting high-social-value, ie they are taught to select brands, and to prefer ideas which are identified in the media to possess high-social-value.

Nonetheless, try to be rational and independent

How to describe the stable physical systems?

Should any such "new description concerning physics" be about a new structure for interaction, or new ways in which to deal with sets, so as to also have new types of operations defined on new types of sets, or can it be about the properties of the containment-set?

The claim of these books, is that "the new types of ideas" which can be used to describe the properties of stable spectral-orbital physical systems, which exist at all size scales, are new ideas about the properties of the containment-set.

It is a new physical and math construct which transcends the idea of materialism. Such an idea should be exciting to the religious community, but it is not. Again, it is issues of domination and authority about which the public and their so called religious leaders are most aware, not (religious) curiosity and they are certainly not trained to be curious about the nature of existence so that both stars and planets and life-forms exist. Rather they are told this is all a result of the properties of elementary-particles and the anthropic-principle (where the anthropic-principle is the idea that elementary-particles have the particular masses and the other particular properties which they possess, and that other physical constants have the values which they possess because then life will exist) this is a true hoax, or obfuscation, since the laws of particle-physics cannot be used to describe the properties of any hadronic-system with "more than one" discernable particle, eg the general-nuclei's stable spectral structures have not been described using the laws of particle-physics, yet, particle-physics is basically a theory about hadrons, SU(3) x SU(2), but since they have failed to describe the observed properties of hadronic-systems, instead they are, deceptively, claiming virtual omniscience, "God's job is to select the properties of physical constants and elementary-particle masses ("whatever God might be?" in this context)," (note: properties which cannot be logically related to the observed properties of physical systems by using the, so called, laws of physics) while, on-the-other-hand, it is "the job of the ruling-class to determine both what is known, and what is to be created within society, by means of the society's very carefully controlled knowledge."

(one could say, in a much more realistic manner than the anthropic-principle's, supposedly, rational claim, that this (ie that the rich have attained a higher-place than God) is a case of Puritanical-Calvinist (devil-worship) of the cold-hard cash . . . , apparently, the high-point of the protestant reformation was to make "possessing cold-hard cash" a great virtue, . . . , the cold-hard cash places the very-rich above God, at least in regard to human-efforts within society,

Though (God) tweaking physical constants is not a very "controlled context" for "using knowledge to create," ie it more-or-less implies a trial-and-error approach to creativity, but, in fact, it is not really an established form of knowledge, it (fundamental constants placed in the descriptive context of particle-physics) leads to neither accurate descriptions nor is it related to practical usefulness).

New alternative ideas about mathematical-containment constructs

(These ideas were first expressed before 2004)

A choice about how to construct knowledge based on physical description

Can the descriptions of the stable properties of physical systems only be a result of formulating partial differential equations (based on physical law) which deal-with material interactions (based on geometry) and/or material structures (based on randomness), or

Do stable physical system come into being because they are stable circle-space shapes, which are in resonance with the finite spectral-set which is a part of (or defined by) the properties of the high-dimension containing-space for all of "material" existence (where within the new descriptive context "material" is defined on many different dimensional-levels)?

Unfortunately, this question does not get filtered by a rational process concerning considerations about precise languages, in regard to finding new physics-and-math patterns;

The question "about the validity of an idea" is, effectively, (in western culture) answered by the ruling-class, the bankers, oilmen, and military monopolistic industries, who dominate all aspects (all considerations which are based on society's communication channels) of what is considered to have social-value, and they have shown that what, they believe to possess "value," is arbitrary, and imposed by institutional violence, but it will always be about that which provides greater power to them.

They are concerned about "into which aspects" of their monopolistic businesses "can the various categories of knowledge" be directed and used.

The social structure of the west

In a society which is a collective, whose collective purpose is to support a very small set of ruling-elites, then one must consider the attributes which such an social-organization causes, first the main attributes of these rulers (and thus, also of the people within the society, since this defines the main set of behaviors of the society) they possess the personalities of obsessive near-autistic psychopaths and sociopaths who seek social and personal domination, ie they possess a very externally-directed attention, and also they possess a very material-centered world-view, which is (actually) built from subjective measures of value (it is clearly a system which not directed by rationality), where through these arbitrary attributes of high-social-value, identified in the propaganda-education system, overly authoritative experts are chosen to direct the knowledge, so as to fit into the creative designs of the monopolistic big-business-people.

For example, the high-wages which are paid to the-experts lead the would-be experts into a narrow dogmatic path through which current knowledge is controlled, and narrowly categorized, so that the experts compete for the top spots associated with these categories of high-social-value. Thus, social domination is defined or expressed by the ruling-class and then it is expressed by their authoritative experts as the dogmas of the universities. These experts are people who are also attracted to socially-dominant positions within society. The bomb industry along with both the propaganda-system and the managers of big-institutions was able to turned university physics departments into nuclear-bomb-engineering departments.

And this is how knowledge gets its structure in our western society, and it is this highly controlled vision of knowledge which the public experiences in the education system.

The methods and actions which have led to the high-social positions of the ruling-few are: violence, stealing, and deception.

Within such a collective society, which is based on barbarity, and opposed to both variety and thought (or rational-ness) one finds a very fixed society, and the social-state which is needed to acquire and maintain one's social power is a social-state which is based on the material world and its relation to monopolistic business interests, so that such a material-based viewpoint stays fixed, and is narrowly categorized (based on monopolistic business interests), wherein such a material-viewpoint the public must possess a limited viewpoint about life-styles and material resources, concerning both the type and their use (the importance with which they are used, or needed, to maintain social-power).

A fixed society filters-out and uses only certain types of technologies associated to violence and control ie military and communications technology, which, in turn, requires that the ruling-intellectual-class . . . , also based on (or led-by) obsessively domineering personalities (which supply the technical, as well as the propaganda, support for the society's Roman-empire-like social structure) . . . , also be very opposed to new ideas, where new ideas would make the dominant-few vulnerable to having others questioning their intellectually-superior social-positions, since their authoritative dogmatism, which focuses on a narrow set of both instruments and intellectual categories, which are associated to a narrow set of products, ie in a scientific-thought framework, which depends on certain narrow ideologies, ie the idea of materialism, and narrow set of categories associated with the commercial interests of the ruling-class.

The population, especially the manager-class, is subsequently also possessed with character-traits similar to those of the ruling-class: obsessive, violent, coercive, mean, manipulative, and a person who possess a very narrow focus, and who has no inclination for rational-thought, particularly the intellectual-class (who represent authoritative-dogmatists, as the church represented authoritative dogmas in the age of Copernicus).

The population are all externally-directed, and very focused on the material-world, and on social-domination, particularly the, so called, religious leaders, where lyrical-spirituality (or propaganda) is all that is considered (as the relevant subject-matter of religion), and the expert authoritative-dogmas come from the intellectual-class who are the only ones who are allowed to have a voice about social issues under the ruse that everything is too technical, but in reality everything is based on barbarity, and not-based on "valid technical description" nor rationality, nor about an alternative to a world-view which is based on materialism, not simply expressions about a need for community and community interests . . . , {since the point of the western community is for the public to support the ruling-class, and to support the high-value which it the ruling class provides to the public, through the propaganda-system. It is this high-value (of the ruling-class) which the liberal-left so self-righteously upholds.}, . . . , but more importantly (a scientific-and-religious interest in new knowledge) for individual people to independently support the seeking of new-knowledge, and the relation of knowledge to new contexts associated to "practical" creativity.

Reconsider physical descriptions

Precise descriptions, new contexts for (practical) creativity, accurate knowledge, and practically useful knowledge

Precise descriptions must be consistent with the observed properties of the physical world, and be able to describe in a sufficiently precise manner the observed properties of the physical world, ie where the physical world is the world we experience.

When a math pattern of quantity and shape is sufficiently confined, then any of the inter-related properties (ie functions) and/or coordinate-properties (which compose the descriptive context) can be found, but such system-information is usually formulated in the context of initial conditions on the coordinate space, ie the control one possesses in the laboratory coordinate-frame.

If a precise descriptive context is confined by properties and context then other properties must also exist, is the basic model of math deduction.

The introduction of physical law (into physical description) causes the physical descriptions to also be considered to be deductive, though physical-law is supposedly developed (or found) by induction.

However, there are many math properties (qualities) whose math context is (or can be) also a physical context, eg the position of an object's center-of-mass in a coordinate system.

Is there really a difference between physical description and math description?

Apparently both languages (math and physics) can languish (or be trapped) within a context of absurdity, as string-theory has shown, where there is great professional concern about vague speculations, for a professional community which only considers a narrow range of (peer-reviewed) ideas.

The context of deduction is to:

Formulate and/or construct various sets of quantities, identify operations which inter-relate the different sets of quantities then identify the confining construct, eg sets of equations, to solve.

Can a complete description of a system's properties exist (or be contained) in a (purely) quantitative context?

If yes, then things (properties of systems) can be measured and placed into a "correct" context (or put in the correct place) for a particular use.

In this scenario, one needs both reliable measuring and stable patterns used for the descriptions.

But are the math professionals providing such a precise descriptive context?

Apparently, not!

For one thing, the domain spaces . . . , ie the system containing-sets . . . , which are being used now (2013) are incorrect.

Review of modern physics and math

The rules which have been identified, in dogmatic (peer-reviewed) physics, are about the operators, which either determine material-interactions, or which are used to try to determine material structural-averages about a system's energy properties.

The material-interaction descriptions, based on geometry, are essentially correct,
ie either 2nd-order force-fields associated to 2nd-order relations of inertia, or the "shape of space" determining inertia (are essentially, correct).

However, small material-components become discontinuously discrete (the vacuum-point), and most classical systems are non-linear, and thus, in both cases (small or non-linear) their geometric inter-relationships are difficult (actually impossible) to discern,
and there are no valid models of the "shape of space" determining inertial properties, especially, since the geometric version of material-interaction provided by general relativity is expressed almost exclusively as a non-linear relationship, and thus, it is quantitatively inconsistent and chaotic (and thus not relatable to describing the stable properties of the solar-system).

As for determining material structural-averages about a small material-system's energy properties . . . , though there is good reason to believe in fundamental randomness for small material-components, and an associated function-space model for a physical system (where the function-space model is based on randomness) . . . . , these probability methods have only been vaguely related to material-components confined by walls of infinite potential-energy, and to the questionable notion of a lowest-energy-point harmonic-oscillator (of an, essentially, classical model which is re-interpreted in quantum physics), otherwise, it seems to be all about questionable mathematical-tricks, and the application of math-patterns which are un-relatable to, actual, descriptions of stable quantum-material-systems (where this failing is due to non-commutative-ness and/or non-linearity).

The many-(but-few)-body quantum system . . . , a type of system which is both plentiful and fundamental to our existence . . . . , whose observed properties are seen to be very stable, go without a valid description, which is based on the, so called, laws of quantum physics.

That is, the property of randomness is not the property which is fundamental to understanding the structure of material-systems, though it (ie randomness) is an observable property.

However, the prominence of material-interactions, in regard to the mathematical descriptions of material systems, where math-based descriptions of material-interactions are, assumed to be, needed in order to determine material properties of physical-systems, leads to an incorrect mathematical context for describing the observed set of very stable spectral-orbital properties . . . , which have been observed to exist for physical systems at all size-scales . . . . , of the vast majority of the observed properties of physical systems.

Though material-interactions are (seemingly so) important for relatively "free" condensed-material systems, these material-interactions have only minor affects on the very stable properties, which are observed for the very-many . . . , and most fundamental . . . , (stable) material-systems.

A new idea

The more fundamental attribute of physical description . . . , in regard to understanding the structure and properties of stable physical systems . . . , is the containing-space itself.

Materialism is the wrong model of existence, and material-interactions . . . , either based on force-fields or based on the "shape of space," . . . are only contributing relatively-small perturbing properties, which only slightly-adjust the more fundamental properties of existence. Whereas, the model of randomness may-work if the functions in the function-space (of a system which is modeled as being random) are, actually, the set of the stable circle-space shapes, based on a finite spectral set, so as to provide a new model for the harmonic-functions.

The more fundamental properties of existence are the stable-shapes, ie stable-shapes which are in resonance with the containing space's (finite) spectral-set, and which are defined in the various dimensional-levels, and on various subspaces, which compose the high-dimensional containing-space of existence.

Defining containment-sets as coordinates which are "metric-spaces," in turn, leads to considering the stable shapes of the discrete isometry subgroups, especially, for the non-positive constant curvature metric-spaces, and, in turn, these stable shapes can be used to organize (or they can be placed in) a many-dimensional context . . . and if they are placed into a context where metric-spaces have properties associated to themselves, and as a result, are also associated to pairs of opposite metric-space states (leading to complex-coordinates, and to the stable shapes of the discrete unitary subgroups) . . . which identifies the math context where the stability of material-systems can be both described and understood in the context of spectral-resonances.

Placing a description of a system within a quantitative containment space, where this containment space is a set of different quantitative types, ie a set of independent (or pair-wise orthogonal) coordinates, which, in turn, can place the system's properties, where such (quantitative) properties are functions defined on the system-containing, independent, coordinates, into a valid quantitative relation (or context), so that the descriptions (or rules based on both the containment construct and the processes by which the system's properties are found) are: accurate, wide-ranging, measuring is reliable, and the description is based on stable patterns, and it is a practically useful description (ie the rules are the math objects which relate the function-values to the containment set, or there are "the defined processes" [ie the rules] which relate a function to its system-containing domain space, where the domain space also has measurable properties, such as metric-functions).

Placing a system, ie a component which is related to other components, into a measurable containment set, whose measuring properties are associated to the measurable properties of the system, may necessitate both (1) certain properties of the containment-set which relate to the identification of the system, and that (2) stable properties of the system are only relatable to certain types of stable shapes.

For example, set-containment coordinates and a metric-function defined on those coordinates . . . (at this level can define a descriptive context where measuring is reliable) . . . . imply that the only stable functions (which have any amount of variety) are the stable shapes of the isometry fiber-group, which when these shapes are related to partial differential equations and differential-forms, have the math properties of being linear, metric-invariant, and continuously commutative (almost) everywhere, ie sufficiently defined to be solvable, where the stable shapes of the isometry fiber-group are the discrete isometry subgroups, which, in turn, are defined by coordinate-space "lattices" (or checkerboards) composed of "rectangular-related-shapes."

Hypothesis: The system properties and the containment spaces are the same type of math-construct, but discretely, discontinuously separated from one another by a dimensional-level (or by dimensional-levels), where the system is usually considered to be a lower-dimensional math construct, eg a lower-dimensional metric-space, than is the dimension of the system-containing metric-space.

Furthermore, the stable orbits of (condensed) material (contained within a metric-space) result from the metric-spaces' stable shapes, which interact with the material components, while the stable spectra of a material-component also result from stable metric-space shapes, but the material-components are at least one-dimension less than the dimension of the metric-space within which the material-components are contained.

The rules for the containment-space and for "material-interaction processes"

1. there are many-dimensional levels, and each of these different dimensional levels (and each subspace for each dimensional-level) are all related to stable shapes, which determine the stable (geometric) structures of the observed spectral-orbital properties which are observed for physical systems at all size-scales.
2. These stable geometries are realized by their being "in resonance" with a finite spectral-set, which is defined for the over-all, 11-dimensional, hyperbolic metric-space, containment-set. This finite quantitative spectral-set, the set upon which the quantitative constructs of existence depend (in this model of existence), means that the quantitative construct does

not depend on "there being defined a continuum" for the models of the quantitative-sets which are used in the precise descriptions (in this new model).

3. Both condensed material, and material components, interact with one another in a local discrete (discontinuous) context (where the discontinuities are determined by very short time-intervals) of stable-shape related geometric properties. However, the context of "pairs of opposite metric-space states" define metric-space states which, in turn, define the continuity of motion which is related to these newly constructed processes associated to material-interactions, where the continuity is identified in either one of these (pair of) metric-space states.

Thus, {there can be a continuously conserved property (or quantity)} in the . . . (ie in low-dimension [up-to hyperbolic-dimension-5]) . . . stable metric-space shapes . . . , which possess the topological property of being closed and bounded . . . , where the topology (within such a metric-space shape) is an open-closed topology, ie so within the metric-space which possesses these stable shapes these stable shapes can be closed-bounded shapes.

It should be also noted that the stable shapes of the (new) interaction-construct result in no singularities being defined by the material-interactions (this is an improvement over the 1/r singularities at, r=0, of the potential energy terms a (non-linear) spherical shape).

Furthermore:

Both material systems and material-containing metric-spaces exist as stable (circle-space) shapes because they are in resonance with a spectral subset of a large but finite spectral-set, where the large spectral-set exists within an 11-dimensional hyperbolic metric-space, which, in turn, contains all of existence.

The size and shape of the metric-spaces . . . , (where material systems are really all stable metric-space shapes which are lower-dimension than the metric-spaces within which they are contained) . . . , is what most determines material organization and material motions. That is, the shape of a material-containing metric-space can interact with, and change the motion of, a material component which is contained in the metric-space ie this is an example of a linear-shape determining material motions (as hypothesized by general relativity, but in general relativity the descriptive context is usually non-linear).

The incorrect model

The system properties of material systems, in particular, the many-(but-few)-body systems are not related to solution functions to (partial) differential equations in a containment context defined by materialism modeled either as inertia or charge, or in a thermal context, or (as in quantum physics) by assuming that all material systems reduce to random behaving point-particle-events in space and time, where energy operators (or sets of operator), which are related to partial differential equations, which, in turn, are to be (formulated and) solved, and which are supposed to determine a system's energy structure, where the random quantum system is modeled as a function space, and the set of operators act on the function space.

What descriptions can be quantitatively consistent?

The range of the set of valid quantitative descriptions may define a very limited context in regard to:

1. set-containment,
2. stable (math) patterns, and
3. measurable reliability.

That is, a containment-set within which stable patterns can be contained, so that measuring the stable pattern within the containment set's coordinates remains reliable, and practically useful, ie so that quantities can be defined and remain consistent, so as to be used to describe a stable math pattern within a set of measuring properties, seems to be quite limited, (in regard to the containment-set and the stable patterns which can exist within such a containment set).

One begins with stable (geometric) patterns of energy (in the form of stable geometric shapes) which can possess measurable properties, which can be measured in a reliable (continuously-stable) context, with stable quantities, where these quantities can be modeled in regard to a uniform unit and stable linear changes in scale (eg fractions) as (on) lines and circles, in a quantitatively consistent manner, ie the real number-line and the complex-number plane (though it may be best to "not" provide these quantitative sets with the property of being a continuum, since a continuum may be too-big of a set to be logically consistent).

The context is that "measured properties," in containment coordinates, are either spatial-displacements or energies (where m=E).

or

Can the (a) system's pattern be contained in a thermal context of set-containment coordinate measures? [but, in this thermal-context, which relies on the existence of differential-forms, how is a metric-function defined?]

Can sets of measured properties define a valid context of (a measurable pattern's) set-containment and measurable description?

That is, "What stable patterns can be both defined, so as to be quantitatively consistent, and be contained in a coordinate space?"

The properties of coordinates require that metric-functions be defined on the coordinates, so that geometric-measuring can be done in a quantitatively consistent manner. For a description to be quantitatively consistent, one requires that (partial) differential equations, which are assumed to model physical systems, need to be linear. These two attributes (ie invariant-metric-functions and linearity) require that the coordinates be in a shape which is continuously commutative. This descriptive context is solvable and controllable, ie it is a stable math pattern.

This relates the description to a fiber group of the Lie-type SO, and the need to define pairs of opposite metric-space states, in turn, relates to a fiber group of the Lie-type SU. These Lie groups have related to themselves the discrete isometry and unitary subgroups, which, in turn, define the stable shapes. Furthermore, the metric-spaces are of non-positive constant curvature, where the metric-functions have constant coefficients, so that the stable shapes are the circle-space shapes (and in the real-case) these shapes are both the discrete Euclidean shapes and the discrete hyperbolic shapes.

That is, the stable math patterns are the very simple circle-space geometric shapes, where these stable shapes can also define metric-spaces. These stable shapes can define metric-space or material-components, where such a shape is either one or the other (either metric-space or material-component) depending on the dimension and the set-containment context.

More generally, but still in regard to stable shapes

Stable geometries, which are quantitatively consistent, can be built from line-segments, and circles, and lines, eg both bounded and unbounded stable shapes, where the stable bounded shapes are related to either rectangular shapes, or shapes based on rectangular-components attached at vertices (where the rectangular-component shapes are distorted, but distorted so as to identify coordinates [of the distorted-shape] which remain continuously commutative) where these shapes are "moded-out" (ie an equivalence-relation which associates opposite sides of the rectangular-components) so as to form circle-space shapes.

Note: These "moded-out" shapes are related to partial differential equations (defined on differential-forms) which are linear, metric-invariant, and defined on coordinates which are continuously commutative everywhere (except at the point where the vertices of the rectangular shapes meet, in regard to the "moding-out" process), ie they are solvable, controllable and stable geometric shapes.

The two main examples of these shapes are the tori (doughnuts-shaped) which are "discrete Euclidean shapes," and the shapes composed of toral components, which are the "discrete hyperbolic shapes."

The discrete hyperbolic shapes are very stable and rigid shapes, with very stable spectral properties, which, in turn, are defined by the (lower-dimension) facial-structures of the (distorted) rectangular fundamental-domains, where these fundamental-domains are "moded-out." to form the circle-space shapes.

Note: The properly distorted fundamental-domains of the different toral-components (or different rectangular components attached at the vertices) implies a continuously commutative coordinate system for the shape.

The number of toral-components in a discrete hyperbolic shape is related to the different stable energy states for the material system which the shape models.

Holes (or different numbers of toral-components) in a stable metric-space shape are normal, and they are not the exception, in regard to the descriptions of stable patterns in a set-containing coordinate metric-space, where measuring is reliable.

A stable energy-shape (ie a discrete hyperbolic shape) resonates to the finite spectral set of the defining spectral-energies of the containment space.

This finite spectral-set is determined by a partition of an 11-dimensional hyperbolic metric-space, which according to D Coxeter, cannot possess a discrete hyperbolic shape, ie the last discrete hyperbolic shape is 10-dimensional (of a hyperbolic metric-space), and a 10-dimensional discrete hyperbolic shape is an unbounded shape.

Note: The last bounded discrete hyperbolic shape is of hyperbolic dimension-5.

The partition of the 11-dimensional hyperbolic metric-space, is composed of sets of discrete hyperbolic shapes (of specific dimensions), so that the partition set is determined by the set of the largest discrete hyperbolic shapes for each subspace of each dimensional level up to hyperbolic dimension-11.

Note: Both the subspace partition elements of a given dimension (where these partition elements define a hyperbolic metric-space for that subspace) and the same-dimensional facial-elements of higher-dimensional discrete hyperbolic shapes in the partition set, together, determine the full set of the allowable resonance-set (ie the finite spectral set) for a given dimension (of spectra), or of a given dimension discrete hyperbolic shape which can be identified by its stable spectral properties.

The partition identifies (hyperbolic) metric-spaces (which possess shapes) which can contain lower-dimension stable metric-space-shapes (ie discrete hyperbolic shapes) which, in turn, model material-components if:

1. They resonate with the over-all finite spectral set,
2. Are contained in the correct subspace (ie the subspace of its containing metric-space-shape),
3. Are small enough to be contained in the metric-space-shape, (where the metric-space-shape may be bounded or unbounded in size),
4. The material-components become trapped in the metric-space-shape (within which they are contained), (where the material component experiences an open-closed topology in regard to its containing metric-space-shape).

Material interactions

When trapped in a metric-space, which has a shape, the motion of a material component will be determined by it following the coordinate's shapes as well as geodesic properties of the shape, ie this is a linear model of general relativity.

Material components can be discrete hyperbolic shapes, or they can be condensed material, eg crystal material.

The form of a stable material-component, which is a discrete hyperbolic shape, can deform (or change) into another discrete hyperbolic shape based on interactions, as well as collisions, where if the energy of the interaction is in the correct range, and if the discrete hyperbolic shapes which are interacting (eg by collision) can resonate (to the over-all high-dimension containing space's finite spectral set) then it can (or so as to) form a new discrete hyperbolic shape, or for condensed material, the condensed material-system can change its (assumed to be fixed energy,) energy representation between kinetic and potential energies, as time changes, and in this regard there are perturbing material-interaction processes (or their associated operators). (*)

The geometry of the new construct for material interactions

These material interactions are governed by discrete Euclidean shapes (or toral-shapes), which act by action-at-a-distance, where the Euclidean context is related to spatial-positions and spatial-displacements in regard to the distant fixed or rotating stars, and this interaction structure is related to (or organized around) discrete time-intervals.

Both material and material-interactions are defined by stable shapes which possess inertia associated to local linear spatial-displacements in time, as well as being associated to material interacting with the shape of the material containing metric-space, ie this is needed to describe the stable motions of the planets in the solar system.

Where there is also, for condensed material, a gravitational-force-field (which in 3- and 4-dimensions may be confined to a 2-plane) and an electromagnetic-force-field, where force-fields are associated to (linear) metric-invariant, local functional (or force-field) relations to geometric-measures, ie differential-forms, where these relation are 2nd-order (partial) differential equations, which relate material geometry to force-fields by means of the discrete Euclidean shapes and the local linear geometries (vector relations) which these discrete Euclidean shapes determine . . . , and whose matrix-representation-properties of the differential-form (matrix since of order-2) . . . , are related to either the vector geometry of the metric-space as a domain space, or to the adjacent lower-dimension metric-space (so as to define the force-field),

(*) where the material-interaction operator perturbs the condensed material component (which exists within a metric-space) and allows the energy structure to change between the energy-function's kinetic and potential energy terms.

The great reach about which the new ideas can be related in a very strong manner

This new viewpoint "about existence" not only provides accurate descriptions of the properties of physical systems,

that is, it not only opens-up physics,
but also chemistry (understanding catalysts etc),
but also life-science,
but also understanding mind, and
also provides a means for traveling throughout the universe,
so that religion has a (new) meaning, and
That meaning is about creating all-of existence, and
religion can also (come to) be about "what we now think of as mythological adventures of exploration."

so as to gain knowledge, so as to acquire a greater creative capacity, so as to create existence, and these ideas provide the context concerning "what might need to be created" in order to expand the extent of existence.

Appendix 1

The range of the set of valid quantitative descriptions may define a very limited context in regard to:

1. set-containment,
2. stable (math) patterns, and
3. measurable reliability.

That is, a containment-set within which stable patterns can be contained so that measuring the stable pattern within the containment set's coordinates remains reliable and practically useful, ie so that quantities can be defined and remain consistent, so as to be used to describe a stable math pattern within a set of measuring properties, seems to be quite limited, (in regard to the containment-set and the stable patterns which can exist within such a containment set).

Can sets of measured properties define a valid context of (a measurable pattern's) set-containment and measurable description?

That is, "What stable patterns can be both defined, so as to be quantitatively consistent, and be contained in a coordinate space?"

The range of valid quantitative description may define a very limited context of how to organize a description based-on set-containment, stable (math) patterns, and measuring reliability.

If one deals with random (well-defined, and finite-event) set properties for a description of random events

Or

Valid identification of a non-linear (partial) differential equation's critical-points and subsequent limit-cycle structures

The domain space modeled as a continuum is too-big of a set, so that its properties may not be logically consistent,

And

Function spaces define even bigger sets

Appendix 2

The differential operator

That is:

1. the derivative, or
2. connection (used for the non-linear geometry of curvilinear coordinates), and
3. the subsequent (partial) differential equation, and
4. the (almost) inverse integral operator; to name some differential operators, and their inverse operators.

The differential operator is a perturbing operator which is associated to (relatively) free systems of material objects which are defined in a stable geometric context, of an object affected by distant material-geometric properties. So that the structure of the process which defines a perturbing-operator is based on the same type of stable geometric shapes of the circle-space shapes upon which it is hypothesized that the rest of the stable properties of existence depend (or of which they are composed), eg stable metric-spaces of certain dimensions and stable material components contained in metric-spaces, but with an emphasis, in this new construct of a local (where if solvable, then it will be linear) perturbing operator on its relation to the formation of instantaneous, action-at-a-distance, discrete Euclidean shapes which connect the distant geometric properties to the object whose dynamics are being perturbed so as to create the local tangent properties of the distant geometry associated to the object.

The operator is discontinuous in (small) time intervals as well as between dimensional levels which are also a natural part of the shapes which are a part of the material-interaction process which perturbs the motions of relatively free-objects in metric-spaces, but it is continuous in the sequence of (discontinuous) time-intervals which exist in a (or in any) particular metric-space state, where there exist two opposite "directions of time" in the two opposite metric-space states (where the two time directions are associated to two opposite dynamical flows, of the object whose free-motion is being perturbed) where the two opposite dynamical flows, in turn, define opposite metric-space states. The point of the relation to Euclidean space to the perturbing operator is that the perturbance of the object is (or can be) identified by spatial-displacements, which naturally are identified in Euclidean space, where locally for each discrete time-interval (which are a part of the process which defines the perturbing operator) the spatial-displacements, in the two opposite metric-space states exactly identify local inverse spatial-displacement relations to one another, ie to the two opposite metric-space states, so that this is done within each discrete time-interval, where the discrete time-intervals are defined by the period of the spin-rotation between the pair of opposite metric-space states in a hyperbolic metric-space, but correlated to the opposite metric-space states of Euclidean space by the spatial-displacements of the perturbing action of the perturbing operator.

Thus, an object's relation to distant relatively-stable geometric structures is to be perturbed and this can introduce both an apparent randomness to the descriptions of small material components which are always being perturbed so as to create for these small material components a Brownian motion and a deep relation to an apparent fundamental randomness, furthermore, many of these perturbing operators are non-linear, so as to make chaos also a result of the perturbing properties of the perturbing operators, thus for free macroscopic objects the perturbing operators can appear to be the dominating set of properties associated to material descriptions. However, these perturbing differential operators only seem to have any relevance at all to the two body material system (in regard to either a pair-of planets or the H-atom) and they have no relation to being able to describe the stable properties which are seen for so many fundamental, many-(but-few)-body systems.

Furthermore, the perturbing differential operators, which are used so often to model "physical law," have no relation to the chemical-molecular (apparently collision) interactions, where new molecular systems emerge from an interaction.

????

The spatial-displacement in one direction of time, at the position each distinct time-interval, is the exact inverse of the local spatial-displacement in the opposite direction of time. ??

Newton's laws of inertia center around a summation of local inverse-functions of "opposite" metric-space states, where the function and its inverse can be identified with each point of the domain space, where the inverse function is the inverse value of the local spatial displacement, ie their related opposite spatial-displacements, which exist separately in each of the two different metric-space states, where these "opposite" spatial-displacements possess an exact inverse relation to one-another for each local discrete time-interval.

It is (or Is it?) the local pairs of inverse functions as a structure for spatial-displacements which allows a summation-process for discrete time-intervals of a Newtonian inertial-system to be "in an inverse relation" to the object-perturbing differential-operator, so as to allow the definition of an inverse integral operator, to be defined on a set of discrete local (differentially) perturbing processes.

This pair of dynamic functions defined on each discrete time interval can be defined as being a function which can have associated to itself at each of its domain points both its value and its inverse value. That is, a bifurcation which contains (or which is composed on a function and) its own inverse function.

the spatial-displacements, in the two opposite metric-space states exactly identify inverse spatial-displacement relations to one another, ie to the two opposite metric-space states, so that these pairs of inverse values exactly define each local opposite dynamical-state in the pair of opposite metric-space states, within each discrete time-interval. This can be easily related to local unitary transformations acting (locally) in complex-coordinate systems.

Chapter 26

Math education

Math education

Here is a sequence of elementary math ideas which develop the structure of math methods so as to provide a global picture of math and this is done by means of identifying a sequence of elementary math constructs.

Math is used primarily in relation to measuring.

1. Measuring
2. The system or pattern (of observed measurable properties)
3. Containment of the measured pattern within a set upon which consistent reliable measuring can be defined.
4. Coordinates (models of many quantitative sets which are modeled to measure length in space by means of curves, but one wants the local tangents to the curves to always be independent of one another, ie orthogonal to one another, ie such orthogonal coordinates which are also quantitatively consistent can be based on coordinates built from lines and circles which are orthogonal to one another, this is possible because lines and circles can be made quantitatively consistent with one another, eg the real-line and the complex-number plane)

    4a. Metric-spaces (metric-functions are often symmetric and they may be represented as matrices so there is a local coordinate system, wherein the matrix of the metric-function is diagonal)

    4b. Geometric measurements on the coordinates (local alternating forms defined on the local linear properties of the coordinate curves (or coordinate functions, functions defined on some simple coordinate structure such as rectangular coordinates (so that the individual coordinate functions define a set of curves within the rectangular coordinates), or on orthogonally structured circle-spaces))

5. Geometric shapes
   [Note: The stable geometric shapes are predominantly the discrete hyperbolic shapes.]
6. Functions (identifying the measurable properties of the observed pattern (or system) which is contained in a set of measuring properties, eg contained in a set of coordinates)

   The system upon which the measured properties, eg position, motion, etc, are defined, are contained, so as to be contained in a "domain space," which is usually (always) a space of coordinates, where the formula which defines the function's values is given in the variables defined by the coordinates.

   There is the domain space of variables . . . , (primary measured values, upon which all other measured properties are dependent, by way of formulas in the given variables [formulas] for the functions, which represent the measurable properties of the system) . . . . , and the function values. So for a region defined in the domain space the function evaluated on these coordinate values results in an "image" in the set of function-values, where functions may be either scalars or they may be ordered n-tuples, in a similar way as coordinates can be n-tuples, eg (x,y,z) etc.
7. One wants to relate the properties of functions (often represented as a graph defined over the coordinates) to algebraic constructs (operator patterns which only act on [or apply to] numbers) where the algebraic constructs are defined in relation to the tangent (or linear) properties of the graph, ie a quantitatively consistent linear relation between function-values and coordinate-values, so the coordinates are in this case best represented as a set of orthogonal lines, ie rectangular coordinates, so if all the coordinates "except one" are held fixed, then the function's graph appears as a 1-dimensional curve (in relation to the one variable coordinate) for which a tangent to the function's graph (which is defined for only one variable) at a particular "domain point" can be determined, where the slope of the tangent line can be found based on the elementary idea of the slope of a line in the (x,y)-plane by the relation, "rise over the run," is determined as the one free variable gets closer to "the given point" of the coordinates.

   This can be done for each coordinate, so as to identify a linear map between the function values and the local linear coordinate values, defined around the given (particular) domain point.

   Furthermore, the rectangular coordinates can be related to the curved coordinates by means of coordinate functions, or by means of local linear relations between the rectangular coordinates and the coordinate curves, ie coordinate functions.

   One would suppose this same method of determining a linear relation of functions slopes and local linear variables for coordinates which are circle-spaces.

8. That is, geometry, ie (5), has come to be defined in the language of functions. Namely, coordinate functions of curves which define the (natural) curved contours of measuring on a shape.

   But must the curves be either lines and/or circles?

9. The process of "getting close" needs to be associated to a particular method of getting close, in regard to both "set structure" and the quantitative structure of small numbers associated to being close. Namely, a region in the domain space is related to the function's values by identifying an inverse-image . . . , of a set of function-values in the function's image-space . . . , where the inverse image is in the domain space.

10. That is, local linear (tangent properties to a function's graph) quantitative relationships which exist between "local linear coordinate values" and "local linear function values" allow (cause) the properties of algebra (the laws for operating on quantities) or algebraic patterns to be applied function values and coordinate values. These local linear, or vector, relations then allow geometric measures to be defined on both functions and curved coordinates (simultaneously). This local-ness and closeness allows linear quantitatively consistent patterns of numbers to be associated to geometric measures defined within the system's containment set.

    But, can this local linear structures, which are quantitatively consistent, only be realized for certain types of global shapes?

11. Essentially, the partial differential equations . . . , used to describe: either equivalent types of quantities (or properties) which have different representations (inertial properties associated to force-fields), or identify the local linear quantitative properties of change of the system or of the system's components in the domain space, for physical systems . . . , can only be solved in a linear, metric-invariant (quantitatively consistent) context, when the local algebra is: either everywhere invertible (globally, throughout the coordinate space of the function's domain space the linear algebraic relations are invertible), or everywhere commutative, ie everywhere the linear matrix operators which are defined between function values and coordinate values are both linear and diagonal, or in regard to an ordinary differential equation the solution function is a product of functions where each function-factor is defined for only one-variable and the differential equation is separable in regard to this factored solution function. Such a (separable) factor-function is related to a coordinate structure which is everywhere independent (or everywhere orthogonal).

Spectra

But measuring is about either measuring geometric properties [and/or local properties of motion and change] or it is about spectral values of physical systems (or of spectral values associated to math patterns).

12. For local measures of functions to be consistent with the geometric-measures of the system-containing coordinate-space the local measuring context needs to be a 2nd order partial differential equation defined in the spatial derivative terms, but in turn, the form of the 2nd order metric-invariant partial differential equation depends on the nature of the metric-function, eg the metric-invariant equation for Euclidean space is the Laplace (or Poisson) equation, while the metric-invariant equation for space-time is the wave-equation (or the vector-potential wave-equation for Maxwell's equations, equated to the current vector).
13. Then there are the 2nd order differential equations associated to Newton's law of force, which are time-dependent differential equations broken-up into the different independent directions associated to a geometric force-field.
14. There are the different types of 2nd order partial differential equations

    (1) elliptic (Poisson) type, and they deal with bounded orbits, the
    (2) parabolic type the semi-free material components and the angular momentum of (small) components, and the
    (3) hyperbolic type which deal with physical-waves and component collisions.

15. Then there are the:

    (a) harmonic equations defined by F=-kx,
    (b) the wave-functions, ie solution functions to the physical wave-equations, and
    (c) the parabolic "energy operator" based on relating the structure of wave-functions to differential operators, where these differential operators form into a parabolic type 2nd order partial differential equation.

    Each of these types of partial differential equations defines a set of operators and a set of solution functions, so as to define sets of spectral values, in turn, defined on a function space, ie a bunch of "spectral-functions."

16. This context has led to the algebraic development of operators defined on function spaces, so as to solve the spectra by algebraically "diagonalizing" the function space by applying sets of operators [or by finding a complete set of commuting Hermitian operators to be applied] to the function space.
17. Harmonic functions can be associated to describing random events in space, of random particle-spectral events in space, and the spectra are both related to the observed spectral values and to the harmonic wave-structures which can describe randomness.

    Are the random properties and the observed spectral-values of a system really different types of descriptive properties?

However,

When these sets of operators are associated to a descriptive structure which is based on randomness then in general the spectral set of either spectral values or the spectral-functions cannot be found.

Yet, the physical quantum systems possess sets of very stable and definitive spectral properties, so that these spectral properties seem to be fundamental to physical description and physical measurement.

18. The set of stable spectra associated to material components can be related to the very stable circle-spaces defined in regard to the discrete hyperbolic shapes. The stable spectra are associated to stable shapes and the significant geometric structure is the hole-structure of the circle-spaces.

Education for mathematics

The following 20 questions form a set of fundamental questions about math, which are of significance to beginners and to seasoned professionals. Grappling with these questions brings one into math and, when answered, they deal with how math is . . . , how precise descriptions are . . . , related to creativity.

Current descriptive languages are not working

The math constructs based on both non-linearity and randomness [particularly indefinable randomness, randomness defined by unstable events and incalculable events], do not lead (have not led) to a context of practical creativity, nor to accurate descriptions of stable, physically-observed patterns.

For example, the stable spectral-orbital properties of: general nuclei, general atoms, molecules, molecular shape, crystals (the critical temperature predicted by BCS has been exceeded by high-temperature superconductivity), as well as the stable solar system, are all systems, which are not now being accurately described through the application of physical law. For such a wide range of general, but fundamental, systems which are stable and "definitive in their properties," yet which go without valid precise descriptions means that physical law should be reconsidered.

When a precise descriptive language does not work . . . , ie either has very little accuracy or very little practical value, . . . , then the correct interpretation of the Godel's incompleteness theorem is that the language should be revised at the level of assumptions, contexts, interpretations, containment, and definitions. etc.

Furthermore, it is at the elementary level of assumptions, etc, that the language is best related to serve the interests of practical creativity.

If a precise language is not suitable for one's creative desires then one needs to create a new precise language.

In order to do this one needs to be aware of the context of precise description.

Consider the elementary questions:

0. What is measuring? What is shape?
1. What are the axioms of quantity (in regard to the operations of addition and multiplication)?
2. What is a pattern (or system) which one wants to measure or carefully describe?
3. What is the structure of the number system, (eg place-value, small, large, etc)?
4. What is the set of quantitative properties associated to a containment set, and/or associated to the measurable properties of the system, which is being contained in one's quantitative construct?
5. What is the relation between shape and quantity?
   (functions and coordinate domain-spaces, the real-numbers, etc)
6. What is the relation between dimension and shape?
7. What is a derivative? [Must derivatives be defined within a continuum?]
8. What is an equation?
9. What is a physical law?
10. Are metric-spaces associated to physical properties?
11. How does one relate a containment set to a system's measurable properties? eg defining functions on the domain space, where the domain contains the material system,
12. What is a function?
13. Are functions best considered in regard to: functions as geometric properties vs. functions as a set of spectral properties? (Where such spectral-functions define function-spaces (of particular types)),
14. What is the set structure for inversion (of a system of operators acting on sets of functions) and its relation to solution of an equation?
15. What set of properties of a math construct allows quantitative consistency?
16. How can a function's values be consistently measured in relation to its domain space? (in a linear manner?), and

17. How is nearness to be determined?
    For some local operators the property of closeness is needed, but
    "How close does one need to get?"
    "Does one need a continuum?"
18. What is a spectral set?
    Are they averages and/or arbitrary vibrations?
    Do spectra correspond to sets of operators, eg sets of eigenvalue differential equations?
    or
    Do stable spectra possess a geometric structure?
    Are stable spectra related to a space whose structure is defined by its holes?
19. Is measuring about metric-invariant measures defined on coordinates whose shapes are quantitatively consistent, ie with shapes related to lines and circles?
    or
    Is measuring about the spectral properties of material components or systems, and associated to random harmonic properties? [If yes, then why would this be true?]

Are metric-spaces associated to physical properties?

Inertia is defined within a Euclidean space, where spatial position can be determined within a Euclidean space, while stable energetic, and electromagnetic systems are defined in a hyperbolic metric-space (note: a hyperbolic metric-space of dimension-3 is equivalent to space-time).

Note: If one defines mass to be related to stable geometric shapes, so that, m=k/r, where r is the smallest radii of a discrete hyperbolic shape [so that this is a property which relates the spectra of a discrete hyperbolic shape, by resonance, to a discrete Euclidean shape, whose spectra is associated to the same value, r, so that, k/r, is the inertia (or mass) associated to the discrete Euclidean shape].

It might be noted that, m=k/r, is consistent with (or similar to) the relation, p=[h/(wavelength)], where p is momentum.

Stability and quantitative consistency depend on discrete isometry subgroups of classical Lie groups, which, in turn, are based on the general expression of "cubical" (or limited sets of rectangular) simplexes (or fundamental domains). These stable shapes are the very stable discrete hyperbolic shapes, and the continuous, flat, and adaptable discrete Euclidean shapes.

Is a derivative best placed in a context in which it is a discrete operator, defining both discrete multiplication by constants, defined between dimensional levels and between toral components of a hyperbolic space-form, along with discrete angular changes, and discrete changes in material positions?

Math which is stable, and quantitatively consistent, and solvable, and controllable, then it is to be constrained to a "checker-board," or "cubical" lattice.

Can sets be "too big?"

Generalizations concerning geometry, topology analysis, etc, which are based on the belief in a continuum, have not led to valid, or accurate, or practically useful math constructs.

Math papers should not be peer reviewed, rather their assumptions, contexts, and interpretations, etc, should be identified by the author and a computer should sort them out based on the fundamental attributes of the ideas (or patterns) being expressed, and they should be published openly, but the author needs to also identify the practical creative, or practical measurable pattern, or context to which the description (or the math constructs) apply,

and they should be published openly

Education is only valid (useful) when it is related to creativity, but the people of a society can only be creative if they are all equal creators. That is, equality allows creativity, whereas inequality is about: domination, stealing, selfishness, dogmatic authority, and the extreme violence which allows and maintains these destructive forces which are a part of every unequal society.

An example of a set of answers to these elementary but fundamental and yet difficult questions.

Consider the elementary questions:

0. What is measuring? What is shape?
1. What are the axioms of quantity (in regard to the operations of addition and multiplication)? [(+, x), (+) counting things of the same type using a stable unit of measuring, and (x) rescaling, ie the linear functions, and (x) is also related to the changing of number-types],
2. what is a pattern (or system) which one wants to measure or carefully describe? (something continuous in time, or something contained upon (within) a coordinate grid of lines and circles, or an unstable but distinguishable, within a metric-space, set of properties)

3. what is the structure of the number system? (it has an order, numbers modeled as polynomials, it has a set of axioms associated to itself)
4. what is the set of quantitative properties associated to a containment set, and/or associated to the measurable properties of the system which is being contained in one's quantitative construct? (properties of measurability, ie determining relative size but done in a stable and uniform context, the containment of material)
5. what is the relation between shape and quantity? (functions and coordinate domain-spaces; in the new descriptive structure the shapes of coordinate spaces and the properties of functions are essentially the same as the shapes of coordinate spaces, though often defined on different dimensional levels),
6. what is the relation between dimension and shape? (one perceives 2-surfaces in 3-space, and 3-faces in 4-space etc.)
7. what is a derivative? [Must derivatives be defined within a continuum?] (local linear method of measuring formulas, or linear operators on function-spaces, or discrete operator defined between dimensional levels and between uniform, discrete time-intervals)
8. what is an equation,
   (different representations of the same thing, or an expression of how measuring sets (measured values) are changing, or a rule for an operator (function) on a set)
   [Equivalent measurable properties which are contained in a stable (math) structure, which allows and maintains a reliable measuring context.]
9. what is a physical law, (two representations of a system's properties, or "how a system's properties change")
10. Are metric-spaces associated to physical properties? (yes, eg position in space, stable pattern which is continuous in time, eg conservation of: mass, shape, pattern, etc)
11. How does one relate a containment set to a system's measurable properties? eg defining functions on the domain space where the domain contains the material system,
12. Are functions best considered in regard to: functions as geometric properties vs. functions as a set of spectral properties? where such spectral-functions define function-spaces (of particular types),
13. what is the set structure for inversion (of a system of operators acting on sets of functions) and its relation to solution of an equation? (continuously linear, and commutative, and metric-invariant, and invertible, as well as both one-to-one and onto)
14. what set of properties of a math construct allows quantitative consistency? (Discrete hyperbolic shapes)
15. what is a function (a model of measurement within the set which contains the system, which the function is modeling as a measured value of the system),
16. how can it be consistently measured (in a linear manner), and

17. how is nearness to be determined {by a continuum, by (locally) linear geometric measures, by a set structure of a function (function-values and its domain values and the containment within the domain set in regard to an inverse-image set, where the image set is defined by the function-values)}? For some local operators the property of closeness is needed, but "How close does one need to get?" "Does one need a continuum?"
18. What is a spectral set? Are they averages and arbitrary vibrations? Do spectra correspond to sets of operators, eg sets of eigenvalue equations? or Do stable spectra possess a geometric structure? (They are associated to the stable geometric properties of discrete hyperbolic shapes.)
19. Is measuring about metric-invariant measures defined on coordinates whose shapes are quantitatively consistent, ie with shapes related to lines and circles, or is measuring about the spectral properties of material components or systems, and associated to random harmonic properties?

Chapter 27

Pure vs. applied math

Pure vs. applied math and physics

In the AMS Notices December 2013 issue, there was an interesting selection of articles which can be interpreted in the context of both education and technical (creative) development; and thus, also in regard to, the relation that the social organization of both knowledge and its application has to the investor class.

These articles included a brash criticism of "pure" math, the brashness would be appreciated, but it was (or seems to have been) written so as to set-up a straw-man for the intellectual dogmatists, ie virtually the entire math community, to tear-down and to criticize these opinions . . . , let alone the problem with applied-math . . . , which the (straw-man) critic claims is a superior way of doing math . . . , is that they (the applied mathematicians) are following the lead (of the language used within the community of) the pure mathematicians, . . . , and two applied-math articles, one about quantum computing (a highly invested developmental project), and the other about "hearing a shape," ie the relationships between spectral sets and geometries (mostly represented as a problem which would be framed by a pure-mathematician).

Quantum computing

The very applied quantum computing article (which was very poorly written, apparently, so as to "guard" trade secrets from the public, either that, or it should have been better edited) "showed" how applications are getting stuck within narrow states of application (where these narrow states of application are being defined by the considerations of "pure math") . . . ,

Where the ideas followed, based in pure math, have very limited ranges (of application) in regard to the ways of thinking, where this limited way of thinking about applications of quantum computing is being determined by the pure math language which guides thought, where this limitation of pure math is due to the limitations of the language which the pure mathematicians use [eg theorems about transforming many "computing problems" into quadratic binary problems adaptable to the Ising-model, and theorems about computing-speed, as well as theorems about framing discrete optimization problems as a random Ising-modeled set of computing processes etc] . . . , by theoretical math claims, ie the ideas determined by pure-math.

That is, the ruling-class proclaim that they will invest inside "such and such" of a range of thought, and this, subsequently, defines the authoritative dogmas of the intellectual-class.

It might also be pointed-out that the quantum computer, ie based on the Ising-model (a square-lattice defined on a 2-plane with spin-½ properties defined at each lattice-site, where the spin-½ properties interact with other distant lattice-site spin-½ properties), is more like a classical statistical system, which is a random classical probability-statistical construct, than like an actual quantum-system of:

> "the quantum-computer system is supposed to be a probability wave representation of a quantum-process"

The quantum-part is the processor-part of an input-processor-output "machine"

The processor-part of an input-processor-output computing "machine," is a non-local system, composed of a finite number of components [a quantum system is also indefinably random and non-commutative] and the component's properties (of the quantum system) are all to be determined by probabilities, ie it is not controllable, and the way in which distant quantum systems can couple to it in a non-local manner, is not controllable,"

That is,

it is never known when and/or how the, so called, "isolated" quantum-computer-system (ie the processor-part) will (by means of a non-local manner, and in a random way) couple-to (or interact-with) a distant, external quantum-system, where this would cause the collapse of the quantum-system's, supposedly, "causal" wave-function, which is supposed to be functioning as the quantum-computer's processor part, and this non-local interaction-process would give the "computing-machine" a random in-put, by means of some random coupling context, which is not definable,

The quantum computer is supposed to function as follows:

Either, first a state of the processor-part of the system is identified, Or, perhaps the properties of the processor-part are changed, eg applying an external field to the Ising-model, so as to (help) define the input to the quantum-computer, And (then), apparently, an identified "quantum"-state of the processor, would be created by the processor's (supposedly, causal) quantum wave-function [where one is assuming the processor has not coupled-to distant quantum-systems in a non-local manner during the quantum-computer's processing stage], is then the processor would coupled-to, an out-put so as to provide an out-put state, which is identified as an out-put signal.

Apparently, this indefinable descriptive context, identifies "the great fun" for some "applied mathematicians."

The knowledge gained by working in an applied-math-context may seem important, but if one were working on this quantum-computer, one can be sure, that no-one would be making-up stories about how a quantum-computer might-function, or how it should be built, rather, one can be sure that there would be a highly disciplined adherence to the dogmas which are put of the quantum-computing investment scenario, which guides the selection of the personnel used on the project.

However, the author of this article admits that there can be at least a sequence of 100 computations, ie there are 100 input-processing-output events, where the output is not close to the answer to which the machine is being used to try to identify.

This means that indefinable randomness and/or a quantitatively inconsistent model is being used.

If it was truly a probabilistic based descriptive structure (for the quantum computer) then increasing the sample sizes (eg increase the number of computer runs) should solve such a problem, though this suggests a need for a very high-number for the sample sizes, which are being determined in a theoretical context of many samples "taken."

The author of the quantum-computing article provides a list of, supposedly, relevant words associated to the quantum-computers description, and then provides an incoherent text, apparently, communicating in-code to other experts. It is more informative to make-up one's own story about "what this quantum computer actually might be," since the article is written in an incomprehensible manner, and so as to provide (to the reader) confusing ways in which to use words and acronyms (ie initials substituting for the words of the experts).

Hearing a shape

The "hearing a shape" article, which was very close to being a pure-math discussion, was much more clearly written, though its adherence to the dogma of functional analysis, essentially, renders the discussion difficult, at best, to comprehend. That is, its fundamental idea of associating a spectral-sequence to a physical vibrating system, in this article, is a function-space construct, whose validity is based on the relation that these spectral-sequences have to the convergence properties of a (solution) function defined as a converging series, and thus, the actual spectral values for the vibrating-system might (seem to) not have any relation to the system's physical properties (or only a few of the spectral values have any relation to the system's physical properties), that is, arbitrary spectral-values may be used but when they are placed into an infinite series which is supposed to converge to a solution function the arbitrary spectral-values become irrelevant in the infinite series limit process for defining a (solution) function, Thus, there is an inability to make sense out of the problem in the first place.

The problem is supposed to be in regard to a vibrating system, where the system has a shape; Then "the problem is," either Use the shape to find the spectra Or Use the spectra to find the shape.

In eigenvalue problems there are (can be) issues about coordinate directions . . . , where there is an analogous construct about the "directions of functions" in a function-space, eg sets of commutative functions which are (usually) the harmonics of "the principle geometric length" of the vibrating system's shape, eg plucking a string and considering its set of harmonics, . . . , spectral values related to some set of eigenvalues associated to some fixed dimensional vibration, and . . . , in regard to the spatial directions of an eigenfunction's periodic closed wave-pathways, ie a vector direction for the wave-propagation path, in the region where the wave-function is defined, {Note: Such periodic pathways must exist on general regions if a wave is defined at-all (on such a general region, or arbitrary shape) unless for some shapes such as a scalar circularly symmetric wave propagation which is initially defined at the focal point of an elliptic shape.}

The question is: Does an eigenfunction have a relation to the direction of a periodic closed wave-pathway on the region upon which the eigenfunction is defined? This must be true, or the eigenfunctions are not properly defined, ie otherwise their wave motions disintegrate (dissipate), except on a few special shapes, eg initiating a wave at the center of a circular shape.

The issues can be formulated in regard to the local linear approximation of a vector-function's values in relation to (linearly approximating with) the local (spatial) coordinate directions . . . , ie in the domain space of the function . . . , can have their independence, ie the independence of the row-vectors (of the local linear map), determined by the zero or non-zero value of the "determinant" of the (n x n) matrix of the (local) linear approximation. Thus, there is the linear map, Ax=y, where y is the linear approximation of the vector-eigenfunction at x in the domain space.

This determinant function (going from the n x n—matrices to the real-number-field) is a polynomial function (of degree-n) in the local linear coordinate variables. The roots (or zeros) of this polynomial identify the eigenvalues of A, and they also determine the diagonal values for the diagonalized version of A, but not all metric-invariant real linear approximations of the vector-eigenfunctions have real roots.

That is, "independence of direction" (on the local linear coordinate approximation of the vector-eigenfunction) or the consistency of the wave-propagation direction is related to the factorization of this (determinant) polynomial, (related to solving for the zeros of this polynomial expression) where

1. distinct linear (polynomial) factors (or zeros) imply independent directions, and a consistent direction for the wave-propagation,
2. the same set of factors imply an effective reduction in the system's dimension (or reduction in the set of the system's independent directions), and
3. quadratic (polynomial) factors imply a non-commutative dependence of direction, in regard to that particular-pair of linear directions (associated to the matrix representation of the linear approximation of the values of the vector-eigenfunction, in the local linear coordinate directions).

Do 2 and 3 imply dissipation of the wave?

If this is not the case, then there are (may be) rotations defined on the dependent spatial directions, but which may also be interpreted to be properties of vibrations, and/or (still) dissipation occurring between the two local independent (eg imposed Euclidean) directions.

Note: In a context of local dynamic spatial displacement directions, there can be two opposite metric-space states defined in which the local displacements in the opposite states is exactly the inverse of the other state.

The solution structure in the complex number system, in regard to the zero's of the determinant's polynomial (in the local linear coordinate variables) determines the dependence or independence of the coordinate directions in the function's (or eigenfunction's) domain space, where these complex-numbers are the eigenvalues of the matrix, but in the dynamic context (of opposite metric-space states), one wants a real and pure-imaginary structure for the eigenvalue quantities to locally represent a pair of opposite metric-space states (or inverse spatial displacements in the two subsets) defined on a pair of real-shapes, which otherwise has (each have) continuously commutative local coordinate directions (except maybe at a single point). But how does one relate the spatial directions in the real subset to a corresponding set of spatial directions in the pure imaginary set? There is, (-i)(iR) = R (but is it R or (-R)?). So the matrix operation of "complex-conjugate of its terms (or matrix component entries) and then transpose," would invert the spatial displacement so as to then be defined in R, in this opposite metric-space state context of containment.

This matrix operation defines a unitary context, and the context of inverse unitary matrices.

That is, the displacement state in iR would be brought into the "correct direction (?)" in R and inverted to give the correct displacement in R from its spatial displacement in (iR).

Though the writers (of the article about "hearing a shape") use expert techniques, where some of the techniques are described at an intuitive level, while others are simply referenced, they seem to try to describe things so that the technical jargon is not a hindrance to following the ideas they are trying to express. Their outline of the heat-trace and the wave-trace techniques . . . (which are used and taken from analysis) . . . are used without any questioning of the assumptions, or (without) questions about . . . . "if these techniques are related to a wide range of, essentially, valid descriptions, which can be used in a practical context" . . . . are not considered, though they do admit that the techniques have only the smallest set of complete solutions in regard to these general problems about hearing shapes {or about vibrating systems being modeled by means of partial differential equations} (even though the techniques lead to an abstract context within which to pose the questions, which they pose in this article).

Furthermore, these techniques, applied to quantum systems have not led to valid solutions to any of the most fundamental stable quantum systems, whose stable spectral properties have been observed, eg nuclei, atoms, molecules, crystals, etc even though these systems have stable properties, [where possessing stable properties implies that these systems are linear and solvable] . . . and thus, one should conclude that this nearly incomprehensible context {about vibrating and random systems being modeled by means of partial differential equations} should be reconsidered, and all of its assumptions questioned.

This problem is formulated in the context of partial differential equations with boundary conditions and/or with initial conditions.

Then a point-disturbance which causes a wave-vibration to emerge from the given point, is modeled as an infinite series of functions, where the functions are "taken" from a function space. Thus, the discussion is about the convergence of the series of functions.

How the region is related to the definition of such a series' convergence properties can be arbitrary (eg point-wise, or uniform convergence).

And convergence, defined in (on) (such a big set as) a continuum, can be questioned, eg it allows quantitatively inconsistent constructs to be defined on the containment set due to different types of convergence constructs which can be defined.

In general, such a problem is not solvable, unless the shape of the region is a very simple shape, eg cubes, circle-spaces, and conic sections.

1. Why would one accept that there is an infinite spectral sequence for such a problem, if this idea can only be implied by applying (functional) analysis to the problem, but in general there are only a very limited number of actual solutions identifying a spectral sequence, and whereas these solutions are associated to a limited number of shapes; such as the cube, the disc, and the cylinder, etc. in regard to such a descriptive context based on solving a partial differential equation, where the region's shape and the propagating wave can seldom, ie almost never, define a continuously commuting coordinate system, which is needed for solutions to be found.

Instead "what is claimed," (by the experts) is that an infinite spectral sequence, "apparently with arbitrary spectral values," can be defined, so as to, in turn, define an infinite-dimensional function-space, wherein the spectral-functions in the function-space are identified as possessing a function-space property of being independent, so that the solution-functions are defined to be a converging series defined on the set of "independent functions" in the function-space. This is contrived, and without any (or very little) physical (or geometric) motivation.

On the other hand, they do a very geometric approach to the wave-trace's defined-values, essentially, defining them (the wave-trace's defined-values) as the set of periodic closed-curves which are defined in relation to a propagating-wave's reflections from the region's boundaries (ie the region where the vibrations are defined). But, this (the wave-trace's defined-values) is being defined in regard to a set of "functions" (distributions), where each "function" is being defined as a diverging series of "actual functions." These "functions" are used to model the effect of "dropping a pebble into a pond" so as to start a circularly-propagating wave (circular, until it is reflected from the region's boundary), and thus they finally get into a discussion about the identifiable vibrations (or relatively stable vibrations) of a system which are defined geometrically as the region's periodic closed-curves.

Thus, move-around the analysis into a more practical descriptive context for measurable patterns and instead assume that relatively stable vibrating systems are defined by simple geometric shapes

That is, consider:

2. For example, why not try to partition the region into rectangular sub-regions, and try to find rational sloped curves which return to their point of origin so as to be moving (propagating) in the same direction as when they left the original point. In fact,

3. There might only be a few shapes where the vibrations can be identified with an infinite spectral sequence, and where the infinite spectral sequence is, actually, generated by a finite spectral set (and the vibrations have a property in their vibrating context of being-free, eg the vibrating motions are not constrained, which, in turn, is associated to the shape's geometry, most notably the set of discrete hyperbolic shapes, but where in the discrete hyperbolic shapes the spectral structures are very constrained in regard to their vibrating-motions (on the vibrating-medium upon which they are defined, or plucked [or a pebble-dropped] to initiate the wave-motion), where in the discrete hyperbolic shapes the periodic motions would be parallel with the geometric-shape (which can engage in periodic motions) and its geometric-measure, which, in turn, defines the (or its, ie "the vibration's" or periodic-motion's, or a spectral-flow) spectral-value.

The difficulty in understanding the problem, of "hearing a shape," when formulated in the context of a partial differential equation, and if the machinations of the math experts, concerning . . . , functional analysis related to defining "a vibrating system's, so called, infinite spectral sequence" . . . , are not actually valid, then the problem . . . , framed in relation to the existence of an infinite spectral-sequence associated to the region's vibrations . . . , is also not valid, and thus, the expression of the problem is also invalid.

This is a reasonable criticism, because there are only a few simple geometric regions which can be solved: rectangles, circles, discs, cylinders, and apparently even the 1-dimensional triangles are difficult to solve, except for some triangles (identified in this article), but this is to be expected because most (all) triangles coordinates, where the idea "of a solution" is framed (in the article) in the context of periodic closed-curves defined on (within) the triangles, are not-commutative in regard to both the shape's coordinates, and the shape's periodic, closed wave-pathways.

That is, triangles are not solvable in this context, even though the triangle has obvious geometric properties, which should be easily related to a triangular region's vibrations, but apparently the problem, in regard to trying to solve for the (mythological) spectral sequence for many triangles, is that their shapes, as well as the region's vibrations do not form into a commutative set of coordinates (defined on the triangular-region) which are consistent with the periodic closed wave-pathways, as well as the notion of independent-functions, which can exist within function-spaces, is not as a "refined enough of a quantitative idea" to be all that helpful, ie defining an infinite spectral sequence for a triangle is not all that helpful, in regard to determining if one can "hear the shape of a triangle."

To avoid using math structures which are being applied to a set of overly general and abstract contexts, which are not meaningful, the experts need to make the case that this type of overly general and abstract contexts . . . , that such incomprehensible descriptive contexts, such as using function-space techniques to solve for the spectral structure of a vibrating system which is being modeled by means of partial differential equations . . . . , do, in fact, possess some relation to practical (or practically useful) value.

But this is not the function of an expert, rather an expert is to extend the dogmas into which they have been indoctrinated.

As yet, it is not at all clear that they have made this case.

Nonetheless, this mostly useless contexts and dogmas constitute the dogma which govern peer-review, and thus, these difficult (and seemingly irrelevant) constructs are used to identify a, so called, legitimate way of "math-thinking." That is, one cannot justify the dogmas which determine the ideas expressed in peer-review science and math publications (about the attempted extension of useless dogma) any more than one can justify the useless dogma of racism, or religious fanaticism. That is, the failed intellectual dogmas (of science and math) are the result of the interests of the investment class, and one of those interests is promoting inequality within society, and subsequently, both the investors and their experts are opposed to

"the requirement that a failed dogma must submit to"
"challenges to its domineering language"
"at the elementary (precise language) level of:"

assumptions, context, interpretation, containment-set, and organization of the descriptive patterns of the language etc etc, to which (any person addressing the concerns about a measurable descriptive language which is supposed to describe (stable) patterns, which in turn, are supposed to be, or expected to be, of practical useful value) every "reasoned voice" has validity, and should be expressed in a meaningful manner within the propaganda-education system.

eg "How can the idea of a "quantity" have meaning in a descriptive and measurable setting?"

It might, very well be, that the only shapes, where such a relation between finding a spectral set and relating that spectral set to the region's (the shape's) geometric properties are the very simple "discrete isometry subgroup" shapes, defined on metric-spaces which possess constant curvature and their metric-functions only possess constant coefficients, ie the partial differential equations (which the dogma of the experts requires to be the models for physical systems) are linear, and the coordinates of the system's shape are continuously commutative, ie the partial differential equation is solvable.

The main question has to do with the constraints of the vibrations of a shape, which has geometric relations to a natural vibratory structure, eg periodic closed-curves.

Do naturally vibratory shapes, which are free . . . (ie free to vibrate in a geometric-way which is normal [perpendicular] to their natural shapes as objects), . . . , really define vibratory systems, or is the free-vibratory phenomenon about an unstable property, and thus, it will always be difficult to describe?

Or

Is the better question:

Can the set of "naturally vibratory shapes" absorb the energy of particular (external) vibrations (particular spectral-values) by means of their resonance properties, due to their natural (geometric) vibratory properties?

Thus, there is also the question:

Do the set of "naturally vibratory shapes" dissipate (vibratory) energy, or can they store (vibratory) energy?

* * *

Some of the main concerns about energy dissipation deal with:

1. the periodic, closed, wave-pathways which result from the shape of the region which is vibrating (or caused to vibrate)
2. The vibrations being realized in a periodic wave-oscillation which is normal to the region's (usually, assumed to be bounded) shape so that the vibration is immersed in a bigger-set or a higher-dimensional system.
2. (b) Are the normal vibrations the main cause for the different wave-lengths (of the wave-functions associated to the spectral-sequence) to travel at different wave-speeds, so as to dissipate the energy (of the initial impulse-cause of the vibration) more quickly?
3. Wave dissipation due to both the arbitrary values of the spectra which are selected (in the infinite-dimensional spectral-sequence setting of analysis) and these spectral-values are due to the deformability characteristics of the medium (material composition) of the shape, so as to attribute the sudden pulse (and the material's deformation reaction to that impulse) to sets of spectral-values which are also assumed to identify an instantaneous relation to the assumed set of eigenfunctions and their associated eigenvalues.
4. The relation of the wave-lengths to both (a) the size of the region, perhaps dividing the region into sub-regions (of rectangles) associated to different sizes, and the relation of the wave-lengths to (b) the small components which make-up the material, which vibrates in its confined local-regional-shape, eg the crystal-lattice ie the size-scale of the discrete components which compose the material which is modeled as a continuum in regard to the frequencies (or wave-lengths) as well as wave-amplitudes, which the vibratory material-medium can accommodate, without either failing or (without) the material decomposing (from the wave-oscillating-actions).

5. And finally, the main mathematical issue concerning quantitative consistency, and its relation to local (linear) approximations of function's (or coordinate's) values [ie linear approximations of local function-values made in terms of the local linear coordinates] the local continuously commutative-ness of such linear approximations, so that locally, the wave-directions (of periodic closed pathways) are independent of one another so that different wave-lengths can be associated to independent directions in the region. Without this independence the wave-lengths would be arbitrary due to the interference of normal-vibration properties as well as the quantitative inconsistency of non-independent coordinate directions.

This need for independent coordinate directions seems to be very limiting where one should consider either the conic sections where some of these shapes can be placed in sets of (almost) always locally orthogonal coordinates in the context of "sets of:" elliptic, parabolic, and hyperbolic coordinates, so that all pairs of local directions (in regard to the "sets of" these conic-section shapes) are orthogonal to one another, and the fact that the reflective properties of the conic section-shapes are well known, eg the focal-points of an ellipse define periodic-closed-pathways in the ellipse, etc, and a set of rectangular partitions of regions can be used to approximate various shapes, where rational sloped lines (in regard to the rectangular regions of the partition) define periodic closed pathways, on the rectangles; but only cubes would be able to carry (accommodate) continuously commutative coordinates.

Thus, the vast majority of regions (and their bounded shapes) will only vaguely sustain a vibration, and rather quickly dissipate the wave's energy, while only a few shapes, which have been struck at the correct-point (upon or within the shape), would be able to demonstrate a substantial capturing of energy due to resonances which a vibration (of the strike) might define on a shape.

The vibrations of a relatively stable system (or object, or stable material-component) when struck (with a mallet) should dissipate the energy of the strike, unless the position that is struck (and, perhaps, its energy) coincides (on the object) with a periodic closed wave-pathway on the object (where it is assumed that wave-propagation defines spherical shells which reflect from the boundary of the bounded object's boundary-shape) so a resonance within the system forms, and thus, there is vibratory energy which is absorbed within the object, so as to either subsequently dissipate that energy, eg heat flows, or store that externally derived energy, ie energy from the initial striking of the object.

Perhaps the better context of inquiry . . . , instead of vibration and dissipation . . . , is about the context of continuous stability, eg conservation of energy (in a stable descriptive context).

"What shape?" (A set of discrete hyperbolic shapes). and "What condition of set-containment?" allows for a stable set of systems, and subsequent conditions for the energy, to possess [associated with such a] stable shapes? ie to possess the property of stable continuity of both shape and energy, but so that different (stable) material-components can change and/or form?

That is, the dissipation of energy in a physical system is about coupling the dissipated energy . . . , eg the energy associated with the striking of an object with a mallet in order to initiate the object vibrating . . . , to smaller components a smaller context of lower-dimensional stable shapes (with their own spectral properties, and which the original set of vibrations are not in resonance) or a higher-dimensional (eg contain what is normal to the struck object's shape) stable set of larger shapes (or shape) which, in turn, possess stable spectral properties.

The context of dissipation can be:

both

1. Heat and collisions of smaller components, and
2. sets of waves with different wave-lengths, which move at different speeds within the common medium of wave-propagation, so that (or where)
3. the waves, or vibrations, can be normal to the object's shape or boundary, and thus the vibrations can couple-to a greater inclusive system, or region of space, or the greater freedom allows the different wave-lengths which compose the system's assumed spectral-sequence to move more easily at different speeds so as to allow quicker dissipation of vibratory energy within the shape.

Whereas the context of conserved energy for discrete components is the context of:

1. (elastic) collisions of waves with geometric boundaries or of smaller components which compose the larger vibrating system, as well as
2. electromagnetism for both charged material systems and electromagnetic-waves propagating discrete energy packets across space at a fixed speed (in a vacuum).

But one needs both

1. periodic closed wave-pathways, and
2. continuously commutative coordinates . . . , to be used to describe the properties of the region's shape . . . , which is needed to describe the vibrations and their periodic closed-wave-pathways, in order to sustain a continuous energy content.

The (descriptive context of) indefinably random (events are not stable or the assumed spectral values cannot be calculated, or there is no list of stable events which composes the (an) elementary event space of the random system), and non-stable-geometric shapes (eg non-linear shapes), and dissipative (many) spectral-value (associated to vibrations defined which are normal to the system's shape) contexts for either general descriptions of vibrating systems, or for non-periodic wave-pathways defined on the vibrating system's shape, as well as non-commutative coordinates and the different coordinates of the different functions in a function space, which are all descriptive properties which can interfere with the uniform descriptive structure of a local coordinate structure, of a vibrating-system's descriptive context, and these (various descriptive) properties take the quantitative properties of the description out of a valid (measurable and quantitatively consistent) context and result in the descriptions of these physical properties to be related to unstable dissipative processes, and instead of being waves they are (dissipative) properties which are related to the heat-equation that is commutative functions in a function space may define coordinate shapes which are not compatible with the shape upon which the vibrating-system is defined.

For stability there needs to be defined a stable set of conditions, or set of component-shapes, upon which the dissipative processes can be considered, or identified, or modeled, as well as the conditions of the containment set which allows for such a set of stable shapes to exist in a common containing space.

One sees [two] different sets of assumptions:

1. That the constructs of analysis can be used to identify a vibrating system's spectral-sequence
2. That the vibratory system can be modeled by means of the heat-equation
3. That the vibratory system can be modeled by means of the wave-equation, which is supposed to identify a relatively stable construct, but which dissipates due to the assumptions of analysis that there is a spectral sequence of many different wave-lengths so that each different wave-length is associated to a different propagation speed in the (mechanical) medium of vibration, yet there is also an overtone relation of the wave-lengths, which are so often modeled as instantaneous wave-constructs so as to model the striking-initiator of the wave (or mechanical vibration) furthermore many of the values in the analytic context may very well be arbitrary, especially, if there are no periodic closed curves for the defined pathway of wave-propagation associated to the point on the object which is struck, so these many spectral-values are arbitrary and seem to come from (be based on) both analysis and the way analysis is related to the limit processes in which one might define convergence, where the explicit spectral values are not needed

Instead of allowing math to descend into its own chaotic descriptive context of:

Indefinable randomness
Quantitative inconsistency

Non-commutativity of functions and/or coordinates (or functions which are commutative within the algebraic constructs of functions spaces (or of analysis) but whose coordinates are not commutative, eg 2-dimensional spectra)

Non-linear ordinary differential equations, and non-linear partial differential equations, and

Where convergence is defined both on too-big of a set, eg the real-numbers modeled as a continuum, and in a structure of many contexts in regard to how to define convergence (uniform, point-wise, etc).

Rather

Give some definitive definition of a stable component, and their context for existence, within a containing space, ie resonances which define stable components.

So that the description is:

Linear

Metric-invariant, with non-positive constant curvature so the metric-functions only have constant coefficients

And

Continuously commutative everywhere, except maybe at a point

And any randomness is about an elementary event space built from events which are defined by their stable properties,

So that the math descriptive context is based on quantitative consistency.

What math contexts can be related to practical creative efforts?

Whereas, descriptions of patterns, in relation to quantity and shape, which possess a relation to practical value, need to be within a descriptive context where:

"measuring is reliable" and

"the descriptive language is based on stable (and usually controllable) patterns."

Some shapes which do have a strong relation to their spectral properties are the many discrete hyperbolic shapes

However, there is one shape which has a very close relation to its spectral properties, the discrete hyperbolic shapes (in the context of discrete isometric subgroups) for the generalized space-time metric-space, eg R(n,1) becomes an equivalent n-dimensional hyperbolic metric-space, ie a hyperbolic metric-space, where the spectra of the many discrete hyperbolic shapes of various dimensions are determined by a direct relation to the geometric measures (in a metric-invariant space) of the faces of the (discrete) shape's rectangular simplex structure, called its fundamental domain.

Mathematics is not being done correctly, particularly in regard to understanding stable (math) patterns, ie the stability and reliability of quantitative sets and shapes.

Vibrating systems or periodic systems . . . , or relatively stable and bounded systems which are 'struck" or its geometry displaced . . . , are of central concern for: spectral-geometry, quantum physics, and macroscopic electromagnetic systems. However, the stable many-but-few-body gravitating systems may also need to have a relation to stable periodic-motions.

What are the stable shapes which can possess an ability to store energy? Is the main math pattern, in regard to stability, the discrete hyperbolic shape?

Consider that though the ordinary differential equations and the partial differential equations may be the natural local linear measuring math structures which are useful for describing material-component interactions where solution functions are (or represent) observable properties. But Vibrating-systems and periodic structures (or patterns) of physical systems need to be distinguished (or classified) by the properties of stable and/or natural vibrations, ie resonances, where a natural vibration is about a shape's natural set of "periodic closed-pathways for wave-propagation" related to the propagating-wave's reflections off the (bounded) region's boundary,

And

Unstable, un-calculate-able, and/or dissipating vibrating structures.

A displacement of a semi-rigid body (object) can cause sounds but not necessarily stable vibrating structures rather temporary unstable or dissipating vibrations, emanating from the struck object, however, some stable geometric relations, eg periodic closed-curve wave-pathways defined on the object's shape, may or may-not be activated by the object's initial displacements.

There is the natural property of a bounded object which is struck can identify "local" (or nearby) periodicity, but its regional bounded geometric (or system) properties define relatively stable properties, in regard to a periodic closed wave-pathway.

That is, one may want to consider a stable (vibrating) structure vs. a local arbitrary vibrating structure (or system, or shape) whose vibrations dissipate.

Then there is the more definitive question in regard to:

"What vibratory systems can store energy?" eg the vibrations are not normal to the object's shape, vs. Those systems which dissipate energy.

That is

The conservation of energy

vs.

The relation a vibrating system has to an increase in entropy to the higher-dimensional region which can couple to the given (vibrating) object, ie a local region which is normal to the object.

This is related to the question as to:

"What is a stable math pattern (or stable form)?"

And

"How, in turn, can it be related to dissipation of energy and/or sending energy to smaller stable (material) components?

Basically, the shapes with stable properties are related to rectangular simplex structures and subsequently, to continuously commutative coordinates.

The descriptive contexts which are:

1. indefinably random, eg the elementary events are not stable-events, (algebraically)
2. non-commutative, and
3. infinite dimensional,

where the always "locally orthogonal" geometric coordinates associated to a bounded shape, ie independent coordinate directions, cannot be related to an algebraically analogous math patterns defined on an infinite-dimensional function-space in a consistent context of independence.

That is, the dimensional and quantitative (algebraic) structures of coordinate spaces and function spaces cannot be compared unless the domain of the functions of the function space is continuously commutative, essentially, everywhere, and even then the infinite dimensions can hide patterns which result in quantitative incompatibility for solution-functions, especially, if one's main idea about modeling a measurable system is the derivative, for example the definition of a function as an infinite series and the various ways in which convergence can be defined, eg point-wise or over the entire domain, etc etc.

All of these issues contribute to the context of a descriptive language which has a very limited relation to a meaningful context.

A system's vibrations are geometric and free-dissipative vibrations are related to the region's periodic closed-curves which might be followed by free-waves dependent on both the displacement point and the shape of the object.

An idea very clearly expressed in this article about "hearing the shape."

Weyl-angles, shape, and discrete operators

The discrete hyperbolic shapes have a standard shape of "linearly" arranged finite sequence of toral components, which can be adjusted by folding the shape's lattice structure, ie the metric-space's discrete partition so as to form the fundamental domains of the stable shapes, by the discrete Weyl-angles, which can change the shape, by changing the angles defined between the shape's toral-components, but which preserve the shape's spectral properties.

This explains the reason that biological molecules can change shape, and thus, they can mechanically function in different ways at different times.

The spectra would remain the same but the energy, due to the introduced folds, can change in, usually, small discrete ways, and where it is to be assumed that conservation of energy is defined in regard to the continuously stable spectral-shapes of the discrete hyperbolic shapes.

Thus, the folds between the shape's toral components (of the stable shapes), or of the lattice, can be described by the Weyl-angles, which are defined between different maximal tori of the shape's containing-metric-space's fiber group, SU(n) [or SU(n,1)], so that the different toral components of the folded shape would be related to different toral components defined in different conjugate maximal tori in the fiber group, thus the Weyl-angles are defining conjugations between the different conjugation classes of maximal tori in the metric-space's fiber Lie group. These conjugations are defined in a discrete context, where the fold action is a discrete Lie-fiber-group conjugation operator, which can easily fit into the "discrete" operator structure defined in regard to the (small) time-intervals defined by the (time) periods of the spin-rotations of metric-space states, so as to define such a fixed discrete structure associated to the Weyl-folds for a (relatively) stable shape.

In the standard position, for a discrete hyperbolic shape, there is a return to the distinguished-point (the vertex of the shape's fundamental domain) for a continuously commutative path (on the shape within the metric-space) defined by an "orbital" group action by a maximal torus which is acting locally on the shape, and at this distinguished point a (further) continuation of the "orbital" group action can cause an orbital-path which is also continuously commutative on another toral-component of the discrete hyperbolic shape, but where the other toral-component is oriented [in the shape's containing (real) "metric-space"] in the same orientation as the other toral component. Thus, the continuously commutative orbital path (on the new toral-component of the shape) can still be defined by the group action defined by the same (orientation) maximal torus of the fiber group (ie the shape is really in complex coordinates, but so as to have two real-shapes, which are contained in the pair-of real and the pure-imaginary subsets of the complex coordinates).

Thus, for a new shape, where one of the toral components is changed by a Weyl-angle, then the maximal torus which is associated to defining the orbital-paths on this toral component is also changed by the same Weyl-angle (by group conjugation, ie where the conjugation is defined by the Weyl-angles) so that the new maximal torus, which is defining a continuously commutative orbital path on the folded-shape (in the metric-space) is allowable since both the orientation of the folded-toral component and the orientation for the new maximal torus are consistent with one another, so as to be able to define an orbital pathway [due to the group action by the conjugated (new) maximal torus] which is continuously commutative, and it also returns to the shape's distinguished point.

This type of a transformation is allowed within the context of operations being defined in a discrete sense, so that the fold defines a discrete context, in regard to "that upon which the operators can act" in the discrete relations which can exist (or be described) between quantity and shape.

This is defined in regard to a torus, T, where T^(n-1) is contained in R^n, or the corresponding condition in complex coordinates, ie C^n, where two toral shapes are contained in the two {real and pure-imaginary} subsets (in complex coordinates).

Thus, in C^n there can act SU(n), whose maximal tori are (n-1)-dimensional, and which act on the T^(n-1) (toral) components of the pair of discrete hyperbolic shapes (ie the pair in the real and pure-imaginary subsets). Note: Locally all the operations . . . , which are defined on (in) the two opposite metric-space states . . . , are inverses to one another, but globally, on the complex-coordinates, there is mixing, but which projects down to the two shapes in the real and pure-imaginary subsets, and where the local structure is defined by the two shapes in the real and pure-imaginary subsets (of the complex coordinate space).

* * *

The two sets of local inverse properties of the two opposite metric-space states can be (naturally) related to the inverse matrix structure in a (special) unitary group, when the two opposite metric-space states are defined on the two subsets of the real and pure-imaginary parts of the complex coordinates (which contain the two opposite metric-space states).

The local transformation of the coordinates is given by a unitary Lie algebra element.

However, when, in turn, (the Lie algebra matrix is) acted-on by complex-conjugation, . . . so as to change the direction of the dynamically defined local-time in the pure-imaginary subset's metric-space state, . . . , and since the reverse transformation would be related to the transpose of the original local matrix, then the inverse of the original unitary matrix is defined.

Now there are two unitary matrices, where the original unitary matrix moves the system "ahead" in both metric-space states, while the complex-conjugate transpose of the same matrix brings the two metric-space states back to their beginning state.

For each point in space, the "original," or the moving ahead dynamically (special) unitary, matrix is determined by the local tangent properties of the dynamical interaction torus, which is defined between the two interacting toral-approximations (of the two interacting-material system's discrete hyperbolic shapes) and these tangent's geometric relation to the local (special) unitary (or real SO) Lie algebra matrix, where in turn, the SO-matrix is related to a unitary matrix by its local inverse dynamic (spatial displacement) properties.

In fact the, n, in SU(n), may be the number "eleven," even for a much lower-dimensional shape, due to the nature of the over-all containing, 11-dimensional hyperbolic metric-space (where sub-tori of the maximal tori would define the group actions), so that there is a discrete conjugation (determined by Weyl-angles) for each distinct toral-component of the discrete hyperbolic shape, so that the energy changes due to the folds defined on the discrete hyperbolic shape's standard-shape. The idea of conservation-of-energy is defined in relation to a set of continuously stable discrete hyperbolic shapes.

Social issues associated to communication

. . . . When one has actually won the game which is being set forth by the math and science academic community (as the author has so won the game) . . . {namely, the solution to the problem of the many-(but-few)-body systems [both classical and quantum] which possess an unknown capacity to be stable} . . . but the, so called, "winning idea" is outside the intellectual dogmas, which are related to commercial interests, and which subsequently, define high-intellectual value within society,

There is a grave mis-perception (held by the people, which is due to the effects of the propaganda-education system) concerning the domineering intellectual authority of the society, namely, that one must start with the assumptions (precepts) of the domineering authority and then prove their system wrong (within their own system of assumptions), but to show how mis-conceived "this idea actually is," one can consider the elementary example which exists in the history of science. Namely, that this would be similar to requiring Copernicus start with the assumption that "the earth is the center" and then prove that the sun is really the center (of the solar system), ie different assumptions lead to "incompatible" descriptive languages "incompatible in the sense that the languages describe different sets of patterns."

Note: When a precise descriptive fails to describe the observed stable properties of the world then new assumptions need to be considered as remedies for such a failed language.

The conceptual dogma (which defines the intellectual-class, who the technical people who best help the investment bankers)

One does not need to begin with the idea of materialism and the partial differential equation (where it is assumed that partial differential equations define physical law), instead one can examine the rather small percentage of valid descriptions which are a part of this "old context." Namely, the linear, metric-invariant, and continuously commutative (except at one point) context, or equivalently, the linear solvable context (of the materialism and the partial differential equation context for a descriptive language), and then examine what characterizes these solvable systems . . . , ie they are based on very simple shapes, eg cubes, circles, lines, cylinders, tori, various ways of combining these shapes, and discrete hyperbolic shapes, ie stable shapes built from toral-components, as well as the shapes of conic sections of: parabolas, ellipses, and hyperbolas, and the orthogonal coordinate systems which an be defined by using these (conic section) shapes. . . . , and then build from these ideas,

Where, in particular, the ideas about the simple stable shapes of the discrete hyperbolic shapes have been extended, by D Coxter, to higher-dimensions, . . . , where these stable shapes (ie discrete hyperbolic shapes) are good models of both stable material-systems, and stable metric-spaces, though the dimension of these two constructs are most often different when a consistent description of a system's material properties are being determined, ie they (material and its containing metric-space) are usually spaces (or shapes, which are also metric-spaces) which possess the properties of being adjacent dimensional structures, etc etc. the discrete hyperbolic shapes possess a very strong relationship between the geometric measures of the faces of these shapes' fundamental domains, and the spectral properties to which theses shapes are very strongly related.

Within this context "the fundamental reason" as to "why things exist" is that these stable metric-space shapes (of various dimensions) are in resonance with a large, but finite, spectral-set defined over five-dimensional-levels, where this spectral set is a property of the containing-space, which has been partitioned into these stable shapes (in the various subspaces and in the various dimensions of an 11-dimensional hyperbolic metric-space) (note: the rank of SU(11) is 10, so this spectral set can be defined on a set of conjugate 10-dimensional maximal tori in the (fiber) Lie group SU(11).

Since it (the new solution to the many-body-problem) is outside the accepted dogma, this means that the professionals cannot "see (or accept)" that a superior containment-context has been put-forth, and thus, the professionals cannot acknowledge that a better context for the descriptions of existence has been found.

Such ideas cannot be peer-reviewed since they lie outside the authoritative dogmas of the technical, or intellectual, social-class, the social-class who follow a technical dogma which the bankers believe best serves their (the investor's) interests. Thus, the "mythological peer" would see that the expressions do not fit into their authoritative viewpoint.

Apparently, as much as anything else, this (peer-review) is about defining inequality, and confining thought, so that there is a "stable of intellects," like a "stable of show-horses," who can be trotted-out (into the media) to express narrowly confined expert-ideas which validate the narrow interests of the elite ruling-class.

Nonetheless, in the new descriptive context for science and math, the ways in which to "build things" will be seen to be more like piecing-together components, in a way which is similar to how Venter has pieced together DNA (in a {[his]} surprisingly capable methods), but nonetheless, the complexity of the chemical context of life's functions is (clearly) not understood (by Venter).

But now with new ideas, and with much deeper understanding of the relations which can exist between the components which comprise a physical system in regard to: nuclei, atoms, molecules, crystals, life-forms, and the stable solar-system, and the types of components which can be put-together, in each of these different contexts, so as to comprise a system, eg a system to heal; a system to travel; a system to generate cheap clean energy; etc etc.

But it is also a higher-dimensional construct, so this gives problems in regard to the comprehension, and formulation, and use of these ideas (or contexts), but it is a geometric based description, and we are geometric living-systems too, and where we are also higher-dimensional systems, ie life might be "the best instrument" which one might use to traverse the "entire-world (entire-universe)" which exists (we can do this independently, within our very-own living-system), due to it (our living system) being built from components which are (relatively) stable and put-together in a many-dimensional context.

Furthermore, it might also be noted that the dynamics of the higher-dimensions are more related to rotations than translations, in regard to spatial position, which in turn, is defined in relation to the fixed-stars, ie translational position might not be so well defined in dimensions 4 (or 5) and higher.

. . . . Then (if one follows the new, precise descriptive language) one sees the total collapse of the language, in regard to the language's relation to the development of practical creative projects. The language has been distorted within society by selfish commercial interests, which define the endeavors of the, so called, professional class, and subsequently, define the creative range of our society.

Thus, all dialog is mired in a confining language which identifies, what is considered to be, "very high cultural value" within all of society.

This is because, we live in a very flawed society, where the key point of social-organization deals with the violence which is focused within the institution of justice, ie a nation based on law, but where the main law is property rights, which, in turn, necessitates (at least within western culture) that law will be determined by lying, stealing, and murdering (thus, this point of enforcement of stolen property is also the corner-stone of the national security state (NSS)) . . . . apparently in the same way as it was so determined in the Roman Empire, but where today the emperors are, instead, the small set of bankers, apparently 10-banks and their oilmen-military cohorts.

This social condition was described by the highly decorated US Marine S Butler back in the 1930's, where Butler explained how the military worked to support the "banking interests," in regard to lying, stealing, and murdering.

However, the propaganda-education system controls the language and the thought within society, where the focus of the media is on both inequality, and money, since the main mechanism for the slave-like servitude (to the banker-investors) imposed on the public, is the requirement that the public are to be wage-slaves, ie one must have money in order to live (survive).

* * *

However, the idea of money is about the quantitative structure of social domination.

The ideas of "value-added product," and "supply and demand," is all a bunch of hooey.

Rather,

It (money and the, so called, economy) is about:

"how 'what is created' is determined by investment," ie it is all about social-domination by the investor-class,

Thus, education is about business risk management, and defining a servant (or worker, wage-slave) social-class.

That is, we are educated to build certain types of things, namely, those particular things which are related to the way in which the bankers invest, in regard to the society's creative efforts.

Thus, science and math come to be based on very restrictive dogmas, if one wants to compete then one must strive to extend the dogmas which identify the narrow creative contexts into which the investors have invested.

In this context there is a lot of baloney about unobtainable high-value, the, so called, myths of intellectual excellence, and an imperative to extend the dogmas, and then there is also the narrow competition between wage-slaves to achieve high-paying positions, so as to be attached to the instruments of empire, so as to be putting-forth a "product," so as to maintain the socially domination positions of the bankers.

The high-value of extending the socially-allowed dogmas (allowed by the propaganda-education system), ie the dogmas used to support banker's investments.

This is true for all the professions: law, education, military, medicine, and the media (and economics) etc etc, all are governed by dogmas which have been both narrowly defined and confined so as to serve investment and big business interests.

Consider that the AMS the math-science communities have failed in their social-intellectual construct, since they have become dogmatic, where this is because, within society, they only serve investment interests.

But it is clear that these dogmas are failing to solve the fundamental problems, which they (the experts, who are supposed to be consistent . . . , in their intent . . . , with the subject-matter's purpose) claim to be trying to solve, ie describing why so many fundamental systems . . . , eg nuclei, atoms, molecules, crystals, living-systems, and the solar system . . . , have such stable properties.

But they really only claim that "they (the experts) define the superior intellects of society," and subsequently they are observed to be the, so called, superior-intellects who, so reverently, serve the investor's interests.

Yet, many, if not most, of their problems are about "describing the stable properties of observed physical systems."

But they claim these properties are associated to systems which are too complicated . . . , and to be a part of a set of systems which cannot be . . . , to be described (at least not described in the context of the dogmas which define these, so called superior intellectuals, of all of society (ie either self-proclaimed, or, more realistically, proclaimed "to be society's superior-intellects" by the investor-class).

That is even though the professional math and science communities cannot describe the observed stable properties of some of the most fundamental physical systems, nonetheless, these professional communities maintain their same dogmatic discourse in relation to its failed authoritative dogmas, and they (these professional intellectual communities) continue to apply these dogmas to the usual set of activities defined by commercial interests.

* * *

A relatively audacious opinion is allowed for the dogmatic contexts of peer-reviewed math, as exemplified by the AMS, as expressed (in an opinion) in a journal-section of AMS Notices 12-13, in the section of Opinion, which has been expressed by D Zeilberger, where there is a contrasting between pure and applied math, however, the expression does not touch the deeper relation that exists between the investor class and knowledge and creative development in our society and thus is quite uninformed as to what is driving or evaluating the social condition of knowledge the belief expressed is that pure is too fixed and far too formal and for this it should be greatly criticized, but the criticism should be deeper when the set of three articles which are presented in the journal of AMS Notices December 2013 issue, and thus considered, where the other two articles are examples of applied math. The thing which Zeilberger does not express (and the social condition about which Zeilberger is likely unaware) is that the formal language . . . . of the pure math community (vs. the applied math community) {is that the formal language} . . . . is almost entirely consistent with the research contexts, which are related to the applied side, but the "applied side" "wants results" which are related to both investments and to some observed patterns about which the applied community is sure that there is a technical-capacity which exists (which the tech-community possesses) so as (or is there) to be able to use these patterns in a practical context, eg this is like a manager developing product based on putting technical capacities together.

That is, Zeilberger is concerned that the pure-side of math is too constrained, so it cannot "get things done." (This is a good criticism)

But what is being done is a set of applications which are directly related to the pure descriptive languages.

Consider the expressions of R Warren in the article (in the same 12-13 Notices) about a quantum computer; this is an exceedingly bad article, and it shows the irrelevance of peer-review. The article is incomprehensible, and the writer seems to not know English, yet it is accepted for a peer-review journal, apparently, since it is done by a person working in the industrial side, ie developing an "actual" quantum-computer.

Nothing is clear in the article.

It mainly is a process of listing words associated to a computing process, and then discussing the ways in which one can make sense of the "machine's" output, so as to obtain a "solution" from the very limited outputs, which the (a) quantum computer can provide.

It seems that it is more valuable to simply try to make-up a story associated to the list of words, which "the author provides the reader" than trying to report any actual contextual information provided in the article, which is written in an incoherent manner.

This is how the applied-side (of the failed dogmas) of math and science is represented to the public.

So when N Chomsky, who is supposed to be a linguist, gets-up and says that "one can only trust the peer-reviewed 'consensus' of the science community," or the (Justice) court claims that it must rule "based on proven science and math ideas," this is a way of validating the actions of the investment class, based on a failed dogma of the science and math "consensus."

Where the "real" basis for such dogmatic narrow favoring of the ruling-class is the violence, which is expressed in the justice system.

The media is all about identifying the set of

This is typical of the "highly narrow" specialized descriptive language which is associated to a technology, so a technical community possesses its own "special set" of technical-words. Thus, this author of an article about quantum computers, is talking only to other experts and he does so in their own "coded language." But in regard to this tech-talk, this expert seems to not be capable of describing what he is doing, so, apparently, he does not understand what he is doing, yet the author is well-versed in the machinations of the technical lab-setting, ie the language forms an outline for lab activities (but the relevance of the language is not questioned). This is a natural problem with all narrowly defined languages. This is an important issue, in that highly specialized and narrowly focused languages leave the emphasis on the narrow context, and it does not attempt to explore a wider-range of patterns, which might be used to better organize the subject, and which might also be used to relate these patterns to a larger context, within which the language can be better organized in relation to other categories of quantitative descriptions, thus, making the information more accessible to more people. But this is exactly what the investor-class does not want to happen, since it implies more risk and more competition.

This can be an example of the professional technical math and physics communities "prancing around" with the bankers, reveling in their superiority (higher-wages) granted to them by the bankers (the investor class).

This same phenomenon of the "limited reach" (limited range of applicability, or of the, so called, correct descriptive capability) of a fixed precise language, also persists in the other applied article about "Hearing the shape of a tri-angle," though the authors of this second article are quite polished writers, and they explain the complicated context of their fixed language very well, yet Zeilberger's complaint about the limitations of a formal language again (still) "rings true," but Zeilberger's criticism "misses the mark" the real point is, is that the "pure and the applied" languages are consistent, in regard to the applied contexts of the investor-class, and that this consistency is far too narrow, since the models "do not quite work" (where, according to "Godel's incompleteness theorem," this will always be true for precise, technical languages).

That is, it is the narrowness of the formal language, and the narrowness of the set of things, to which math is actually-applied, which is far too narrow. This narrowly defined professionalism, is being controlled by the investor-class, and the way in which the propaganda-education system causes the population's interests to conforms to the investor-class.

The point of a professional (or academic) superior-intellectual science-class is all about the control of the activities of this, so called, intellectual-class by means of investment interests.

This is done, by testing students by the, so called, authorities, in relation to some arbitrary set of highly desired intellectual contexts, where the authorities are defined as those who focus their efforts on a narrow way in which to use language and this narrow viewpoint is built into institutions which serve the interests of the investors (where, inside institutions, all these people are, essentially, controlled "by force" through the justice system and a wage-hierarchy [a justice system based on property-rights and, a subsequent, minority rule]). The authorities have to maintain the fiction that they are always capable of solving all problems, but simply reviewing the failure of the technical languages today, ie the most fundamental stable systems are described as being too complicated to ever describe yet the stability of these fundamental systems implies they are linear and solvable and thus controllable. This means that the great ability of these so called superior intellects to solve the technical problems of the world does not exist, it is a fraud perpetrated on the public by the propaganda-education system which serves the interests of the ruling-investor-class.

The point of Socrates is not that he could guide a student to a (specific) geometric proof by means of inquiry, but rather that the human intellect naturally puts order onto the world, and this order can be guided by (any) one's own intent, in particular on (any) one's creative intent.

That is:

Education is about exploring the plasticity of language in regard to the motivating creative intents of the learner.

There is no right and wrong which can be tested, except in specialized cases, but then the language fails when it is applied to other cases.

It is rectifying this failure of a formal language which is central to education and the relation that education has to both creativity and to creating new creative contexts.

This problem, technically, is about having a reliable context for measuring and to have a descriptive language which is based on stable patterns.

In math, the idea of a stable pattern means: either stable shapes or the stable and consistent quantitative and algebraic context of the real and complex numbers (where these number systems have shapes associated to themselves), and the relation which these "quantitative structures" have to their stable uniform unit of measurement, ie the relation of quantitative consistency to both a uniform unit of measuring and to the shapes of: the line, the cube, and the circles, and the subsequent, the stable circle-space shapes, as well as cylinders, and to continuously-commutative coordinate systems, which are used to describe (contain) these stable shapes.

Note: The prevalence of the (non-linear) spherical-symmetry in physics, where physics has been based on the idea of materialism, ie based on a local spatial-displacement description of a material component confined to 3-space, is a result of the relation of (local linearly-modeled) material-interactions (for relatively free material components) to the 3-sphere shape of SO(3), which is the algebraic-geometric basis for the derivative-connection which is being related to translational and rotational spatial displacements (for a sequence of fixed [very small] time-interval related spatial transformations) in Euclidean 3-space, where inertia is a Euclidean phenomenon, ie spatial-position is a property of Euclidean-space.

In the article on quantum computing it is not clear if a qubit . . . , ie the main attribute of a quantum computer is the qubit, where the qubit is related to input and output and it is the thing which is processed, . . . , is either (1) a pair of "separated" lattice-sites (apparently, in a 2-plane crystal, and apparently each site possessing an attribute of a spin-½ property), or if it is (2) the superconducting domain characterized by the magnetic field tangent to its domain boundary. (the author acts as if these are equivalent representations [or models] of the qubit)

Furthermore, the energy formula for the Ising model of spin-½ coupled 2-plane crystalline lattice-sites is identified (note: spin-½ coupling would be magnetic-dipole coupling).

Since the author does not explain anything, it does well to make stuff-up, even if wrong it provides a context for clarification, or it may also provide a new set of ideas,

Apparently, the process of the processor is the formation of superconducting domains in a superconducting ("fluid") material within which there is a 2-plane lattice, where the lattice sites possess spin-½ properties before the superconducting state is imposed, (there might also [need to] be an external field imposed on this simple structure) and the lattice is quite small, apparently, composed of only a few atoms (apparently, otherwise quantum randomness interferes with the descriptive context, so that coupling to such a system is not "controllable"). Since a superconductor cannot have a magnetic field within itself, apparently, the pairs of spin-½ coupled lattice sites, (which exist before the superconducting properties [or superconducting-state] is realized), form into superconducting domains, which couple the separated spin-½ lattice sites, and whose boundary-surfaces are defined by their possessing a tangent magnetic field (during the superconducting-state of the 2-plane lattice and its containing pool of "fluid" which possesses superconducting capacities).

It is not clear that quantum properties are actually being identified and used in this process, especially, when the lattice is so small, and if the computing process is an, essentially classical, set of properties associated to the superconducting-state {which apparently, by some resonance properties with the containing space, form in an instantaneous and discontinuously discrete manner, into toral superconducting domains which pair distant lattice sites eg the same answer concerning resonances of the system's newly formed components (or superconducting toral-shaped domains) with an, apparently, external context, but which defines the true containing space, (where the same answer is to be given in regard to the question of; "Why do Dehmelt's isolated electrons form into semi-stable quantum states of a circle-space?"}. This provides deep insight into what they are doing in both cases, ie Dehmelt's-electron example and this superconducting-state "computer processor" model [if, in fact, this made-up story is the computer set-up, which this particular Notice's quantum-computing article is about?].

AMS Notices December 2013 issue, in Opinion, by D Zeilberger, namely, that claim that; the useful math, which is not the math of pure mathematicians, ie their (pure-and-rigorous) math, but (the useful math is) the much more effective and efficient non-rigorous mathematics practiced by theoretical physicists, ie in QFT.

To claim that "pure math is too formally rigorous and that applied math is non-rigorous and very useful" is a disingenuous claim, since both bodies of work are using the same fixed language; the same assumptions, contexts, interpretations, containment sets, models through which interpretations are made, and organization of language, and they are both motivated by the funded and invested interests of the oligarchical (US-European) society.

If one reads the physical literature one finds the same language and methods as one finds reading (either pure or applied) math books. The categories are slightly different, but the basic subjects are: quantity, measuring, shape, geometry, and randomness.

The physical literature is all about framing, ie identifying a context, and then applying the so called laws of physics, ie identifying either a single partial differential equation or a set of partial differential equations which supposedly model the physical system (so as to also be associated to the system's dynamic properties), but it is done in the same math context as the pure math people consider, and that context is about: either non-linear and indefinable randomness, or simply indefinable randomness and a set of non-commutative operators, which in principle one tries to select "so as to commute."

That is, it is the same language and the same context and the same interpretations, where "in measurable physical descriptions" it is always assumed that materialism is true, though the descriptive structures can push to the edge of this materialistic viewpoint, but the allowed interpretations, which try to push to this edge, always revert (or refer) back (to observed material processes) so as to be consistent with the idea of materialism, and that the material-system's properties are to be determined by the (non-existent) solutions to partial differential equations, yet there are both the Van der Waal's forces and there is the mystery of the, so called, chemistry in living organisms, as well as enzyme chemistry; where hidden properties seem to manifest not from a material source.

Even though if one claims that the math techniques of quantum physics and particle physics are not rigorous, as does Zielberger, this does not mean that these techniques are not given their "own set of axioms" and used very formally. That is what peer-review is "all about," ie making sure the particular assumptions are known and that the discussion must also conform to the accepted greater containing set of formalism, so as to lead to a certain type of math formalism, where that formalism is still the same context of partial differential equations and materialism and either non-linear or random (indefinably random) or both but almost always non-commutative, and the applied-people follow this context too.

(where non-commutative can be either locally [ie non-linear geometry], or globally [ie randomness (the unfound operators) and point-wise (ie point-particle-operators) non-linear]

ie in regard to the random models of quantum physics there exist either vague quantitative bounds, eg the shape of the potential energy term, or a particle-collision model).

Peer-review is all about staying within the fixed formal language which is getting funded, or which identify technical projects which have been invested-in.

Truly new ideas require new ways of organizing language and thus, cannot enter the peer-review literature, which requires its special formalized language.

Thus, N Chomsky's apology for the banker's interests in peer-review, and his (Chomsky's) support for peer-review . . . , which he phrases as a demand . . . , is that "new ideas need to be made 'official' by being 'peer-reviewed' (or the public can only trust the new ideas as being valid if they are peer-reviewed, and, thus, a scientific consensus is formed), otherwise they are the work of a nut,"

So as to institutionalize the development of knowledge so that knowledge supports the ruling-class so that knowledge is controlled and knowledge remains consistent with narrow commercial (or investment) interests.

but this is exactly the idea, or the expression of "superior intellectual domination," which was used in the age of Copernicus, so as to not allow the ideas of Copernicus to be expressed by anyone who is in the community of authorities.

Chomsky is a typical propaganda-hound, his basic message is that "people must trust the elites" (thus, he is allowed to have a voice on the media) and as with most media-hounds (or equivalently, politicians) who need to ingratiate themselves with the public (the typical ploy for the personalities who represent an elite domination over the public within the propaganda system), for this (need to ingratiate himself with the public), he expresses "an implied belief in equality," at least, apparently, implying that equality is preferred over arbitrary elitism, where, in regard to this, his main statement is that, "there are a set of very rich few who control so much of society, that their arbitrary demands must be met, or the society will collapse,"

{Note: This is the basic collectivist social structure of the Roman-Empire, where the public collectively supported the emperor.}

yet he does not follow the logical path of this statement to its conclusion, and that is, that by satisfying these arbitrary demands, then one is also placing science and math into a fixed dogmatic context (whose primary function is to support the investment interests of the investor-class) so that within the narrow context which represents the investor interests both science and math have come to fail.

That is, Chomsky's claim is not a belief-in (or a seeking-for) equality but rather, a command to trust the elites, and to trust their dominion over the public, where Chomsky is a servant within the domineering-social-class.

Note: Chomsky is part of the "aristocracy of intellect," and to "the elites" he remains loyal.

He seems to play the role of the southern-gentleman at the time of the civil war, ie keep the slaves (now (2013)wage-slaves) in their place, but where for Chomsky it is the domain of the intellectual-elites to which he is so loyal (but intellectual elitism is as arbitrary as is racism).

Thus (according to the intellectual elites):

Their command to society is that: Do not base education on equal free-inquiry,

{so as to consider knowledge in its many forms, And, subsequently, to consider technical language in many types of different sets of assumptions,} and do not consider an individual as an equal-creator, who requires equal access to both information and to a truly free-market

Where one might want individual equality (in both creativity and in trade)

So as to break-away from the idea of society as a "collective of people" who support the few in the ruling-class, ie those few very rich and powerful whose needs must be met or the society will collapse, where this is the so called, model of capitalism and "democracy."

[support the ruling-elite so they do not destroy society, since I have a nice social-position]

So, Zeilberger goes on to proclaim that the great triumph of applied non-rigorous quantum-math was to detect the Higg's Boson. (ie where the applied community is non-rigorous but they act in the exact same way as does the rigorous math pure-math community)

That is, Zeilberger, like Chomsky, is loyal to the elite-intellectual community, the authoritative community within which they have won the contest (passed the tests), so as to be allowed entry into the institutions, which house those with superior-intellects, and it is these institutions which guide the creative activities into which the bankers invest, so as to have small risks, since no other ideas exist which can challenge those activities into which the bankers invest.

Exactly of "What value is detecting a Higg's Boson?" since it fits into a math context which has no relation, or only remotely relatable, to an applied context, where such an applied context would define an extremely narrow instrumental context, eg the explosion of a nuclear weapon.

Yet the main problems of quantum-physics go without valid descriptions, eg describing the very stable properties of a general many-but-few-body quantum system, eg nuclei, atoms, molecules, crystals, life, and the mystery of the stable solar-system, etc, etc.

Zeilberger is wildly brash in his criticisms.

This is desired in an equal free context of using language since many new languages need to be developed and each such language should be brash.

Though there can be sets of intellectual communities which stay fixed and dogmatic, but just as in the case of monopolistic business take-over of society, these fixed ways of considering intellectual things are far too limited, and result in an oppression "as bad as" is the oppression of racism.

The real point is, is that "there is no set of correct answers through which an intellectually superior set of people can be judged," thus the one's which do compose the aristocracy of intellect are the more dogmatic and competitive, ie seek their sense of value from their external relations. But this is not the natural intellectual context, where intellectualism depends more on a belief in an inner (mental) capability.

Many of Zeilberger's criticisms are, in essence, valid, but since he is making them in a context of "he himself" being an active member of the aristocracy-of-intellect . . . , ie the tried and tested dogmatists . . . , he espouses a state of freedom which is very limited, intellectual considerations and efforts which are corralled into elite institutions, ie institutions which exist within the (already built by means of the investment-class, and the educational institutions which serve the interests of the investment-class) narrow context within which he would naturally function, in regard to his being a contest winner.

Yet the criticisms at a deeper level, ie a need for equal freedom of intellectual expression, are ever-more relevant.

However, it is the so called pure side of the math presentation of the applied, ie "Hearing the shape of a triangle," where the need for much greater freedom, in regard to the use of a precise context for a descriptive language, where the need for greater freedom for many forms of "a disciplined expression" appears. And it is (often very much) related to Zeilberger's criticism of infinity. But it is more easily identified by the occasional very clear phrases which are used in the "'hearing a shape' article."

In the "Hearing a shape" article the problem is clearly identified, the problem (the partial differential equation) D^2u(x,y) = w(i)u(x,y) defined on a bounded 2-plane region in Euclidean space, where w(i) are the eigenvalues, and where an initial condition function, f(t,x,y), (which begins the wave-function solution-function, ie that which plucks, or strikes, the region, or shape) is defined on this region, and where f is zero on the boundary of this region, is solved by (1) defining a function space (ie eigenfunction-space), of {u(i): where i is an element of N}, by means of an associated set of eigenvalues, w(i), so that $0<w(1)<=w(2)<=w(3)<=$ . . . where this sequence is called the problem's spectral sequence, where some of the w(i) can be repeated, ie for some i, and j w(i)=w(j), so that the number of eigenfunctions whose eignevalues are repeated is called the dimension of that eigenfunction subspace.

The article is honest, since it points out soon after presenting the problem of "finding a region's spectral sequence" that the spectral sequence cannot be calculated except for in a few cases: rectangles, discs, and a few triangles (etc).

It is not common for the expressions in the peer-review process to express such humble integrity. Namely, that their viewpoint is most often an expression of failure.

Yet it (the formalism which has failed, nonetheless) expresses the dogma of the intellectual-class, which is "considered to be necessary," since it is "this formalism" which most conforms to investment interests.

Nonetheless

The idea that math should be thought about as an experimental activity . . . , which is looking for ways to organize math patterns . . . , so as to define a new method, or a new context, within which to contain and describe the observed patterns of stable measured properties, so that the observed properties can be used in a practical context. (This is a good criticism)

Unfortunately, the so called applied professionals always are placed in the context of a commercial interests, the investor interests which fund the instruments of the lab within which the, so called, "applied personnel 'work.'"

The point of a description (or an endeavor) being put into an applied context (which is [or might be] different from a commercial interest) is that there usually (or often, or sometimes) exists (or there is needs to exist) an assumption of a system's properties being bounded, but is there really a deeper "assumption, about existence itself," to which such an assumption of "a system's properties being bounded," actually applies, ie that spectral properties only fit into simple stable shapes so that the shapes resonate with an external finite-spectral structure or to which the but the mechanical vibrations of a shape which has been molded into "any geometric form whatsoever" are not likely to need an external spectral set to which the new shape's natural vibrations must resonate, but the idea that "the stable vibrations must exist in a simple shape" seems to have its own merit, ie arbitrary shapes usually do not have a vibratory property which allows resonances to be realized within (or upon) the shape.

How could one determine an arbitrary shape's capacity to carry a set of vibrations to which it can resonate? The way in which this idea about a shape's natural vibratory set to which it can resonate would be related to the set of closed paths which repeat their vibratory path's so as to repeat the path in the same direction in which the path began at the point where the entire period is identifiable. Thus, the representation of the arbitrary shape's possible natural vibrations (natural spectral set) can be considered in regard to the arbitrary shape's relation to rectangular partition of the region, where this is because the rational sloped lines defined in regard to the rectangle's orthogonal axis also define the type of repetitive periodic lines (paths) which also define a natural vibration pathway for the rectangle and thus an approximation of such a pathway for the region where the wavelengths of the periodic pathway defined in the rectangle must be smaller than the sides of the rectangle. One can consider a rectangular partition wherein the re are various size-scales associated to the rectangles in the region's partition. One might not think that there can exist periodic pathways which return to their beginning-point in the same direction in which they left the point but which are very long pathways so that the vibration might dissipate during traversing the long path-length.

Note: Drums, and striking the drum, can define a beat, but not so much a well defined vibration.

There are other shapes such as the conic sections which have a natural definition in relation to the properties of reflection of the conic sections in relation to focal-points and directrixes.

Otherwise the methods of analysis, which are used to determine the region's spectral sequence, might only identify arbitrary values for the spectral sequence which only have meaning in regard to these values being used to define a converging sequence within the context of the values which the analysis can identify in a context of converging sequences.

Existence's properties . . . , ie using math patterns to describe the properties of the physical world so that the description both contains a wide range of physical attributes and these attributes are described so that they are easily accessible and thus usable in a practical context, . . . , are not about identifying arbitrary spectra . . . , where it is assumed that spectra identify fundamental properties of physical systems . . . . , in an improperly formulated context for description (improperly formulated since the information needed for the problem's solution is not available in the problem's [abstract method of] solution)

Rather

Its about how stable spectral structures emerge, ie enter into being (or in some cases condense), for no apparent reason from an unordered state or from a higher-dimensional order an order, the spectral system can emerges from chaos and into a stable patterns. Order emerges from stable patterns which are defined in the containing space itself, ie an intrinsic part of existence.

One way in which to realize the origins of a vibrating system's spectral properties is to form a cubical partition which approximates the vibrating region's boundary shape where there is a size-scale hierarchy for the partition an increasing size-scale sequence so that there is a set of bounds (or approximations) on spectral wave-length size-scales beginning with the smallest wave-length size, where geometric properties of the boundary geometric structures whose size-scales are below the size-scale of the vibration's smallest wave-lengths are not seen by the waves so as to not respond to these small geometric properties in regard to the incoming-large-scale-wave reflecting in a way which is sensitive to these small shapes that is there will be a smallest cube size-scale about the size of the smallest wave-lengths of the vibrating system. Then there can exist other wave-lengths which are not related to the smallest wave-lengths of the vibrating system. These other wave-lengths might be estimated by geometric properties of the region

* * *

The first comment is that this function space context, may be over general, and based on too limited a sense about modeling vibrations when they are defined within the context of a continuum, so that the infinite structure of convergence, upon which this analytic math structure is based, requires only an arbitrary set of quantities, but when placed into a construct of convergences, which is defined on a continuum, can identify, by convergence to, any possible vibratory state, but "are these spectral sequences, upon which the convergence processes depend, truly related to the region's natural vibratory resonances?"

When defining a wave-equation's partial differential equation, the model depends on the pluck or strike, which puts the vibrating medium into an oscillatory state do the pluck's position within the region, where the medium is defined, determine the vibratory properties . . . . , ie do the vibratory paths (defined when the region's boundary exists and cyclic paths can be defined when the (amplitude, or displacement from the medium's equilibrium condition) value of the wave-function (and/or initial conditions) is always zero on the boundary on the region's medium) . . . , depend both on the shape of the region and on the position on (in) the medium, which is struck (the position where the medium is struck), so that these propagation-paths (of the wave-propagation in the medium) are cyclic (periodic)?

How does this depend on the natural coordinates of the region upon which the medium is defined?

Are the eigenvalues, w(i), of a spectral sequence (defined in the given problem) arbitrary numbers, which are only relatable to vibratory resonances (or cyclically defined displacements of the medium), by means of a convergence or limit process?

That is, is this problem so poorly formulated, as a partial differential equation, that it provides only the most limited amount of information about the natural vibratory properties, or cyclic (reflecting) paths, of the region's shape?

There are the other questions about,

(1) where does the set of eigenvalues come-from, in regard to natural geometric considerations about the region's shape within which the vibrations are defined?
or
(2) where does the set of eigenvalues come-from, in regard to other (related) types, of partial differential equations?

In regard to (2) there are two other types of partial differential equations which could be applied to this problem, namely,

(A) the heat equation (parabolic case identifying the "bounded vs. free" context for identified properties) and
(B) the wave-equation (the hyperbolic case the context of being independent, eg independent paths of wave-propagation)

(B) models the most natural physical construct for a struck region's vibrations.

The heat equation is related to mechanical processes, since mechanical energy can couple to the medium which is vibrating, but this would be small, for the resonating vibrations identified on the region, and larger for the dissipating, or unstable, vibrations.
Mechanical resonances depend on stable shapes and on the positions upon the shape which is struck. For example, the center of the circle (or disc), the focal-points for an ellipse, etc.

Part of the set of difficulties which emerge from "understanding the problem" is in relation to the mental images of the vibrating structures that are considered.

Also

Another question is (the question about which this article is concerned):
Does the spectral sequence determine the shape of the region?
No!

There is the very important question:

Can the spectral sequence be found by math methods?
No! (at least not by analytic [or function space] means)

Thus, this math problem is similar to quantum physics but it is more geometric, where the real question is:

Are arbitrary spectral problems, which are formulated (as partial differential equation boundary value problems whose solution functions are dependent on their relation to being placed in a function space) in the context of function spaces, actually formulated in a logically relevant context, ie do their hypothesized solutions provide any valid information?

Where both of these questions are made clear and answered with yes and no answers in the article.

This article is about:

What aspects of the spectral sequence can be determined by the region's shape?

Is the dimension of an eigenfunction subspace relatable to a given wave-function (ie solution-function) having a "congruent wave-function" with a spectral sequence defined on 2-dimensions, as opposed to the 1-dimensional spectral waves associated to the given spectral sequence?

That is, For a 2-plane region are there many independent directions which can define the same eigenvalue, and are these direction that which determines the dimension of such an eigenfunction subspace?

One can see that there are assumptions, and hence other questions, built into these questions.

If this problem is not actually numerically or functionally solvable then how can it be applied?

One answer might be that the physical systems, to which it might not be applied, have bounds in regard to the waves (or wave-lengths) which exist.

The article then proceeds to provide the analytic framework through which finding the shape . . . for the given region (with the unknown shape) based on a spectral sequence . . . might be realized.

This analytic structure is full of assumptions and constructs concerning the ability to define "functions" (or distributions) which, in turn, depend on a convergence process, defined on sets which are continuums, and whose physical basis is quite questionable (and depend on using models of infinity).

That is, finding experimentally a physical system's spectral bounds (a physical system to which one is trying to determine this problem's solution) might lead to a better quantitative model, which can be used in a practical sense, but such a physical model does depend on stable patterns, for the model to be of (any) practical use, and these patterns will be shapes, which are (most often) going to be related to circle-space shapes.

* * *

Nonetheless, at all points of concern the descriptive context of the article can help lead one to other questions, since the expression in this article is quite clear.

That is, even though the context and assumptions of this problem seem to be quite questionable at a very fundamental level, nonetheless good exposition can lead to some useful considerations. That is, any expression of ideas can be useful when considering knowledge and creativity. However, the problem with the given exposition is that these types of questions and math formulations have been around for nearly 100 years without all that much success at systematically being able to use the properties of vibratory systems. So "why is this type of discussion the main focus of academic institutions?"

But

Zeilberger's, and Chomsky's, criticism leads to a context, which is essentially the same as the pure academic context, where the labs, within which people such as Zeilberger would work (as a superior intellectuals), are mainly doing things which are of interest to the investment class (and which are hyped by the propaganda-education system) and not to a wider range of ideas in regard to how language and knowledge can be better related to practical creative efforts

It should be clear that when math models are applied to practical creativity one wants (reliable models of measuring and the language to be based on stable patterns) descriptions which can actually describe observed properties and one wants the context to be about controlling these properties, but, where the observed stable patterns of physical systems imply that these systems are controllable and if formulated as a partial differential equation then they are solvable and controllable.

Consider that if one looks-at electromagnetism, the geometric patterns are:

1. spherically symmetric, ie non-linear,
2. cylindrical,

3. toral, and
4. cubical, and
5. magnets are dipoles, and
6. the useful coordinate context is continuously commutative, ie locally always orthogonal, almost everywhere;
7. (consider) where the multi-pole contexts for charge distributions are toral, or almost toral, where the coordinates are mostly continuously commutative (always locally orthogonal).

An important problem about the abstract considerations in regard to the problem of hearing a (general) shape, (eg can one hear the shape of a drum?) is related to the obvious fact, that drums are basically cylinders (or have symmetric geometries), and this, apparently, is because irregular shapes do not have a clear type-of-sound (thus not useful for musical [or beat] sounds), ie they are non-commutative. That is, placing the problem in too general of a context leads to a discussion of properties which cannot be identified in the non-commutative math structures of this general context.

Thus one might better study the eigenvalue (or vibrating space) problem by considering the stable shapes, which are primarily the stable circle-space shapes. These shapes have obvious symmetries: spheres, cylinders, cubes (especially if discretely defined) and the discrete hyperbolic shapes where a relatively few minimal subset structures fit into mostly a set of orthogonal hyperbolic coordinates, so that periodic pathways can be identified in regard to these symmetries, and where, perhaps, the most rich shape, for many vibratory possibilities, would be on the cube (or the rectangle), where the many rational slopes define a closed cyclic pathways, upon which waves can propagate, ie move, (within an integer, or a discrete rectangular lattices within Euclidean spaces, ie the (Euclidean) spaces where spatial positions and spatial displacements can be defined).

Since the problem (of hearing a shape) is Euclidean, both the heat equation and the wave-equation, supposedly, can be relevant, where the wave-equation, of course, seems to be the more natural context, and it is the wave-equation's description of a function-space's association to the problem's physical properties, which are, in turn, related to analytic constructs called the "heat-trace" and a "wave-trace," where the wave-trace is easiest (or the most geometrically motivated in its description) and they describe a set of closed-curves which relate to the initial position, ie regional-point, where a wave might be begun with a pluck (or by dropping a pebble).

In a context of stable circle-space shapes (identifying a sustainable [or measurable] vibration) these would identify the rational curves, which, in turn, are related to the circle-space shape's relation to cubical (or square) fundamental domains. This would mean that solving this problem is related to partitioning the region into cubes where the size of the cubes depends on a small set of eigenvalues [eg where one of these being the smallest eigenvalue, ie w(1)], where this small set of eigenvalues can be associated to the region upon which the vibrations are defined, and then selecting rational numbers related to this small set which defines the scale-sizes of the regions vibrations related to the vibrating-system's wave-length size-scales, which are consistent with the spectral sequence.

Perhaps the region can possess a finite sequence relation, ie a finite sequence of different size-scale cubical partition structures, to which the full set of rational numbers can be related so that the spectral sequence of the vibrating-system can be identified.

Zeilberger promotes (what he states to be) the RSA-algorithm without any clarification yet he scolds the expert community for their lack of clarity for their technical presentations. He points out, correctly, that communication in math is dysfunctional yet he is a dysfunctional practitioner himself, when he refers to mathematicians he calls them talented, where incompetent would be better but apparently (or clearly), Zeilberger is an elitist criticizing the elite. If one wants both relevance and clarity then one should be about creating new language structures which challenge the assumed containment context, where the language of a professional talk is elementary about assumptions interpretations etc, but this would open-up the floor to the non-expert, rather than the already indoctrinated undergraduate.

Zeilberger talks about the "religious" fanaticism of the professional mathematicians, as if the professional physicists are not also a bunch of narrow dogmatic "religious fanatics" too, where they are being paid to focus on such a narrow context of thought, since the wider context easily reveals the utter failure (of that over-all viewpoint or set of professional assumptions), to which these assumptions lead, where Zeilberger criticizes math rigor "based on 'axioms' which are completely fictional" . . . , (really there needs to be a rational discussion about sets of assumptions as a basis for language and fantasies; "How can assumptions best be organized so as to get-at a descriptive language which can describe the observed patterns?" is the main problem.) . . . , where his example of a fictional axiom is the idea of "infinity."

To his defense the main attribute of the physical sciences should be the type of observed properties which are bounded, and can be placed in a lab and studied, so that the main attention which is now in the professional physical journals are about the "infinitely small point-particle Higg's particle" and the claimed to be bounded, but there is no valid reason for such a descriptive context other than a particular interpretation of the empirical data which is interpreted in a very absolute and dogmatic theoretical context, in fact the true context of an assumed to be expanding universe (in the relation to the observed red-shift) might be about the conjugations of SU(10,1) Lie groups in regard to the structure of galaxies, rather than the, assumed relation (in a material based language) to properties, which are interpreted to be related to general relativity, of the big bang, ie the focus is on particle-physics and general relativity, an essentially unbounded idea which cannot be studied in a lab. (Whereas many data are taken from an infinite context and then interpreted in a very narrow dogma and then it is claimed to be a measurable support for an idea but the idea is assumed in the interpretation),

Both (particle-physics and the, so called, "practical side" of general relativity, ie the big bang and black-holes) are math models with dogmatic interpretations which determines their, so called, descriptive relevance, but in both cases they are dysfunctional descriptive languages.

However, it is really only the bounded shapes . . . , (or both closed and bounded stable shapes, depending on one's containment viewpoint, and on what stable properties, (of, apparently, bounded systems) one is trying to understand) . . . , which can be related to stable properties, ie the systems which actually do fit into labs.

Zeilberger then states that for "the research and teaching we get paid-for" "we should adopt a more open-minded attitude towards math-truth" (now the absurd statement), "similar to the standards of the "hard" physical sciences," and

This, he claims, to be a wide-open way of doing math, without any sense of a realization about how commercial interests are dictating the context of professional dialog and the research which the paid-professional is doing.

The issue of AMS Notices, Dec 2013, goes on to provide two articles about the application of math to math-physics problems which are heavily funded:

One would (might) be considered pure math and the other applied math.

Then Zeilberger goes on to laud "experimental math" without providing the reader the context of "experimental math." but, apparently, Zeilberger's experimental math is all about working with instruments which are placed into an interpretive context of a very fixed precise descriptive language,

Whereas experimental math might be better described as doing thought experiments, made popular by Einstein, but where one considers the basis for building a quantitative and measurable language and asking "what one wants to create?" and then considering "how such a creation would be relatable to a measurable context, where shape is central to the metric-space (with an invariant metric-function) descriptive language of the space within which one is measuring."

The two types of application articles

1. "Can one hear the shape of a triangle?" (apparently, pure math, continuous geometry) And
2. "Using a Commercial quantum computer" (apparently, experimental math, discrete geometry, supposedly, in a random context)

In regard to

(1) this is a problem in continuous geometry in a metric-space, apparently, Euclidean 2-plane, where the model for geometric problems is about local linear measures are used to identify physical law. The continuous geometry is usually either the shape of the 3D structure or the shape of the 2D boundary, but this is a 2D problem but the 3D drum spectral problem is also considered. The physical law is the second order partial differential equation of a vibrating shape where the vibrating medium is bounded, ie the spectra are related to the shape of the boundary, ie the vibrating medium's bounding shape.

While
In regard to

(2) this is typical of a highly funded and narrowly defined research project in that it is essentially completely incomprehensible, but apparently, at least he image provided is that it requires a team of people in order to couple a quantitative problem to the instrument's processes so that the quantitative pattern can result, from the processes, in (with) a related quantitative property which is the solution.

Vaguely in the article there was mentioning of:

Ising's formula which represents the coupling of spin-states (as magnetic-dipoles) between all the different distinctive crystalline lattice sites, in a lattice which is contained in a 2-plane, so as to model a crystal's spin-property relation to ferromagnetism, where Coupled lattice sites, are A natural part of Ising's model.

Superconductivity is a property, apparently, imposed on the instrument during the computing process,

The existence of a magnetic current between coupled lattice sites

The coupling of pairs of spin-states between distinct separated lattice sites of the 2-plane crystal would normally be a magnetic interaction, so if two sites do happen to spin-couple together during the process, then the spin properties either align or anti-align, but in a superconducting context would (could) this define loops of magnetic-fields where the magnetic-field (or vector-field flow) is confined to be parallel with the loop's surface. (?) Is this a physical attribute or an abstract way of talking?

This seems to be a classical interaction between stable lattice sites but some lattice sites do not interact, they are independent and do not see one another, ie so they do not interact. But when many of these interactions between many lattice-sites occur, then the randomness, apparently, keeps the counting "associated to combinatorics" from being distinguished, and the quantities (in the structure of the process) lose their order (or lose their meaning), thus, the need for only a few lattice sites in this machine.

The super-cooled instrument, ie a 2-plane crystal lattice of spin-properties, apparently, is further perturbed, and time is allowed for the spin-lattice to achieve some low energy-state, in which the Ising sum over spin-coupled lattice-sites of the crystal becomes a minimum.

It is called adiabatic which means that during the process the heat energy is a constant.

Apparently, there are formal theories in which it is formally shown that combinatory problems defined on the 2-plane can be optimized with this process.

Perhaps due to non-local quantum-interactions, and the subsequent collapse of the system's wave-function, ie the processor-part of the computer fails, and apparently, this can happen for 100 machine actions in a row, so that the output has no relation to the solution which is desired.

The important question, which current physical theory has no answer, is "how and why does a crystal form with well-ordered lattice positions for the components (which compose the crystal)?" "What is the condensation process?" and "How and/or why does the lattice form?"

This seems to not really be a quantum system based on random distribution of components about the "form" which is possessed by an energy operator, rather it is about a highly ordered system where distinct lattice sites seem to interact or couple in a random context

this is a discrete problem also in a Euclidean metric-space, where (in the new descriptive context, the rigid discrete structure is brought about due to the rigid micro-spectral properties associated to stable shapes which make-up the components which in turn), discrete components compose the discrete crystal lattice. The N components of the crystals are modeled as bar-magnets (or possess spin properties) and are in "up" or "down" states, at each lattice site, and a sum over the orientations of the individual bar magnets in the discrete crystal determines and a linear term which is also a sum are used to identify the system's energy function.

18. Diagrams

This set of diagrams represents a symbolic-map which can be used to help identify a set of analogous higher-dimensional diagrams (or an analogous set of higher-dimensional constructs), where in lower-dimensions these diagrams are consistent with the observed material patterns, though now the ideas of either materialism . . . . , or existence which is contained within a greater set of higher-dimension so that the higher-dimensional analogous constructs possess the properties of macroscopic geometries . . . . , is given a new interpretive context.

(in the Diagram section) The diagrams provide a succinct outline of the simple math, which is based on stable discrete (hyperbolic) shapes, and which is the basis upon which the stability of measurable description . . . . , of the observed stable, definitive properties of physical systems . . . , depends.

The diagrams provide a clear picture (or clear analogy, or clear map in which to think about moving into the higher-dimensions) of the context within which these stable discrete shapes (of non-positive constant curvature) are organized, so as to form a many-dimensional context (whose higher-dimensional properties should be thought of as being macroscopic) of both component containment, and component interaction.

The many-dimensional containment set, possesses a macroscopic and stable geometric context, which is composed primarily of "discrete hyperbolic shapes," which, in turn, are contained in hyperbolic metric-spaces, (wherein it is true that each dimensional level, except the top dimensional level, has a discrete hyperbolic shape associated to itself).

The finite set of stable discrete hyperbolic shapes, which model both the different dimensional levels, as well as the different subspaces of the same dimension, is a geometric foundation upon which the construct of a finite spectral set depends (a finite spectral set for all existence, contained within a high-dimension containing metric-space, ie an 11-dimensional hyperbolic metric-space).

Each dimensional level (and each subspace of any dimensional level) is associated to a very stable "discrete hyperbolic shape," and each subspace (within the many-dimensional set) is characterized by a size-scale (determined in relation to the finite spectral set), where the size-scale of a dimensional level of a particular subspace is also determined by a set of constant multiplicative factors, which are defined both between dimensional levels and between different subspaces of the same dimension.

The fundamental properties of the high-dimension containing space are determined within an 11-dimensional hyperbolic metric-space.

Hyperbolic space is analogous (or isomorphic) to a general model of space-time defined for various dimensions, eg R(3,1)=[space-time], while generally, R(n,1), is a "general space-time."

However, there are also various other "metric-function signature" "types of metric-spaces" of the various dimensions and metric-function signatures, R(s,t), which are involved in the description {where s=space dimension, and t=time dimension, where s must be less than or equal to 11 (it seems (?)), and s+t=n}, most notably the "discrete Euclidean shapes," in R(s,0), which possess properties of continuity (of size), which is needed in the interaction process.

These diagrams identify the context in which "material" components exist within each dimensional level, and they identify the context of both "free" components (associated to both parabolic and hyperbolic second order partial differential equations in regard to inertial properties), as well as orbital components (associated to both elliptic and parabolic second order partial differential equations in regard to inertial properties).

These diagrams also (pictorially) show the basis for "material" interactions, and the relation that a new material system, which is emerging from a material interaction, has to being resonant with values of the "finite spectral set" which is defined for the total containment space.

These diagrams provide a context for the emergence of new stable systems from material interactions.

These diagrams of "'material' component interactions" can be identified at any dimensional level (dimension-2 and above).

Note: Interactions are constrained by:

1. the process itself,
2. dimension,
3. size,
4. subspace, and
5. a finite spectral set,

where the basic form for such "material" interactions has an analogous structure (or is "the same") for each dimensional level, though there are differences, in regard to the properties of material interaction, between the different dimensional levels, which can be due to dimension, subspace, and size.

The diagrams give low-dimension pictures of the very simple, quantitatively consistent, geometric shapes which are stable, ie most notably the discrete hyperbolic shapes, and it is these stable shapes upon which stable mathematical patterns can be described in a context where measuring is reliable, and because the description is geometric this means that the description can be very useful.

Note: Following "the diagrams themselves" there is a section in which the descriptions associated to the diagrams are re-written in reference to the number of the diagram, eg 1, 2, etc. given on each page of the diagrams.

If one cannot read the words on the diagrams, they are here (below), where they are associated to the numbers of the diagrams.

1. If the circle is rigid in its shape then the complex plane defines a commuting number field, as does the real-line.

Linear measuring directions are perpendicular (or they are independent of both one another's measured values as well as measuring directions) a complex number, z, is represented as, z=x+iy=r(cosW,sinW)=re^(iW), (W is a measured angle)
Line segments and circles are quantitatively consistent shapes.

2. The following (above) shapes are quantitatively consistent shapes, and locally their directions are independent, and form commutative algebraic constructs at each point over the entire shape, ie global commutative algebraic constructs, locally linear and invertible [one-to-one and onto] and this is true everywhere on the shape.
3. Cubical (or rectangular) simplexes are related to circle-spaces by means of "equivalence-relation topologies," or equivalently, by a "moding-out" process.

   On such shapes local geometric measures are based on either measuring rectangular shapes or (equivalently) by a measuring process based on tangents to the circle, which is used as a basis for measuring along a circle's curve, eg rdW=dx+dy along a circle.

4.* Lattice in the "hyperbolic circle"
This lattice is more restricted than would be rectangles attached at vertices

5. Discrete hyperbolic shapes are composed of toral components

The number of holes in a discrete hyperbolic shape is called the shape's genus

2-holes are surrounded and caught by 1-curves, and defined by 2-dimensional discrete hyperbolic shapes
3-holes are surrounded and caught by 2-surfaces, and defined by 3-dimensional discrete hyperbolic shapes etc

The faces on the fundamental domain (which result from the faces of a hyperbolic shape's rectangular (or "cubical") simplex) form very stable spectral measures on the very stable shapes of the hyperbolic space-forms.

Discrete hyperbolic shapes, or equivalently, hyperbolic space-forms, have open-closed metric-space topological properties, and they may be bounded or unbounded shapes, but all existing hyperbolic space-forms which are 6-dimensional or greater are unbounded shapes, and the dimension of the last known hyperbolic space-form is hyperbolic 10-dimensions (Coxeter).

Orbits on discrete Euclidean shapes and discrete hyperbolic shapes
Subsystems (or sub-metric-spaces) either occupy spectral orbits or they are "free"

6. The shapes obtained from this rectangular simplex for a 3-dimensional hyperbolic space-form . . . , which contains a 'free" 2-dimensional hyperbolic space-form . . . , are contained in a 4-dimensional hyperbolic metric-space.
7. The separation of two hyperbolic material components is, r, where, r, is defined between the two vertices. Take smallest toral component of each hyperbolic space-form, average the sizes of the two toral components, as is also done in center-of-mass coordinates. Represent the average value as a pair of equal oppositely positioned (rectangular) 2-faces, so as to define a right rectangular volume whose separation, r, is the distance between the vertices of the original interacting hyperbolic space-forms.

Then define an interaction differential 2-form on the geometry of this 3-dimensional Euclidean torus, which is contained in Euclidean 4-space. This determines the force-field, defined between the interacting material components.

The local vector geometry of the differential 2-form is relatable to the local geometry of the fiber SO(4) Lie group, since the 2-forms in Euclidean 4-space have the same dimension as SO(4) the geometry of the spatial displacement is determined in SO(4) by it geometric relation to Euclidean 4-space which is given by the 2-forms so that a local spatial displacement occurs due to a local coordinate transformation with SO(4) acting on the positions of the vertices [(in relation to center-of-mass coordinates) of the original pair of interacting hyperbolic space-forms] in Euclidean 4-space.

If the force is attractive and if the interacting (charged) material (the interaction structure) is contained within either 2-dimensional or 3-dimensional, or 4-dimensional Euclidean space then the force is radial and attractive, and r is made smaller. {Note: If the material is of a new type (oscillatory) and contained within Euclidean 4-space then the force-field has a new geometric structure contained in a higher-dimensional Euclidean space.}

Then the same type of process repeats, for time intervals determined by the spin-rotation period of the spin-rotations of opposite metric-space states (about 10^-18 sec).

In this process the Euclidean torus which forms for each discrete time interval, forms in the context of action-at-a-distance.

Classical partial differential equations are defined within very confining and very rigid sets of both discrete hyperbolic shapes and action-at-a-distance material interaction Euclidean toral components which link the hyperbolic material together, so that the force-field differential 2-form is defined on the torus. (7)

The above interaction for material contained in 3-space results in a 4-dimensional descriptive context, but it can be symbolically represented in 3-space.

8. It should be noted that in this new descriptive context eigenfunctions would also be both discrete hyperbolic shapes and discrete Euclidean shapes (tori).

   Forming new stable hyperbolic space-forms from a material interaction. Assume an attractive (or repulsive) interaction in 3-space, then the interaction of "free" material components would be similar to a collision of components, if the material components get very close during the (collision) interaction then if the energy of the over-all interaction is within (certain) energy ranges and the closeness allows resonances (with the spectral set of the over-all high-dimension containing metric-space) to begin to form, thus forming a new state of resonance for the interaction simplex, so that the over-all energy, as well as the resonances, allow a "new" stable "discrete hyperbolic shape" (in the proper dimension of the interaction) to form, so a new hyperbolic space-form emerges.

9. Weyl-transformations between two maximal tori within a Lie group (rank-k compact Lie group) Two intersecting circles of the two maximal tori may be angularly related to one another by Weyl group transformations, where the Weyl group defines the conjugation classes of the maximal tori which "cover" the compact Lie group.
10. Forming angular changes between toral components of a discrete hyperbolic shape by using Weyl-transformations, which change the angular relations between circles which compose a toral component of a hyperbolic space-form. These Weyl-transformations allow Envelopes of orbital stability for "free" subsystems (or sub-metric-spaces) to be defined.
11.* But rectangles attached at vertices and then moded-out, "without expanding the vertex," shows a model of a discrete hyperbolic shape's toral components, represented as separate tori attached at separate vertices. (This is to emphasize an apparent toral component structure of discrete hyperbolic shapes, which is an important aspect of these discrete shapes.)
12. Various types of unbounded 2-dimensional discrete hyperbolic shapes
13. The figure titled "The mathematical structures of stable physical systems," represents information similar to figures 9. and figure 10.
14. The figure titled "Partitioning a many-dimensional containment space" represents two different dimensional levels, where the 2-dimensional level is identified as an un-deformed rectangular lattice, where a deformed lattice shape (in 2-hyperbolic-dimensions) is given in figure 4. Whereas in this 2-dimensional figure there are contained representations of 1-dimensional shapes. The other (larger) representation of the 3-dimensional partitioning structure are the un-deformed "cubical," or right-rectangular, 3-dimensional shapes, which contain within itself the stable 2-dimensional discrete hyperbolic shapes, this process of partitioning space can continue up into higher-dimensions, where a deformed

3-dimensional lattice shape can be moded-out to form into a geometric-shape, which would exist in 4-dimensional hyperbolic metric-space.

15. The figure titled "Perturbing material-components on stable shapes:" shows an atomic orbital structure which is a stable geometric structure with electrons in the outer-orbits of concentric toral-components (folded into their stable shape) and the nucleus in the center small orbital shape, where the electron's orbit is mostly held stable by it (the electron) following the geodesic path, which is defined on its toral component, and the electron is also interacting as if in a 2-body interaction with the nucleus, so that this 2-body interaction (most noticeably) perturbs the orbit of the electron, eg perhaps causing the electron to possess an elliptical path, where these orbital deformation may result in the variations in the details of the atom's discrete energy structure, where the stable orbital shapes, perhaps related to various Weyl-angle shapes, are the basis for the atom's stable discrete energy structure.
16. The figure titled "Describing the dynamics of 'free' material components in higher-dimensions" is a diagram quite similar to figure 8.

Chart of the face structure for rectangular simplex geometry

2-rectangular-simplex
1-face
vertices

3-rectangular-simplex
2-faces
1-face
Vertices

Diagrams

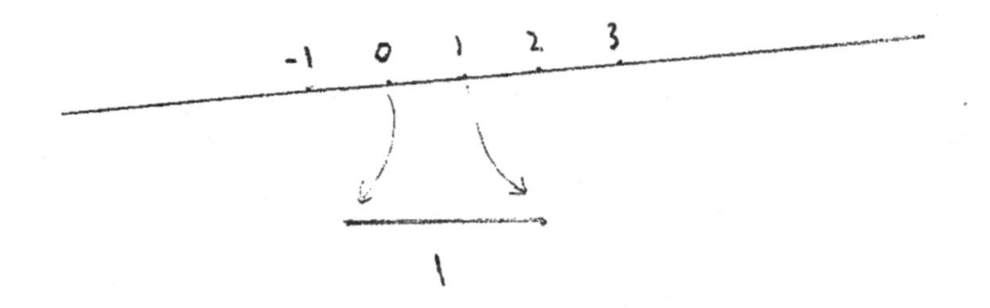

Uniform unit of measurement
(modeled on real-number line),
This unit must remain stable and consistent

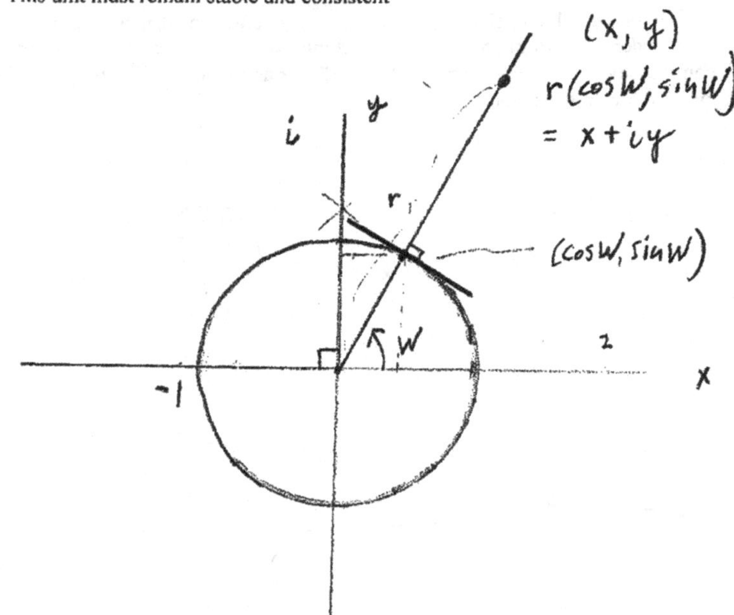

If the circle is rigid in its shape then the complex plane defines a commuting number field, as does the real-line

Representation of rectangular coordinates; (x,y)
as well as the complex-number plane,
Linear measuring directions are perpendicular
(or they are independent of both one another's measured values as well as measuring directions)
a complex number, z, is represented as, z=x+iy=r(cosW,sinW)=re^iW

Line segments and circles are quantitatively consistent shapes.

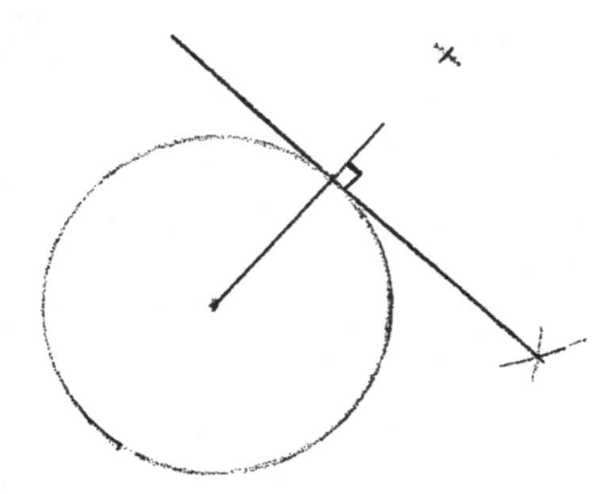

2-dimensions

The following (above) shapes are quantitatively consistent shapes, and locally their directions are independent, and form commutative algebraic constructs at each point over the entire shape, ie global commutative algebraic constructs, locally linear and invertible [one-to-one and onto] and this is true everywhere on the shape

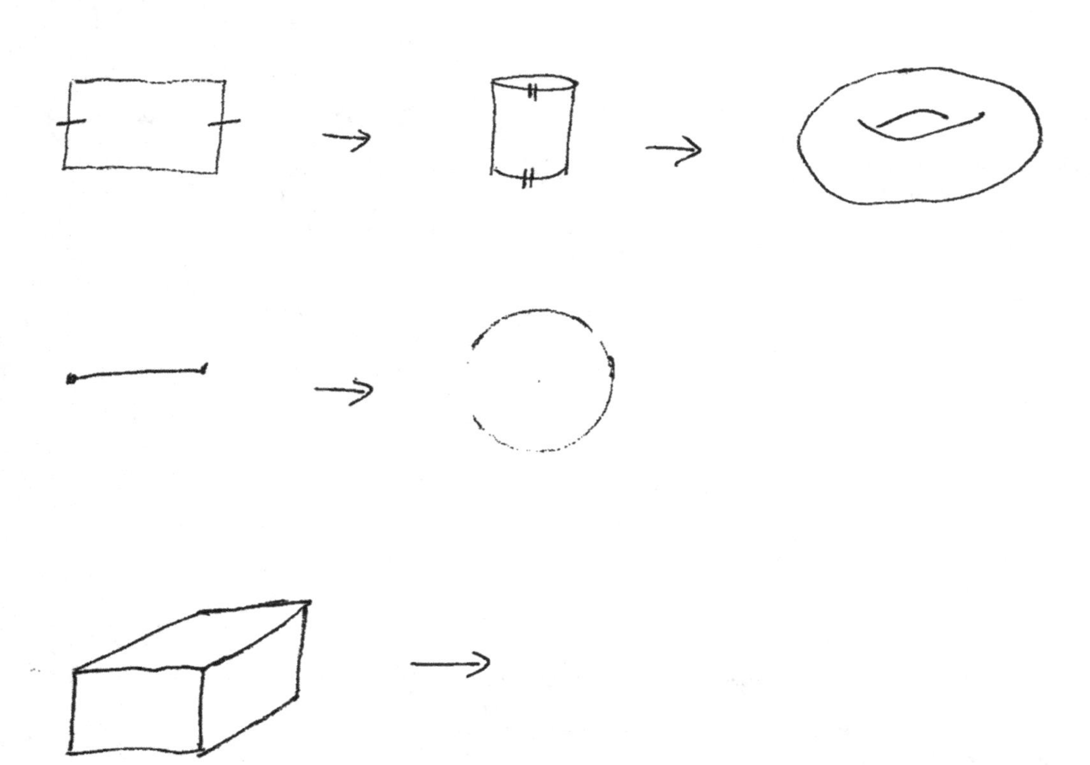

Cubical (or rectangular) simplexes are related to circle-spaces by means of "equivalence-relation topologies," or equivalently, by a "moding-out" process.

On such shapes local geometric measures are based on either measuring rectangular shapes or (equivalently) by a measuring process based on tangents to the circle, which is used as a basis for measuring along a circle's curve, eg rdW=dx+dy along a circle.

Euclidean shapes

Euclidean lattice

Rectangles
Fundamnetal domains

moding-out

tori

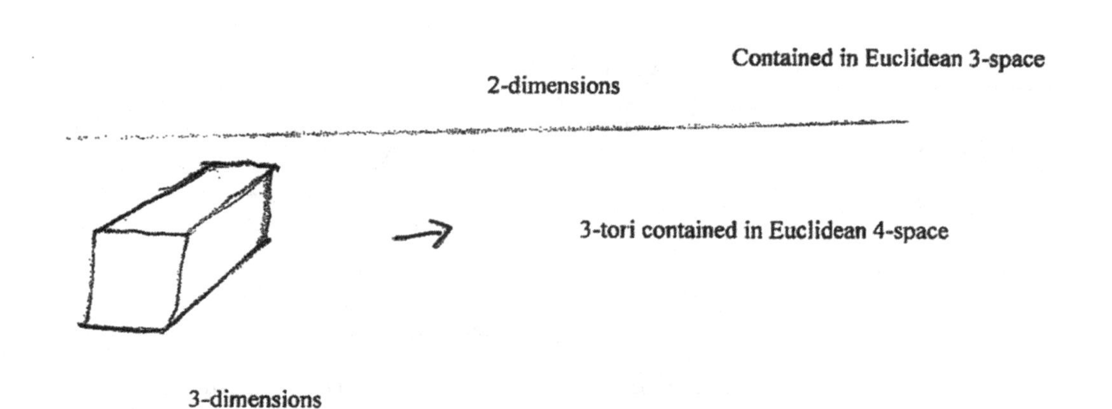

Hyperbolic shapes

Lattices (in hyperbolic 2-space)

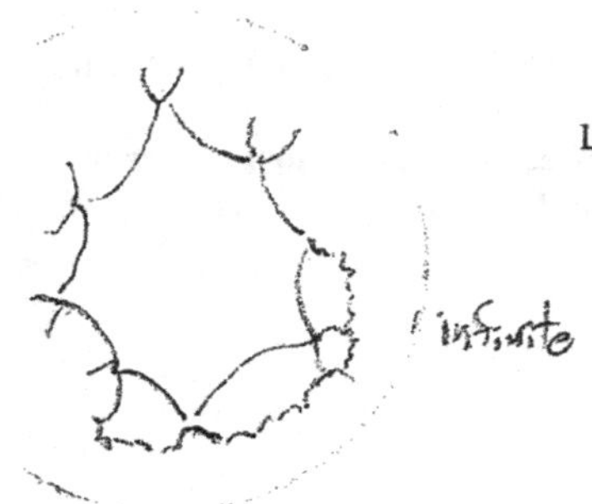

Fundamental domains

moding-out

discrete hyperbolic shapes (contained in hyperbolic 3-space)

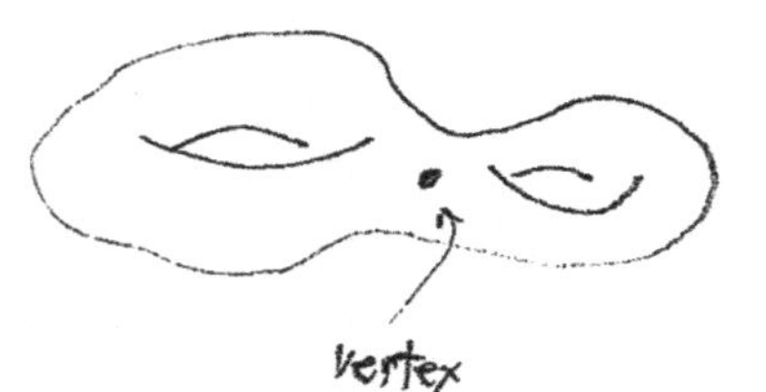

Rectangular simplexes

fundamental domains

discrete hyperbolic space-forms

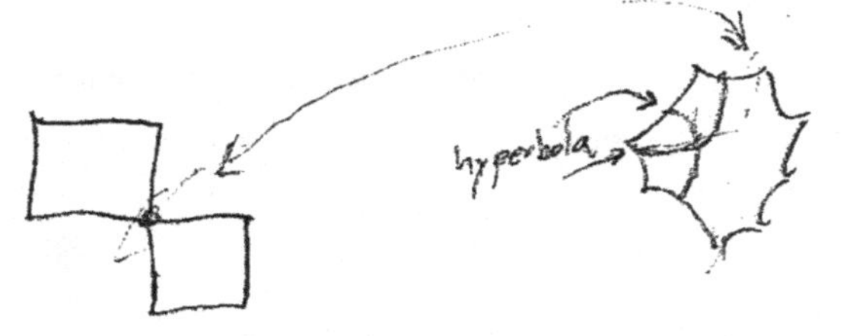

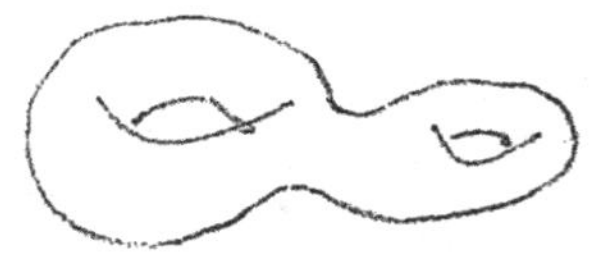

Edges of fundamental domain

The vertex is pulled apart

orthogonal pairs of hyperbolae

Discrete hyperbolic shapes are composed of toral components

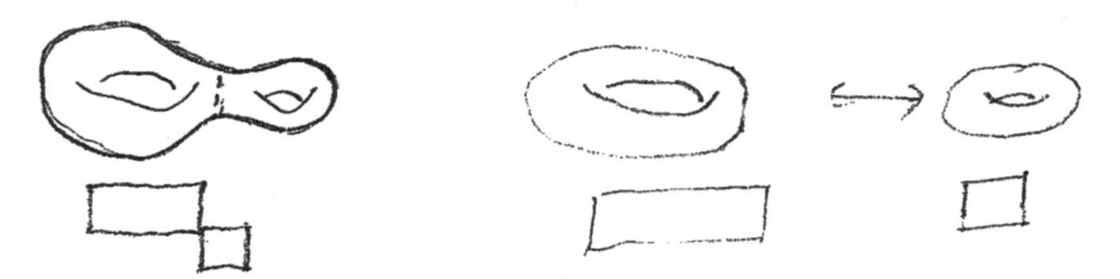

The number of holes in a discrete hyperbolic shape is called the shape's genus
2-holes are surrounded and caught by 1-curves, and defined by 2-dimensional discrete hyperbolic shapes
3-holes are surrounded and caught by 2-surfaces, and defined by 3-dimensional discrete hyperbolic shapes
etc

The faces on the fundamental domain (which result from the faces of a hyperbolic shape's rectangular (or "cubical") simplex) form very stable spectral measures on the very stable shapes of the hyperbolic space-forms.
Discrete hyperbolic shapes, or equivalently, hyperbolic space-forms, have open-closed metric-space topological properties, and they may be bounded or unbounded shapes, but all existing hyperbolic space-forms which are 6-dimensional or greater are unbounded shapes, and the dimension of the last known hyperbolic space-form is hyperbolic 10-dimensions (Coxeter).

Orbits on discrete Euclidean shapes and discrete hyperbolic shapes
Subsystems (or sub-metric-spaces) either occupy spectral orbits or they are "free"

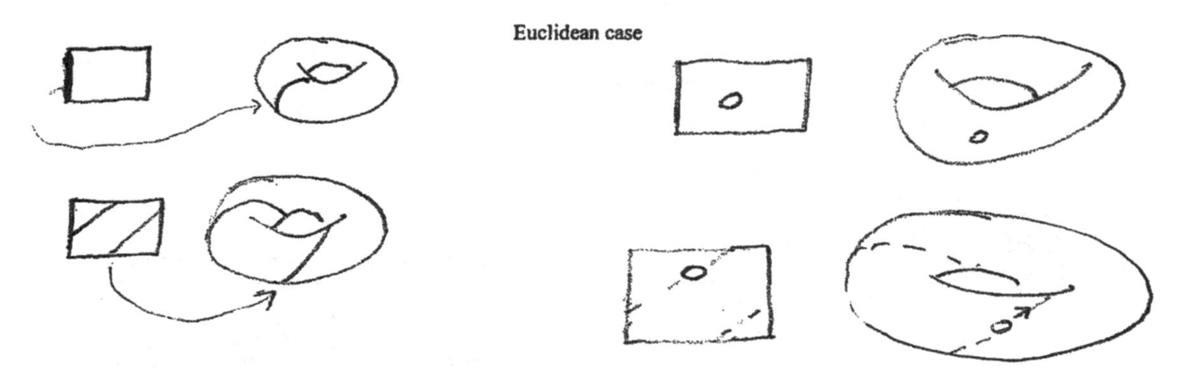

Stable orbits "free" subsystems

Hyperbolic case

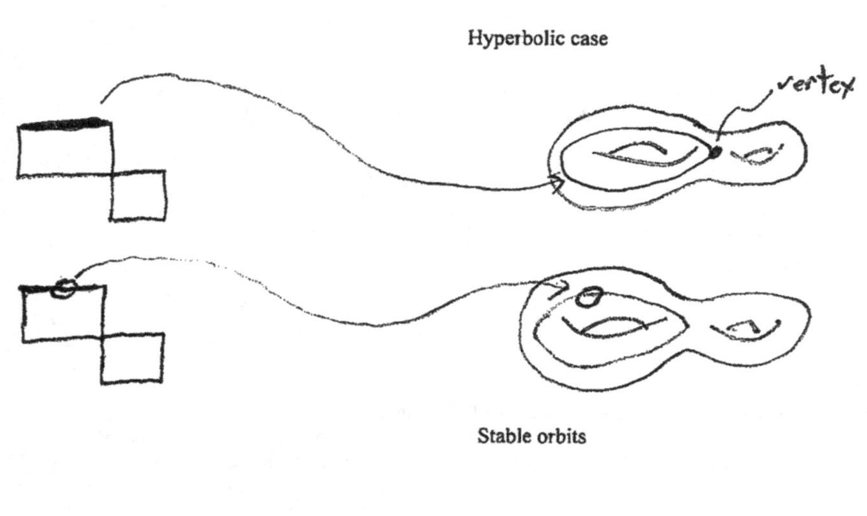

Stable orbits

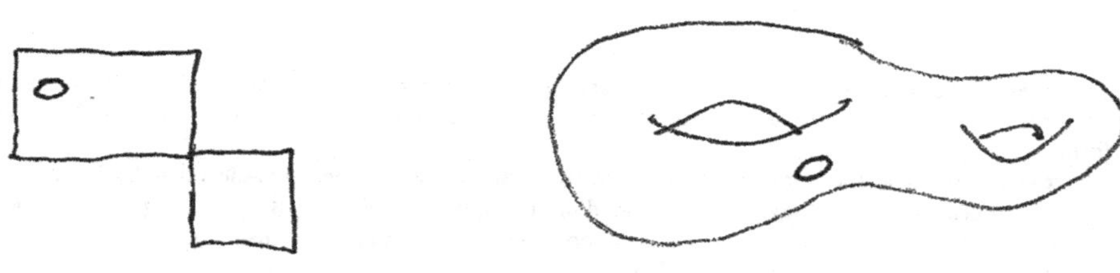

"free" subspaces (or "free" subsystrems)

The shapes obtained from this rectangular simplex for a 3-dimensional hyperbolic space-form. ., which contains a 'free" 2-dimensional hyperbolic space-form...., are contained in a 4-dimensional hyperbolic metric-space.

2-dimensional "free" components (or subsystems)
are contained in hyperbolic 3-space

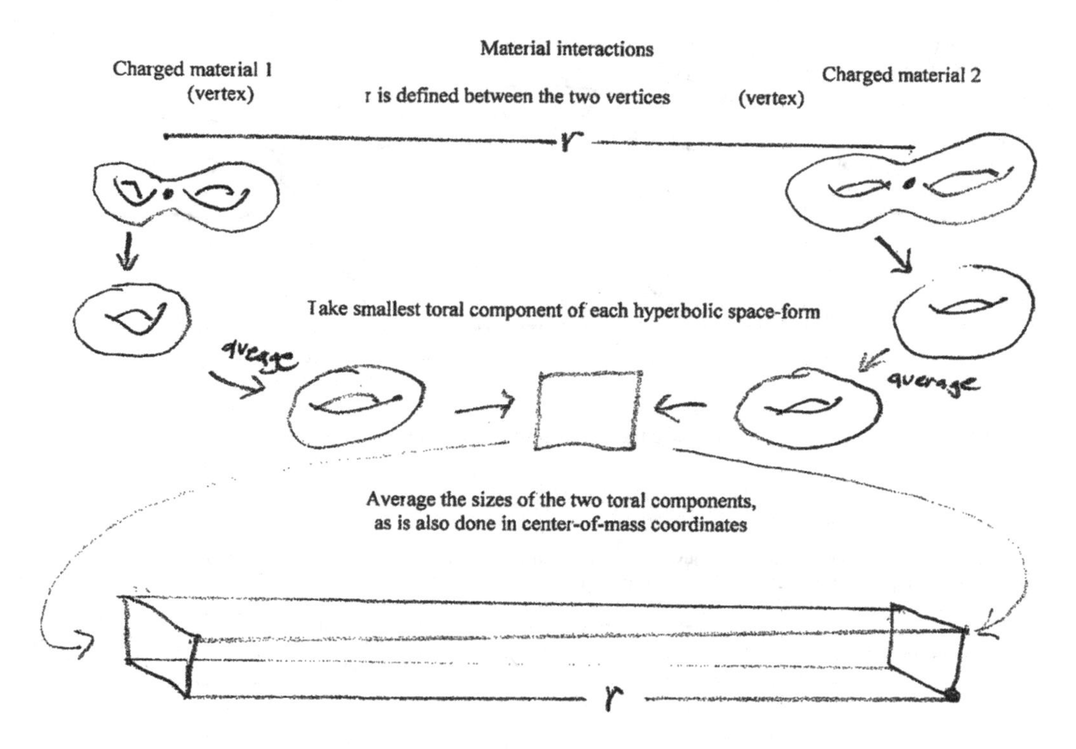

Represent the average value as a pair of equal oppositely positioned (rectangular) 2-faces
So as to define a right rectangular volume whose separation, r, is the distance between the
Vertices of the original interacting hyperbolic space-forms

Then define an interaction differential 2-form on the geometry of this 3-dimensional Euclidean torus, which is contained in Euclidean 4-space This determines the force-field, defined between the interacting material components.

The local vector geometry of the differential 2-form is relatable to the local geometry of the fiber SO(4) Lie group, since the 2-forms in Euclidean 4-space have the same dimension as SO(4) the geometry of the spatial displacement is determined in SO(4) by it geometric relation to Euclidean 4-space which is given by the 2-forms so that a local spatial displacement occurs due to a local coordinate transformation with SO(4) acting on the positions of the vertices [(in relation to center-of-mass coordinates) of the original pair of interacting hyperbolic space-forms] in Euclidean 4-space.

If the force is attractive and if the interacting (charged) material (the interaction structure) is contained within either 2-dimensional or 3-dimensional, or 4-dimensional Euclidean space then the force is radial and attractive, and r is made smaller. {Note: If the material is of a new type (oscillatory) and contained within Euclidean 4-space then the force-field has a new geometric structure contained in a higher-dimensional Euclidean space }

Then the same type of process repeats, for time intervals determined by the spin-rotation period of the spin-rotations of opposite metric-space states (about 10^-18 sec).

In this process the Euclidean torus which forms for each discrete time interval, forms in the context of action-at-a-distance.

Classical partial differential equations are defined within very confining and very rigid sets of both discrete hyperbolic shapes and action-at-a-distance material interaction Euclidean toral components which link the hyperbolic material together, so that the force-field differential 2-form is defined on the torus.

The above interaction for material contained in 3-space results in a 4-dimensional descriptive context, but it can be symbolically represented in 3-space.

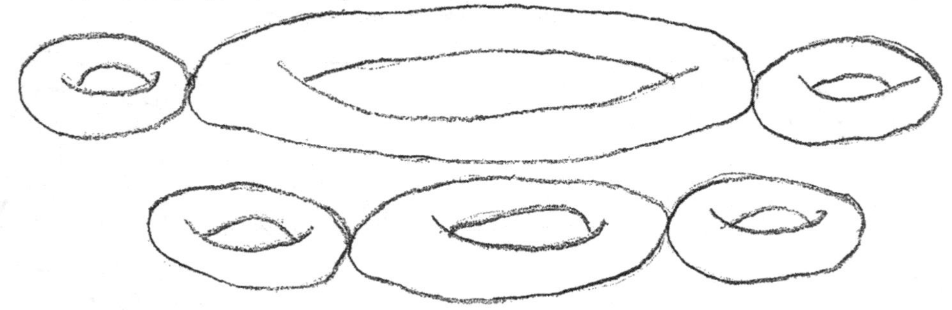

For an attractive interaction in 3-space

It should also be noted that in this new descriptive context eigenfunctions would also be both discrete hyperbolic shapes and discrete Euclidean shapes (tori)

Forming new stable hyperbolic space-forms from a material interaction

Assume an attractive (or repulsive) interaction in 3-space, then the interaction of "free" material components would be similar to a collision of components,

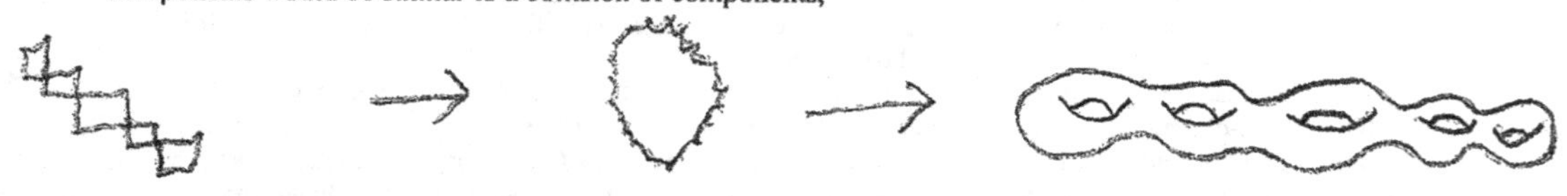

State of resonance new hyperbolic space-form

if the material components get very close during the (collision) interaction then if the energy of the over-all interaction is within (certain) energy ranges and the closeness allows resonances (with the spectral set of the over-all high-dimension containing metric-space) to begin to form, so that the over-all energy, as well as the resonances, allow a "new" stable "discrete hyperbolic shape" (in the proper dimension of the interaction) to form.

Weyl-transformations
Representing two maximal tori within a Lie group
(rank-2 compact Lie group)

These two intersecting circles of the two maximal tori may be angularly related to one another by Weyl group transformations, where the Weyl group defines the conjugation classes of the maximal tori which "cover" the compact Lie group

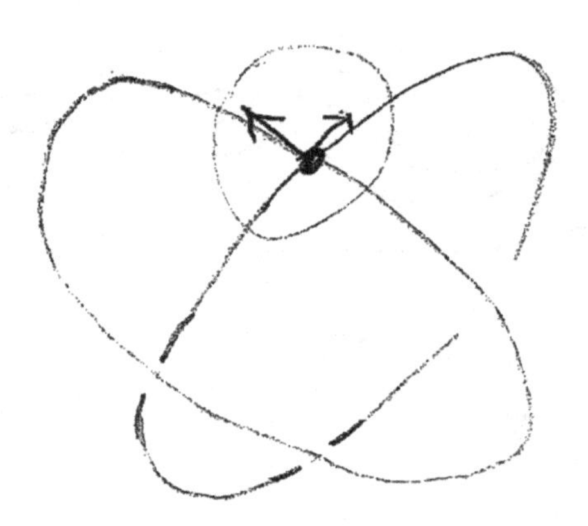

Forming angular changes between toral components of a discrete hyperbolic shape by using Weyl-transformations, which change the angular relations between circles which compose a toral component of a hyperbolic space-form.

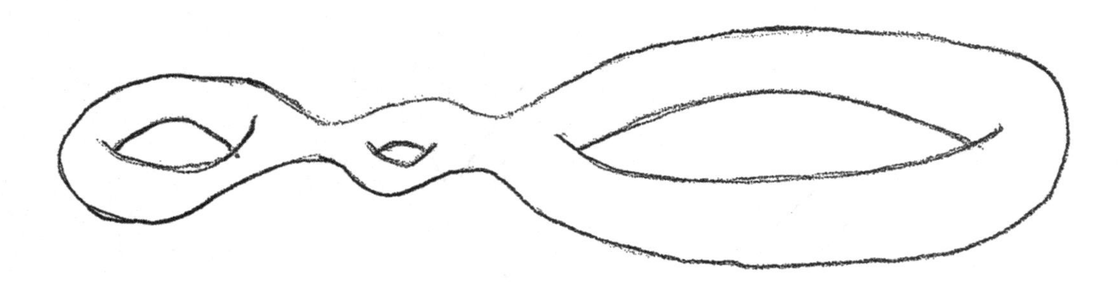

Envelopes of orbital stability for "free" subsystems (or sub-metric-spaces)

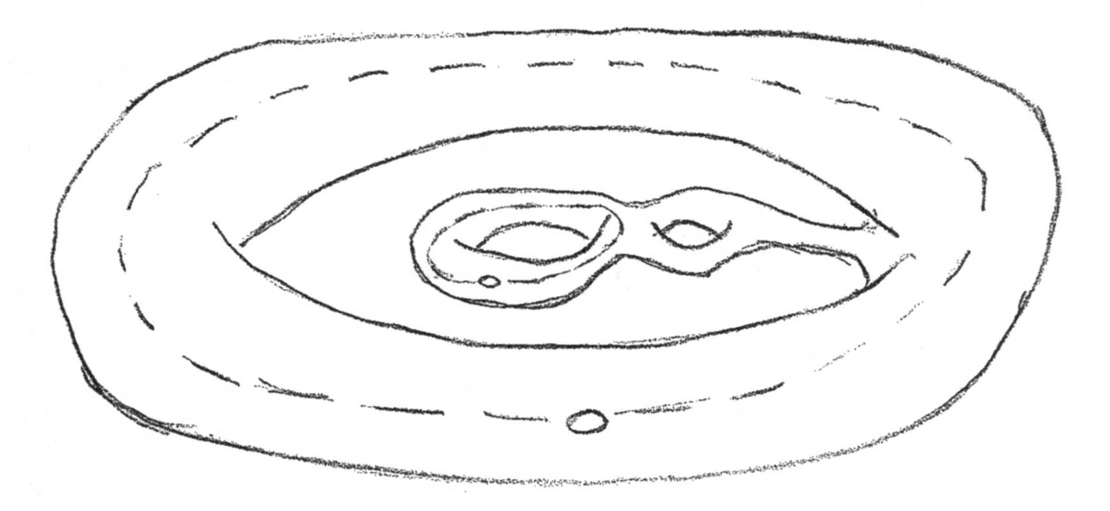

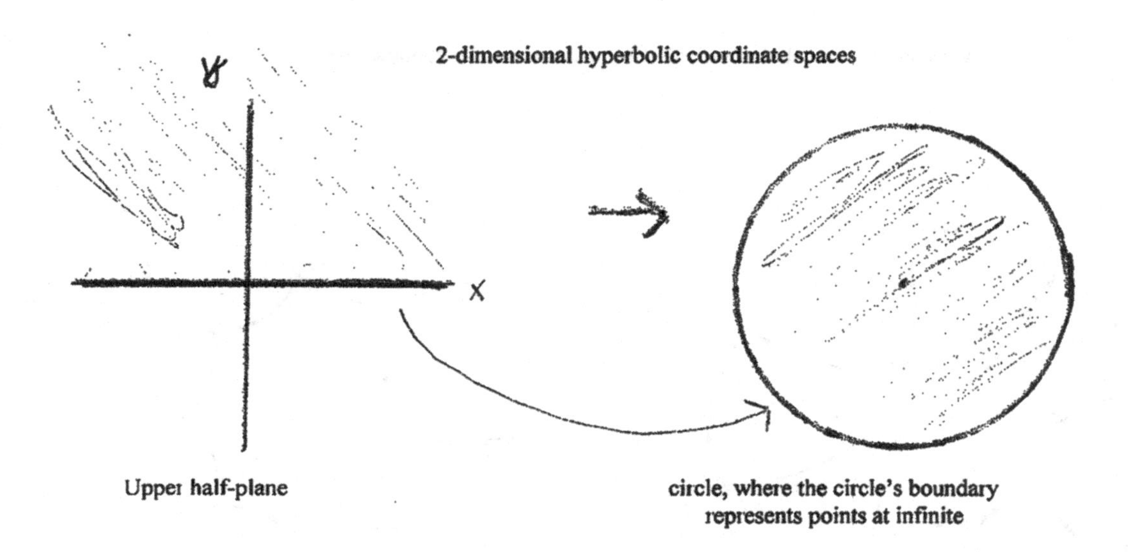

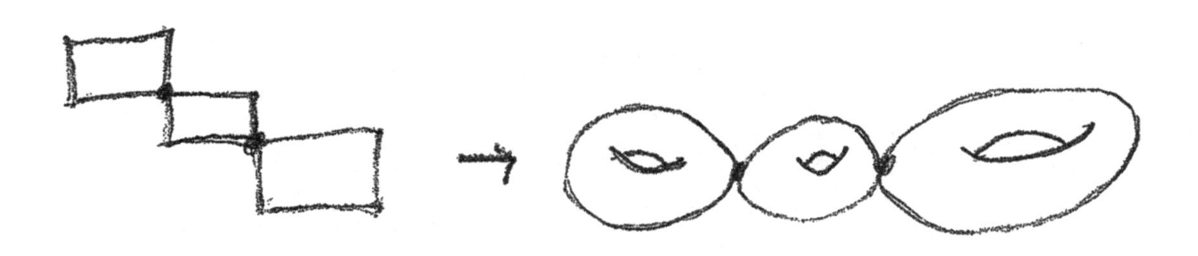

Rectangles attached at vertices and then moded-out, without expanding the vertex, shows a model of a discrete hyperbolic shape's toral components, represented as separate tori attached at separate vertices

Various types of unbounded 2-dimensional discrete hyperbolic shapes

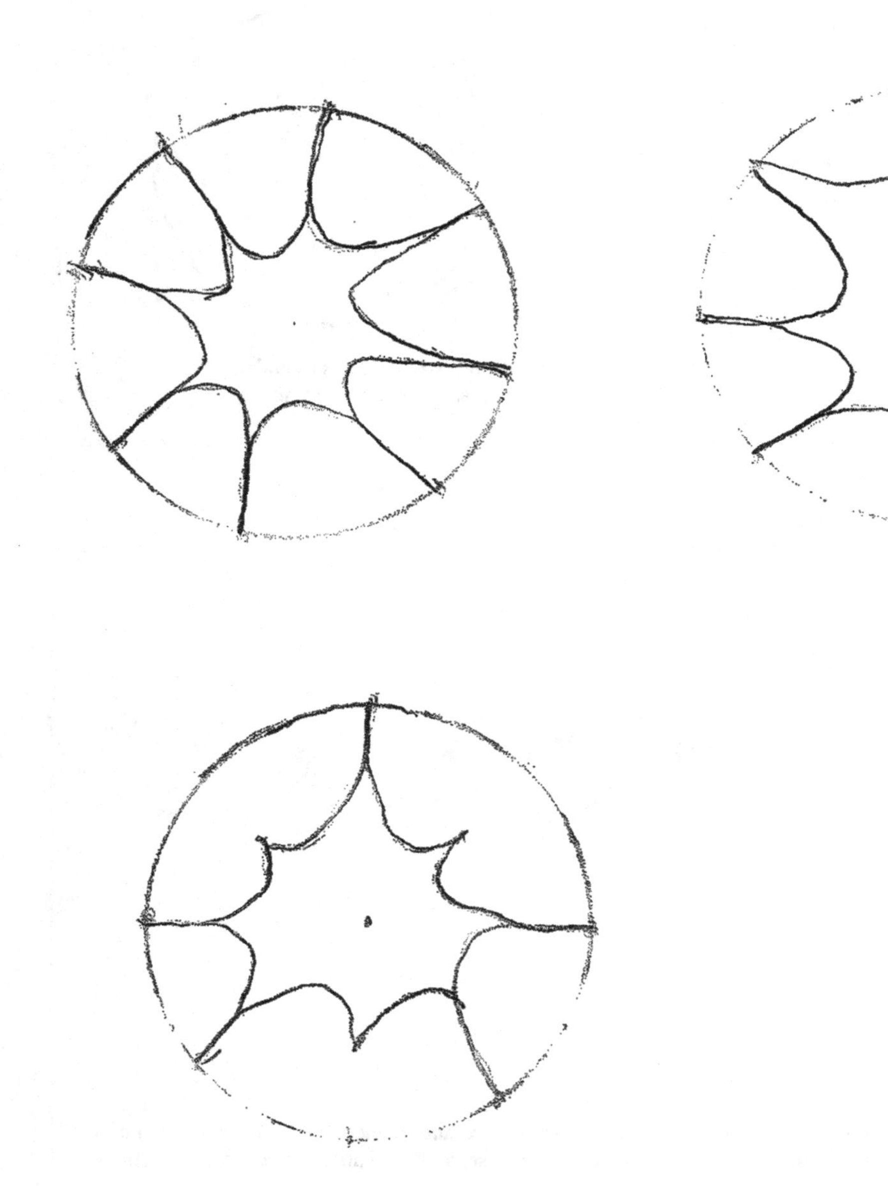

The Mathematical Structure of Stable Physical Systems

The stable geometric shapes

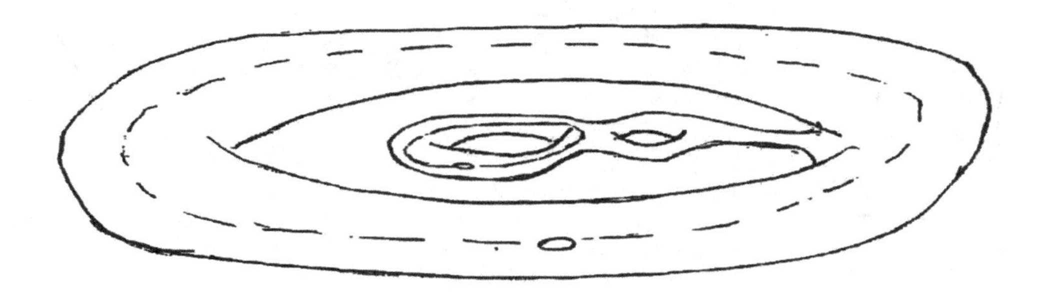

The same shape "folded" by Weyl-angles

Partitioning a Many-Dimensional Containment Space

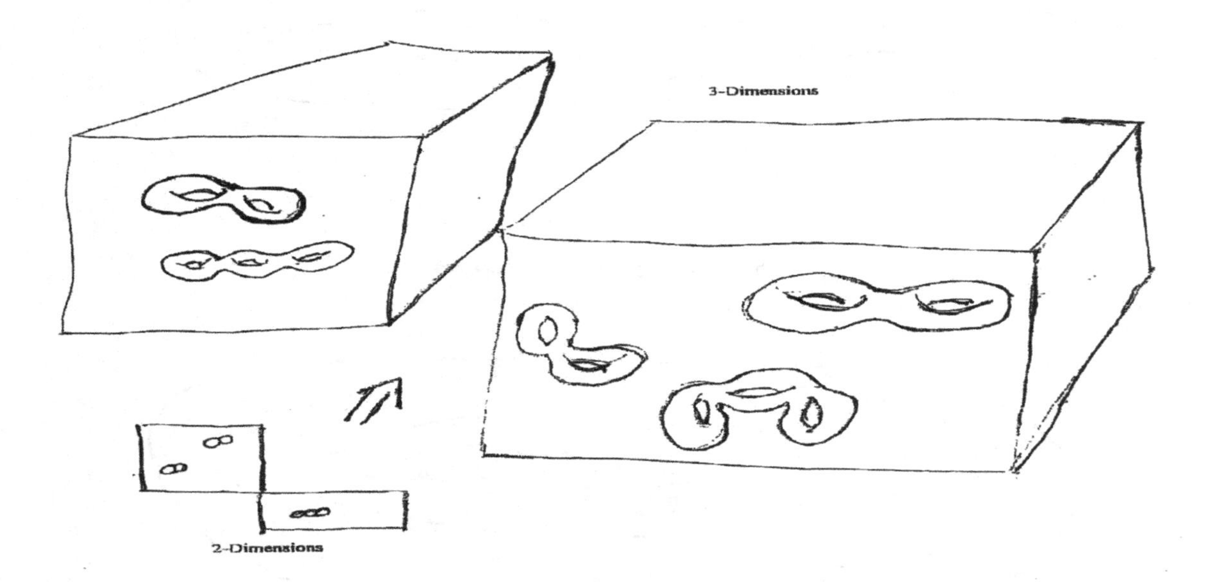

Perturbing Material-Components on Stable Shapes: How Partial Differential Equations Fit into the Descriptions of Stable Physical Systems

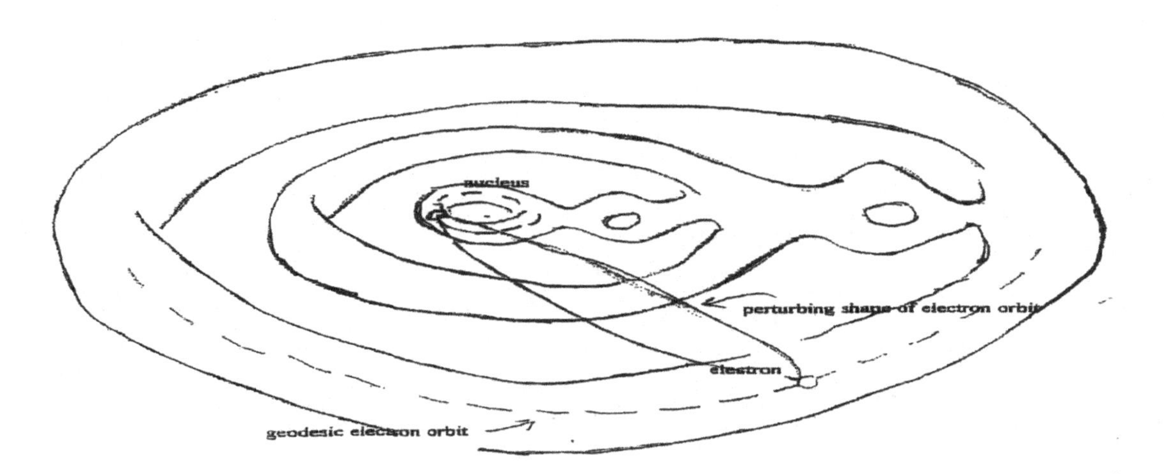

Describing the Dynamics of "Free" Material Components in Higher-Dimensions

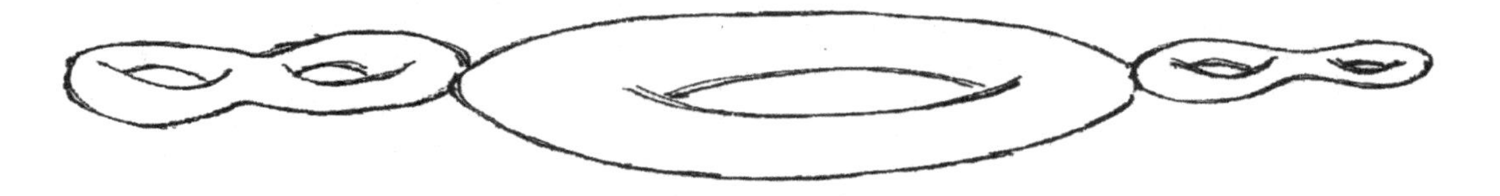

Interaction components and the interaction shape

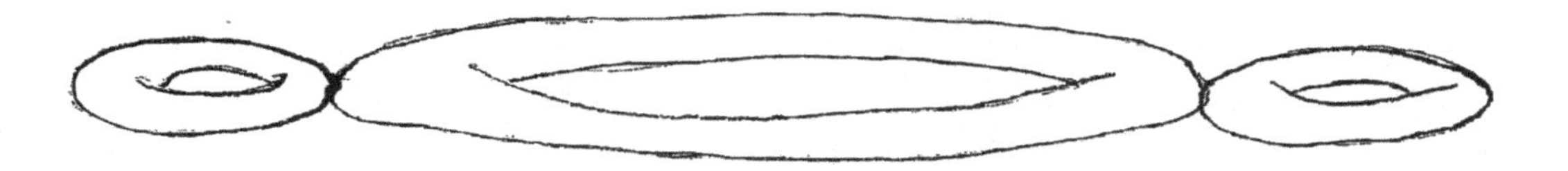

The approximation of the interaction

References

1. A New Copernican Revolution, Bill G P H Bash and George P Coatimundi, Trafford Publishing, 2004. www.trafford.com/03-1913,
2. The Authority of Material vs. The Spirit, Douglas D Hunter, Trafford Publishing,2006. www.trafford.com/05-3038
3. Topology and Geometry for Physicists, C. Nash and S. Sen, Academic Press, 1983.
4. The Infamous Boundary, David Wick, Springer-Verlag, 1995.
5. Function Theory, C. L. Siegel,
6. Three-dimensional Geometry and Topology, W. Thurston, Princeton University, 1997.
7. Gauge Theory and Variational Principles, D Bleeker, Addison, 1981.
8. Geometry II, E B Vinberg, Springer, 1993.
9. Spaces of Constant Curvature, J Wolf, Publish or Perish, 1977.
10. Contemporary College Physics, Jones and Childers, Addison-Wesley, 1993 (High School text).
11. I M Benn and R W Tucker, in, An Introduction to Spinors and Geometry with Applications in Physics, 1987, (Chapter 2)
12. Representations of Compact Lie Groups, T Brocker, T tomDieck, Springer-Verlag, 1985.
13. Dynamical Theories of Brownian Motion, E. Nelson, Princeton University Press, 1967 (1957).
14. Quantum Fluctuations, E. Nelson, Princeton University Press, 1985.
15. Algebra, L Grove.
16. Electron magnetic moment from gonium spectra, H Dehmelt (Nobel prize winner) et al, Physical Review D, Vol 34, No. 3, Aug 1, 1986.
17. Newton's Clock, Chaos in the Solar System, I Peterson, W H Freeman and Company, 1993.
18. Quantum Mechanics, J L Powell and B Crasemann, Addison-Wesley Publishing, 1965.
19. The End of Science, J Horgan, Broadway Books, 1996.
20. Riemannian Geometry, L. P. Eisenhart, Princeton University Press, 1925.
21. Reflection Groups and Coxeter Groups, J Humphreys, Cambridge University Press, 1990.

22. Partial Differential Equations, J Rauch, Springer-Verlag, 1991.
23. The Foundation of the General Theory of Relativity, A Einstein, 1916, Annalen der Physik (49).

D Coxeter
Katok

Just as Copernicus, Kepler and Galileo provided a quantitative-geometric context for the properties of the solar system, which were then precisely identified by the solutions to (the) solvable differential equations of Newton; Martin Concoyle now provides the stable geometric structures which fit . . . , both macroscopically and microscopically . . . , into a many-dimension containment set (hyperbolic 11-dimensional), so that these shapes are the solutions, ie the geometries of the stable spectral-orbital properties, of all the fundamental stable systems which have stable spectra and orbits, and it is the basis for a quantitative system (the spectral set of a measurable existence) which is finite, and These ideas are discussed in the following books: (available at math conference, 2013)

1. A New Copernican Revolution (p286), B Bash & P Coatimundi, Trafford, 2004.
2. The Authority of Material vs. The Spirit (p483), D D Hunter, Trafford, 2006.
3. Introduction to the Stability of Math Constructs; and a Subsequent General, and Accurate, and Practically Useful Description of Stable Material Systems, Concoyle, and G P Coatimundi, (p262), 2012, Scirbd.com.
4. A Book of Essays I: Material Interactions and Weyl-Transformations, Martin Concoyle Ph. D., (p234), 2012, Scribd.com.
5. A Book of Essays II: Science History, and the Shapes which Are Stable, and the Subspaces, and Finite Spectra, of a High-Dimension Containment Space, Martin Concoyle, (p240), 2012, Scribd.com.
6. A Book of Essays III: Elementary Topics, Martin Concoyle (p303), 2012, Scribd.com.
7. Physical description based on the properties of stability, geometry, and quantitative consistency: Short essays which are: simple, "clear," and direct Presented to the Joint math meeting San Diego (2013), (p208), Martin Concoyle, 2013, Scribd.com.
8. Describing physical stability: The differential equation vs. New containment constructs, Martin Concoyle, 2013, (p378), Scribd.com. (also equivalent to, VII 3, at Scribd.com)
9. Introduction to the stability of math constructs; and a subsequent: general, and accurate, and practically useful set of descriptions of the observed stable material systems, Martin Concoyle Ph. D., 2013, (p70), Scribd.com,

See scribd.com put m concoyle into web-site's search-bar

As well as in the following (new) books from Trafford:

1. The Mathematical Structure of Stable Physical Systems, Martin Concoyle and G. P. Coatimundi, 2013, (p449) Trafford Publishing (equivalent to 3. And 5. Above, Scribd)
2. Partitioning a Many-Dimensional Containment Space, Martin Concoyle, 2013, (p477) Trafford Publishing, (equivalent to 4. And 6. Above, Scribd)
3. Perturbing Material-Components on Stable Shapes:
How Partial Differential Equations Fit into the Descriptions of Stable Physical Systems, Martin Concoyle Ph. D., 2013, (p234) Trafford publishing (Canada) (equivalent to 7. And 9. Above, Scribd, and new material)
4. Describing the Dynamics of "Free" Material Components in Higher-Dimensions, Martin Concoyle, 2013, (p478) Trafford Publishing (equivalent to 8. Above, Scribd, and new material)

Alternative title to any of 1-4 Trafford:
The Unbounded Shape, and the Self-Oscillating, Energy-Generating Construct

Copyrights

These new ideas put existence into a new context, a context for both manipulating and adjusting material properties in new ways, but also a context in which life and creativity (practical creativity, ie intentionally adjusting the properties of existence) are not confined to the traditional context of "material existence," and material manipulations, where materialism has traditionally defined the containment of material-existence in either 3-space or within space-time.

Thus, since copyrights are supposed to give the author of the ideas the rights over the relation of the new ideas to creativity [whereas copyrights have traditionally been about the relation that the owners of society have to the new ideas of others, and the culture itself, namely, the right of the owners to steal these ideas for themselves, often by payment to the "wage-slave authors," so as to gain selfish advantages from the new ideas, for they themselves, the owners, in a society where the economics (flow of money, and the definition of social value) serves the power which the owners of society, unjustly, possess within society].

Thus the relation of these new ideas to creativity is (are) as follows:

These ideas cannot be used to make things (material or otherwise) which destroy or harm the earth or other lives.

These new ideas cannot be used to make things for a person's selfish advantage, ie only a 1% or 2% profit in relation to costs and sales (revenues).

These new ideas can only be used to create helpful, non-destructive things, for both the earth and society, eg resources cannot be exploited to make material things whose creation depends on the use of these new ideas, and the things which are made, based on these new ideas, must be done in a social context of selflessness, wherein people are equal creators, and the condition of either wage-slavery, or oppressive intellectual authority, does not exist, but their creations cannot be used in destructive, or selfish, ways.

Index
(key words)

fundamental domains
Geometrically separable
Hermitian form (finite dimensions)
hyperbola
hyperbolic
hyperbolic metric-space
Independent
infinite extent space-forms
interaction
interaction potentials
Inverse
Invertible
Isometry
Isometry groups
Lattices
Linear
Lie algebra valued connection 1-forms
Lie group
Maximal torus
metric spaces
metric-space states
moding out
non-reducible
Orthogonal
Parallelizable

Physical properties and fundamental invariance's, eg translations and linear momentum,
principle fiber bundles
sectional curvature
signature of a metric,

Solvable
space-forms,
space-time,
unitary groups
Weyl group

Appendix I

Notes:

Double spaces can mean a sudden new direction of the discussion without a new paragraph title.

The *'s represent either favorites (of the author) or (just as likely) indecision and questions about (logical) consistency. Information and discussion about ideas is not a monolithic endeavor pointing toward any absolute truth, the wide ranging usefulness, in regard to practical creativity, might be the best measure of an idea's truth, it is full of inconsistencies and decisions about which path to follow (between one or the other competing ideas) are either eventually made or the entire viewpoint is dropped, but this can occur over time intervals of various lengths.

The marks, ^, associated to letters, eg a^2, indicates an exponent.

The marks, *, in math expressions can have various math meanings, such as a pull-back in regard to general maps which can, in turn, be related to differential-forms, defined on the map's domain and co-domain (or range), but in this book it usually denotes the "dual" differential-form in a metric-space of a particular dimension, eg in a 4-dimensional metric-space the 1-forms are dual to the 3-forms and the 2-forms are self-dual, etc.

The main idea of thought (or of ideas) is that it is about either sufficiently general and sufficiently precise descriptions based on simple patterns, or it is about developing patterns (of description) which lead to particular practical creativity, or to new interpretations of observed patterns, or to directions for new perceptions.